Comprehensive

BIOTECHNOLOGY-5

Comprehensive

BIOTECHNOLOGY-5

M.K. Sateesh

Head, PG Department of Biotechnology and Microbiology
St. George College of Management and Science
Bangalore

NEW AGE INTERNATIONAL (P) LIMITED, PUBLISHERS

(formerly Wiley Eastern Limited)

New Delhi • Bangalore • Chennai • Cochin • Guwahati • Hyderabad
Jalandhar • Kolkata • Lucknow • Mumbai • Ranchi

Visit us at www.newagepublishers.com

Published by New Age International (P) Ltd., Publishers
First Edition : 2006

ISBN : 81-224-1934-8

Rs. 195.00

C-06-07-875

Printed in India at Printo Pack, Delhi.
Typeset at Pagetek Graphics, Delhi.

PUBLISHING FOR ONE WORLD

NEW AGE INTERNATIONAL (P) LIMITED, PUBLISHERS
(formerly Wiley Eastern Limited)

4835/24, Ansari Road, Daryaganj, New Delhi - 110002
Visit us at **www.newagepublishers.com**

Dedicated to my beloved
parents
Mr. B. Krishnaiah
and
Mrs. Susheela Krishnaiah

Preface

Biotechnology is a frontier area of science and technology having a great potential for welfare of humanity. Biotechnology is becoming a dream word for the present-day science students. Large number of universities, deemed institutions have started various under-graduate, graduate, post-graduate and diploma courses in the field of Biotechnology. The University of Pune has even started an innovative and extraordinary new program course namely MBA in Biotechnology. India is becoming an emerging leader in the Biotechnology Field. A huge number of books on Biotechnology are published within and outside the country. Most of these books deal with a few specialized areas in depth and are expensive in nature. This book has been written as a compiled text to meet the requirements of 'Animal Cell Biotechnology', 'Immunology' and 'Plant Biotechnology'—major branches of Biotechnology for graduate and post-graduate courses of various Indian universities, for competitive entrance examinations, and also serves as an insight into the basics of the Biotechnology.

Biotechnology is one of the most exciting areas of natural sciences and its applications have to reach the needy people. After the successful publications of the entire human genome (2002) by National Institute of Health and Celera Genomics, Biotechnology has been a centre of attention worldwide. The claims about the birth of Baby Eve, claimed to be the first cloned human being in December 2002, have only reinforced global attention on Biotechnology. This branch is considered to be a field which can solve the world's hunger in lesser duration. It is no longer a specialized segment of science confined to hightech laboratories. The common people have started talking about Biotechnology in the same breaths they talk about information technology. Newspapers are daily giving information on all fields of Biotechnology to educate normal public. The Government of India framed draft of Biotechnology policy in 2005.

Globally too, Biotechnology is slowly entering the hallowed portals of the 'Knowledge economy'. Biotechnology companies, especially in the areas of pharma, agriculture, genomics, proteomics and bioinformatics, are grabbing the public attention. In India too, the similar trend is noticed in the business field and in academics.

Within a span of 10 years more than 265 Biotechnology companies are established in India working on different fields. These companies are targeting for the revenue generation of US$ 5 billion by the end of 2010. The major companies include Biocon India, Astrazenaca, Metahelix, Mansanto, Bharat Biotech, Shantha Biotech, Bangalore Genei, Reddy's Laboratories. At present, in Biotechnology field the focus should be made on research, development and intellectual property rights generation in the area of vaccines, diagnostics, and new medicines derived from ethnobotany, traditional knowledge, rural practice, value added products through employing food biotechnology, bioethics, increased productivity through pest- and disease-resistant transgenic animals and crops with ecofriendly technologies.

Biotechnology is essentially an interdisciplinary field, students' should get exposure to wide varieties of basic and applied sciences. Owing to its interdisciplinary nature, Biotechnology field appeals to both science students as well as engineering students equally. This book is intended to be an introduction to the subject for students of both the streams.

The book presents an integrated approach to 'Animal Cell Biotechnology', 'Immunology' and 'Plant Biotechnology' encompassing information on the basics of the subject. The subject matter has been written in a stepwise systematic manner to meet the student's requirements. Simple, clearly understandable illustrations, correct and up-to-date informations are the main features of this book. Thus, it can serve as a good reference material for students, policy makers and teachers.

'Animal Cell Biotechnology' part deals with the gene therapy, stem cells and its applications, general metabolism of cultured animal cells, growth factors for culture system, production of commercially important pharmaceuticals and various types of bioreactors and media. Culture of animal cells is a primary requirement for successful isolation and culture of animal cell lines. These cell lines are used for genetic modification and production of transgenic products free from all types of contaminants with higher yield. Thus, emphasis was given on various aspects of animal cell culture experiments.

In 'Immunology' topics such as history of immunology, immune system, its types, antigen and antibody structures, effector mechanisms, immunity to infectious diseases and vaccines are dealt in detail. In each chapter the clear definition and illustrations are expanded for better understanding of the topics. These topics are covered from the basics which help for better understanding of cellular mechanisms in protection. This paves the way for designing an appropriate biotechnology product useful in the healthcare systems.

'Plant Biotechnology' part gives introduction to plant tissue culture, different types of media, culture methods and types, embryo culture and rescue, haploid production, somaclonal variations, somatic hybridization and *Agrobacterium* genetic modification systems, *Arabidopsis thaliana* genomics, plants and industry are given from the basics. In all the chapters their respective biotechnological applications are dealt in thorough details. At present large number of transgenic plants are produced worldwide to solve multifarious problems in agriculture, healthcare and environment. Hence, basic concept of plant culture system, genetic manipulation, plant genomics and production of transgenics with its applications are given insight in this part of the book.

Comments and any type of suggestions, criticism from the readers will be deeply appreciated and carefully considered in the preparation of the next edition.

Dr. M.K. Sateesh
mk.sateesh@gmail.com

Acknowledgements

I am extremely grateful to Dr. T. Prabhu, Chief of Laboratories, Medinova Diagnostic Centre, Bangalore and Ex-Principal, Bangalore Medical College, Bangalore, Prof. H.A. Ranganath, Vice-Chancellor, Bangalore University, Bangalore, Prof. V.S. Ranganathan, Prof. Geetha Bali, Prof. S. Shankara Bhat, Prof. K.A. Raveesha, Dr. S. Shishupala, Dr. B. Vedamuthy, Dr. H.N. Ramesh Babu, Dr. Rajeshwari Ramesh Babu, Dr. Sumithra K. Urs, U.S.A., Dr. K.P. Guruprasad, Dr. S. Umesh, Dr. S. Puttaraju, Dr. S. Manjunath, Dr. L. Rajanna, Dr. Y.L. Krishnamurthy, Mr. M.K. Girish, Mr. S. Satish Kumar, Ku. Samrudhi, Ku. Samskruthi, Mrs. Gayathri Girish, Mr. B. Umamahesh, Sigma-Aldrich, Mr. C.R. Naveen Kumar, Biocon India, my teachers, my colleagues, my students, friends and all my well-wishers for their constant support and encouragement.

Special thanks are extended to my parents B. Krishnaiah, Susheela Krishnaiah, to my wife Mrs. Sowmya Satish, my son M.S. Yajat, my brother Mr. M.K. Girish, my sister Mrs. M.K. Sandhya Sateesh Kumar, my in-laws and other family members who have shown great forbearance and encouraged me during the compilation of this book.

Finally I am also grateful to Mr. Saumya Gupta, Managing Director and Mr. V.R. Babu, Mr. R. Srinath, Mr. M.K. Prashanth, G.N. Mahesh, C.K. Suryanarayana, M. Srinivas and other staff members of the New Age International Publishers, in bringing out this book.

Dr. M.K. Sateesh

Contents

Part I
Animal Cell Biotechnology

1

Gene Therapy

1.1 INTRODUCTION

The advancement made in medicine and genetics has changed the entire scenario of treatment processes in patients. For many years, the study of medical genetics was confined to determine whether a particular human disease had a genetic basis or not. Once the genetic nature of a disease was established, it was sometimes possible to devise a diagnostic test to identify the disease in newborns or prenatal conditions. Biomedical researchers are currently redefining human geography. These modern explorers are elaborating a new human map, based on genes, that is likely to alter our views of the world—and our place in it—even more profoundly than did the maps generated by Columbus and other fifteenth and sixteenth-century explorers in the history. This has made a modern generation of therapeutic agents targeting the genetic diseases.

The human genetic maps are likely to alter our perceptions of self and other, of normality and abnormality, particularly in the area of procreation. This can be seen most clearly in the consideration of the impacts of prenatal genetic testing. Thanks to the researchers involved in work on the Human Genome Project and the media reports it has attracted over the past several years, the public has received a steady stream of information on the latest gene to be discovered by science. Such announcements are generally greeted with enthusiasm, and the hope that genetic solutions will save us from our human frailties. But, this hope may be overly optimistic and ambitious. While genetic technologies may offer a few more keys to unlock the genetic puzzle, they have not necessarily produced the solutions required to solve the puzzle. For example, although hundreds of human genes have been identified and mapped, in many cases, a corresponding cure or treatment regime has not yet been developed.

Moreover, although many common diseases are suspected of being genetically linked, in most cases, the risk must be augmented by other genetic or environmental factors in order for the disease to occur. Nor does the detection of a chromosomal disorder, such as Down's syndrome, predict the severity with which the syndrome will be expressed. Advocates of new reproductive technologies, particularly prenatal technologies, argue that this technology will expand women's reproductive choice and decrease the incidence of disability in society. This argument is over simplistic at best, and belies the significant social and ethical implications underlying such technologies for women's reproductive autonomy and the equality rights of women and men with disabilities. It is possible that prenatal diagnostic testing can offer women more control over the occurrence of the birth of a disabled child by providing relevant information about the genetic status of the fetus during pregnancy.

However, because there are no therapeutic interventions available for the majority of conditions which prenatal testing detects, the options open to women following a positive diagnosis consist of preparing for a life with a disabled child or terminating the pregnancy. Negative perceptions about a life with a disability, combined with the difficulties in obtaining adequate social supports may cause a woman to believe that her only real choice is to terminate her pregnancy.

Disability rights activists harshly criticize the intent of prenatal testing and the more often than not, follow-up procedure of aborting disabled fetuses. They are suspicious of the manner in which prenatal testing has been allowed to outpace therapeutic strategies. They argue that gene location is the means for a strategy of eugenics by eliminating defective fetuses. In addition, some form of therapy, such as medication, diet or blood transfusions could be sought to alleviate the impact of gene defect.

Of the over 5000 inherited of gene diseases catalogued to date and most of these diseases lack fully effective therapies. U.S. National Institute of Health website listed a total of 206 gene therapy protocols. There is one Infant in every 100 that has genetic defect. Large number of genetic disorder is associated with liver and with haemopoietic organs. Genetic defect in many instance results into stillbirth or neonatal death. Those who survive with genetic diseases frequently have significant physical, mental, developmental or social problems.

The researchers have long been imagined to curing heritable diseases by introducing the external healthy genes into the patients. The recent advances made in recombinant DNA technology and medicinal fields have made possible the isolation of defective genes containing cells and reintroduction of these cells into the patient again. The recombinant DNA technology has led to an amazing extension of the science of animal biotechnology.

Through these advanced techniques, genes and genomes of a wide range of different organisms from microorganisms to animals are being manipulated for the benefit of mankind. The key technique in genetic engineering is gene transfer which has revolutionized the approaches of genetic analysis. It has enabled the isolation and cloning of genes from any organisms and introduces the same to virtually any living cell that may be prokaryotic or eukaryotic in nature. The ability to recombine and manipulate genes has made once impossible notion into feasible practice and lot of attention is focused in medicinal genetic engineering.

Hence, at present the treatment of genetic diseases by the introduction of gene or genes into the affected cells to provide gene functions that are deficient because of genetic or acquired deficiency is under focus of the medicine. Large numbers of gene therapy therapeutic agents are under clinical trials and still many more are on the way of development. The day will also come to practice of using the genes as prescribed drugs in the medical care.

Table 1.1: Different cell lines from humans, monkey, dog commonly used in animal cell culture laboratories for various work

Cell line Name	Origin Species	Tissue Type	Cell Type	Characteristics
Normal				
MRC	Human	Embryolung	Fibroblastic cells	Diploid, inhibited by contact processes
3T3-L1	Mouse	Embryonic tissue	Fibroblastoid	Contact inhibited; can differentiate into adipose cells
BHK-21	Hamster	Kidney	Fibroblastic	Contact inhibited transformable with polyoma virus
MDCK	Dog	Kidney	Epithelial	Forms 'domes' in culture condition
Transformed, Solid Tissue				
293	Human	Embryo kidney	Epithelial	Very sensitive to human adeno viruses
A431	Human	Squamous skin carcinoma	Epithelial	High level of EGF receptor on cell surface
A549	Human	Lung carcinoma	Epithelial	Makes surfactant protein A
MCF-7	Human	Breast carcinoma	Epithelial	Oestrogen receptor positive cells on surface
CaCo-2	Human	Colorectal carcinoma	Epithelial	Forms tight monolayer with polarized transport
Hep-G2	Human	Hepatoma	Epithelial	Consisting of drug metabolizing enzymes
Hela-S_3	Human	Cervical carcinoma	Epithelial	Shows rapid growth; high plating efficiency; can be proliferated in suspension cultures
Cos-7	Monkey	Kidney	Epithelioid	Transfection capacity is higher
PC-12	Rat	Pheochromo-cytoma	Neural cells	Undergoes neuronal differentiation with nerve growth factor
Leukaemic				
HL-60	Human	Myelocytic	Suspension cells	Differentiates in response to retinoids, NaBt, PMA
Jurkat E6-1	Human	T cell	Suspension cells	Produces IL-2, stimulated by phorbol esters
K562	Human	Erythrocytic	Suspension cells	Synthesizes haemoglobin when induced with Na butyrate

1.2 DEFINITION OF GENE THERAPY

Gene therapy is a novel approach to treat, cure or ultimately prevent disease by changing the expression of a person's gene.

Or

The treatment of any disease by the introduction of genes into affected cells to provide gene functions that are deficient because of genetic or acquired deficiency is known as gene therapy.

Gene therapy is in its infancy, and current gene therapy is primarily experimental, with most human clinical trials only in research stages.

The development of a therapeutic agent for humans usually passes through four clearly defined levels of study and testing before it can be sanctioned for use in regular practice. This makes sure that any therapeutic agents should not show any harmful effects or side effects on the host or receiving organisms.

1. Preclinical trials: Animal testing of the therapeutic agent to determine a rationale for therapeutic use, toxicity's, dose details, and risks involved.
2. Phase I: A small human trial generally involving dose escalation and focusing on safety and pharmacology of therapeutic agent.
3. Phase II: An expanded human trial with the focus on efficacy in a particular patient population.
4. Phase III: A large human trial intended to establish a definitive role for a drug or biologic.

The same things should also be followed by the gene therapy technology. Even though, there is no new or synthetic organics present in the gene therapy, the cells used for the process may creates unexpected problems.

Current practices of symptomatic treatment for genetic disorder are:

i. Replacement therapy for missing factor. Eg: coagulation factor for hemophilia.
ii. Long term blood transfusion for thalassaemia.
iii. Replacement with immunoglobulins for children with congenital hypogamma-globulinemia.
iv. Hormone therapy for certain diseases. Eg: Growth hormone for dwarfism, insulin injection for diabetics.
v. Replacement of missing enzymes. Eg: Gauchers disease, ADA deficiency.
vi. Diet control prevents accumulation of toxic metabolites. Eg: Phenylketonuria, galactosemia etc.
vii. Bone marrow transplantation to correct blood disorders. Eg: SCID
viii. Cofactor responsive metabolic disorders. Eg: methylmalonic acidaemia, homocystinuria colatamine, multiple acyl-coA dehydrogen, deficiency-riboflavin etc.

Symptomatic disease treatments show very low rate of success. It is completely successful in eight diseases (12% of total), moderately successful in 14% of the total and not cured in rest. Majority of the genetic diseases are not treated by regular conventional therapeutic agents such as synthetic organic drugs or naturally isolated supplements. This makes the treatment processes as only a short term solution or cure. This treatment can also cause a large number of side effects in an accumulative manner. And this will not be able to provide the complete curing effect and makes the patient to be under continuous medication or life long treatment. In such long term treatments regular changes of therapeutic agents should be made as a regular practice. This has a potential to alter the entire physiology of the host and thus leads to unrepairable damage to the various organs of the body.

Disadvantages of the above therapy processes

a. High cost is involved in the method of practice.

b. Need of continuous treatment throughout the life.
c. Danger of transmission of AIDS virus and other pathogenic agents during blood transfusion.
d. High iron contents after repeated blood transfusions in transgenic patients has side effects.

Hence, at present lot of emphasis is given to find out a better and permanent solution for the genetic disorders that are found to be effective in all respect. For this purpose the gene therapy is found out to be the most effective therapeutic method.

Historical Highlights of the Evolution of Gene Therapy

Gene therapy can be described as the introduction of a fully functional and expressible gene into a target cell. This results in permanent curing of the specific genetic disease. The evolution of gene therapy as a new method is followed the different phases as listed below:

Phase I: Emergence and acceptance of the concept of gene event. Period: 1866-1953
Phase II: Technical Implementation event. Period: 1955–1983
Phase III: Clinical development event. Period: 1975 onwards

On 14/9/1990 after exhaustive reviews by various regulatory panels, the first human gene therapy with severe combined immunodeficiency (**SCID**), which lacks enzyme adenosine deaminase was done. In this case cloned gene was introduced into lymphocytes (a type of white blood cell) that had been removed from the patient. After culturing the lymphocytes that were not having adenosine deaminase were transfused with the normal gene segment. This cell thus produced the adenosine deaminase enzyme in all the transfused cells. These cells were later returned or transplanted in to the patient.

Severe combined immunodeficiency (SCID), is an inherited genetic disease and in such individuals the blood lymphocytes lack enzyme adenosine deaminase (ADA). This is the key enzyme in purine nucleoside metabolism of the lymphocytes and thus could not synthesize the purines. In the absence of active enzyme adenosine deaminase (ADA), deoxyadenosine is phosphorylated to yield high levels of dATPs that are 50-fold greater than normal levels. This high concentration of dATP inhibits ribonucleotide reductase and thus preventing the synthesis of the other deoxynucleoside triphosphate (dNTPs). This choking off DNA synthesis and thus cell proliferation of the lymphocytes. Thus the lymphocytes concentrations will become lesser and lesser. This condition leads to the increased fatal infections by the potential pathogens. Since lymphocytes mediate much of the immune response and fatal infections cannot be avoided due to the lack of the fighting lymphocytes special care must be taken and isolated (protective measures should be taken) to avoid all types of potentially fatal infections. The mutations in all eight known ADA variants obtained from SCID patients appear to structurally perturb the active site of adenosine deaminase.

As mentioned earlier first breakthrough in gene therapy was carried out at the National Institute of Health by three medical professionals namely Dr. W. French Anderson, Dr. Micheal Blaese, and Dr. Kenneth Culver in 1990, September. They treated a four-year-old patient Miss. Ashanthi Desilva in the hospital. They took T lymphocytes and transfected with the normal adenosine deaminase gene. Four months later another patient eleven year old Cynthia Cutshall also underwent the similar treatment. Both the patients had been injected with 11 to 12 gene therapy processes. This process effectively treated the disease because even after three years their T lymphocytes showed positive for adenosine deaminase normal gene.

Four major approaches to gene therapy

In gene therapy different methods are available to correct the defective version of the gene in an individual depending upon how exactly the changes have been made. By using gene therapy approaches large number of diseases are considered as potential source for treatment. There are four major approaches that can be made to obtain the proper function of the gene in the human and they are gene addition, gene replacement, gene switching and gene regulation.

a. **Gene Addition:** This approach involves addition of a normal gene capable of expressing a normal gene products thus restoring a genetically defective function. The pre-existing mutant (errant) gene is not altered or changed and continue to be present in the system.
b. **Gene Replacement:** In this method mutant gene sequence is altered that has an abnormal gene product. Here the specific excision or repair of a mutant gene is done and transduction by normal gene sequence is carried out. The pre-existing abnormal (errant) gene is no longer present in the cells that are used for the transplantation.
c. **Gene Switching:** In this type of gene therapy the establishment of alternative pathways to circumvent mutant gene function is carried out.
d. **Gene Regulation or Alteration:** By this method of gene therapy altering the regulation of normal or mutant gene is carried out. Here the repression or activation of the normal or mutant gene will be done.

Table 1.2: The major diseases that can be cured by the gene therapy approaches

1.	Acute myeloid leukemia	12.	Leukemia
2.	AIDS	13.	Liver cancer
3.	Atherosclerosis	14.	Lung cancer
4.	Breast cancer	15.	Macular degeneration
5.	Cardiovascular disease	16.	Malignant melanoma
6.	Colon cancer	17.	Neuroblastoma
7.	Cystic fibrosis	18.	Osteoporosis
8.	Emphysema	19.	Parkinson's disease
9.	Fabry disease	20.	Renal cell carcinoma
10.	Gaucher disease	21.	SCID
11.	Haemophilia A	22.	Sickle-cell anemia.

Types of Gene Therapy

In human body mainly two types of cells are present. They are somatic cells and germ cells. These cells are used for correcting the diseased genes to function as normal cells. Based on the types of cells used in the therapy processes it can be two types in humans and they are as follow:

1. Germ line gene therapy
2. Somatic gene therapy

1. Germ line gene therapy

The term indicates the introduction of transgene or transgenic cells into the germ line as well as into the somatic cell population in the next generation. This therapy should have ability to cure patient but some gametes also carry corrected version of the gene for proper phenotype. Thus, the entire future generation gets cured from the genetic disease.

The protocol is relevant for application in humans in the removal of an early embryo with a defective genotype from a pregnant woman and injection with transgenic cells containing the wild type gene. These cells become part of many tissues of the body, often including the germ line, which will give rise to the gonads. This therapy is more ethically problematic. This approach alters both the somatic cell and germ cell lines of the individual and ensures that the genetic alterations will be passed onto the individual offspring.

The ethical issues involved in this approach to gene therapy are currently prohibitive but provide a lively area for debate and discussion about the future of gene therapy. No human germ-line gene therapy has been performed to date.

2. Somatic gene therapy

The aim of the therapy is only to focus on the body (soma) or vegetative cells. This is intended to alter the individual genetic constitution without altering the genetic makeup of the children of that individual. Presently, it is impossible to render an entire body transgenic organism. In such cases it is likely that not all the cells of that tissue need to become transgenic that can ameliorate the overall disease symptoms.

The method proceeds by removing some cells from a patient with a defective genotype and making these cells transgenic through the introduction of copies of the cloned wild type gene. The transgenic cells are then reintroduced into the patient's body which in turn providing normal function. In addition, various strategies for implementing somatic cell gene therapy are beginning to emerge and can be grouped under the broad categories of *ex-vivo* gene therapy, *in-vivo* gene therapy and antisense therapy.

The gene therapy can be done at two different levels of individual. The first one is *embryo gene therapy* in this method genetic constitution of embryo is altered after zgyote is formed. The second method is *patient therapy*, in this healthy gene is introduced into affected person tissues. The healthy gene overcomes the defect without affecting the inheritance.

The difference between the germ line therapy and somatic gene therapy are listed in Table 1.3.

Somatic Gene Therapy

With the discovery of the molecular basis of DNA transformation in bacteria in which a gene from one strain can be transferred to and expressed in another strain, human genetic diseases might be cured by introduction of normal gene into the appropriate somatic cell type. During the 1980s, the prospects of human somatic gene therapy techniques were developed, eukaryotic expression vectors were generated. With this transgenic experiments with mice became routine practice.

Table 1.3: Some of the differences involved in the germ line therapy and somatic gene therapy

Somatic Gene Therapy	Germ Line Gene Therapy
1. Gene is introduced into somatic cell	Genes are introduced into germ line and will be distributed in both germ cell and somatic cells
2. No ethical issues attached	Ethical issues to be answered and precludes its uses
3. Technical expertise for somatic cell manipulation is developed	There are still many technical difficulties in this transfer
4. Changes are confined to recipient only	Changes will be passed to further generations
5. Genes are tissues specific in most instances in their expression although not location, specific in many instances, also it may not be possible to achieve normal level and tissue distribution	Genes are not tissues specific in most instances in their expression
6. No mutations observed	High frequency of insertional mutations are observed in this process and cause teratogenic consequences
7. No harm to reproductive process	It is only abnormal embryo that is to be manipulated, and to avoid causing harm to normal potential fetus early diagnosis and therapy is required

This has paved the way for understanding lots of events associated with the human transgenic methodologies. By the year 1990 after extensive and exhaustive reviews by various regulatory bodies, the first human gene therapy trial was initiated. This was done on a young boy suffering from the severe combined immunodeficiency (SCID) disease. This disease results due to absence of an enzyme adenosine deaminase (ADA).

In gene therapy various strategies for implementing somatic cell gene therapy are beginning to emerge and broadly categorized into three types. They are *Ex-vivo* gene therapy, *In-vivo* gene therapy and Anti-sense gene therapy.

Ex-vivo gene therapy

In this category of gene therapy as the name itself suggests the gene therapy is done outside the individual body. The required cells are usually collected from the patient (autologus cells) who requires the therapy. This results in the absence of any adverse immunological responses to the infusion or transplantation of the transformed cells or repaired cells. The cells are generally collected from bone marrow, liver cells, fibroblasts, pulmonary epithelial cells or nerve cells for gene therapy.

In-vivo gene therapy

In-vivo gene therapy as the name itself suggests it entails the direct delivery of a corrected version or remedial gene into the cells of a particular tissue of a prospective patient. This therapy includes various viruses to deliver the remedial gene and determining, whether the administration of pure DNA is sufficient for uptake into the target cells, is just beginning to be examined.

Now-a-days this method of gene therapy is also gaining importance. The large numbers of diseases such as cystic fibrosis, haemophilia are considered under this category of the gene therapy.

Antisense gene therapy

In contrast to the basic tenet of *ex-vivo* and *in-vivo* gene therapies, antisense gene therapy is designed to prevent or at least lower the expression of a specific gene.

Antisense gene therapy follows the introduction of a nucleic acid sequence that is complementary to all or a portion of a specific mRNA into a desired target cell. After this the two RNA sequences, both endogenous mRNA and exogenous antisense RNA are produced by the cells.

Later both mRNA and antisense mRNA are hybridized to form a double stranded RNA molecule. The amount of mRNA available for translation is reduced. Consequently, a smaller amount of the target protein is synthesized. It is also possible that antisense sequence will hybridize to the complementary gene sequence thereby blocking gene expression.

Ex-vivo therapy

The source of cells used for the therapy for genetic manipulation can either be autologous *i.e.*, from the patient itself or from a compatible donor. The autologous source would naturally avoid the problems that may arise with adverse immune responses. The compatible donor may be relative or siblings that is generally identified by the tissue typing procedures. Different cell types that are currently used for *ex-vivo* gene therapy are the totipotent embryonic blood stem cells from the bone marrow, hepatocytes, vascular smooth muscle cells, fibroblasts, pulmonary epithelial cells and tumor infiltrating lymphocytes. The totipotent bone marrow cells are a popular choice for gene therapy because these will differentiate into a number of different cell types in mature blood including B-lymphocytes and T-lymphocytes, macrophages, red blood cells, platelets and osteoclasts.

The *ex-vivo* gene therapy approach involves the following major steps (Fig. 1.1):

i. Collect the cells from the affected patient and culture the cells that are to be manipulated.
ii. Transfer the desired gene in to the cultured cells by using recombinant DNA technology.
iii. Select for the cells that have been genetically transformed or genetically corrected (remedial cells) and expand the cell numbers in culture.
iv. Either infuse or transplant these cells back into the patients.
v. Observe for the functions of introduced cells in the tissues or organs.

The gene transfer technology for *ex-vivo* gene therapy should be efficient and the remedial gene should be both stably maintained and persistently expressed in the transplanted or infused persons. The various vectors are used for such gene transfer protocols. These vectors can be classified as viral vectors and the non-viral vectors. The viral vectors are generally retroviral vectors, adenoviral vectors, adeno-associated viruses and herpes viral vectors. The non-viral vectors include naked DNA microinjection, gene gun method and liposomes.

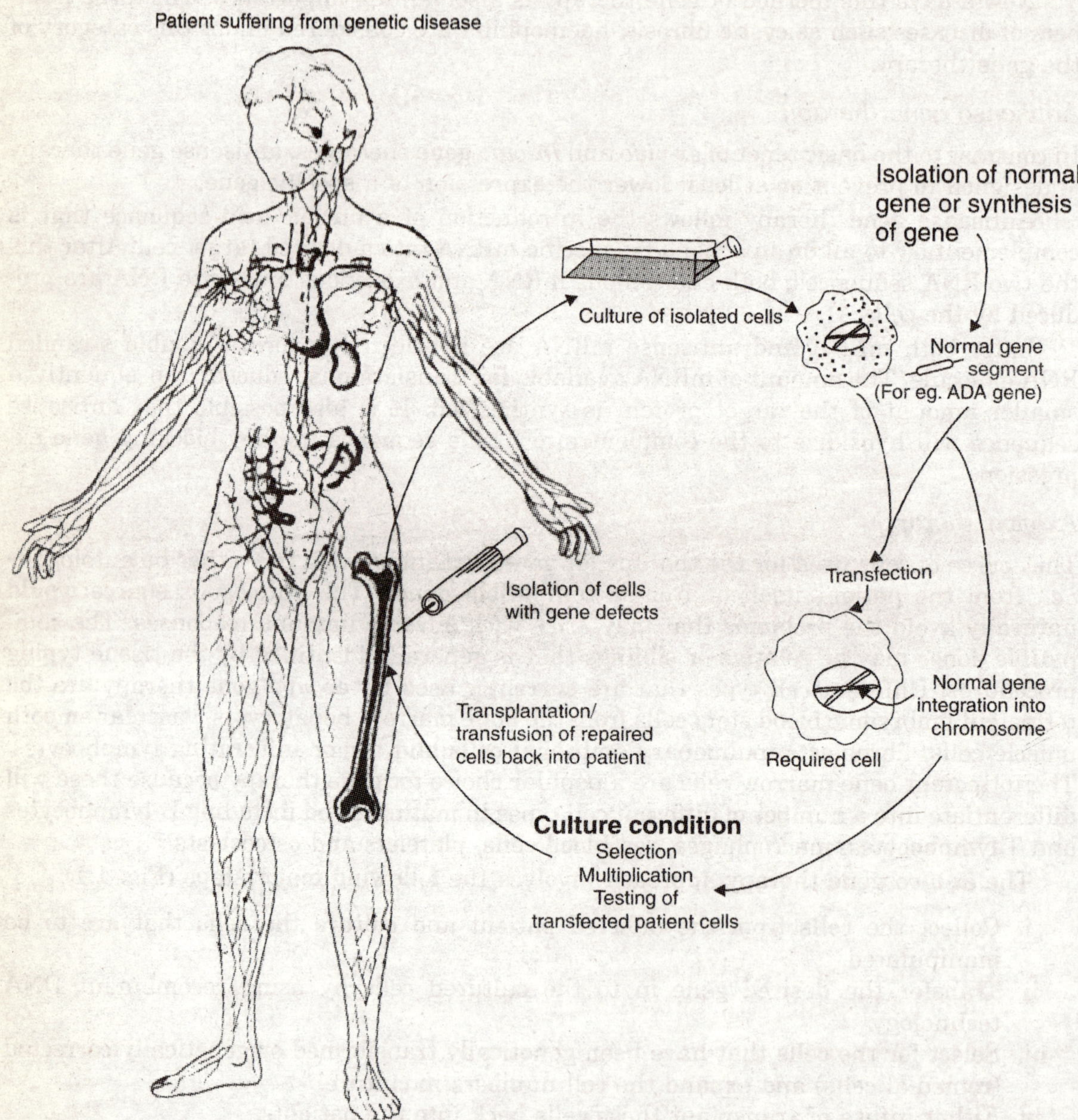

Fig. 1.1. Diagram showing procedure involved in the *ex-vivo* gene therapy in patients suffering from genetic diseases.

Individuals with genetic diseases that respond to the bone marrow transplantation are preferred candidates for *ex-vivo* gene therapy. The reason for that is bone marrow transplantation works as a therapeutic procedure for a diverse set of diseases because the bone marrow contains totipotent stem cells that differentiate into B-lymphocytes, T-lymphocytes, macrophages, red blood cells, platelets, bone cells and others. Thus, for example, when a gene mutation interferes with the proper functioning of macrophages then bone marrow stem cells can be introduced to *ex-vivo* gene therapy and can be easily transplanted back to the patient.

It is important that the gene transfer protocol for *ex-vivo* gene therapy should be efficient so that the remedial gene be both stably maintained and persistently expressed. Currently, for use with human subjects, vectors derived from mouse retroviruses are often used to meet these ends. However, because retroviruses can convert normal cells into cancerous ones, it is essential that the possibility be, at least, diminished and preferably, completely abolished.

Vector system used in gene therapy

Many vector systems are under developing stages for successful usage in the gene therapy. Some of these vectors are listed below:

I. Viral vectors

There are four major types of viral vectors generally used in the gene therapy:

a. Retroviruses
b. Adenoviruses
c. Adeno associated viruses
d. Herpes simplex viruses

II. Non-viral vectors

In this type also there are four major types of non-viral vectors generally used in the gene therapy:

a. Physical methods
Naked DNA injection method
Gene gun method
Microinjection method
Electroporation

b. Chemical methods
Calcium phosphate or dextran or lipids mediated method
Receptor mediated gene delivery

c. Liposome mediated transfer

d. Embryo therapy through *in vitro* fertilization technology

Some of the properties that might be features of the ideal delivery vector include:

1. No limit in the size of the gene that can be incorporated into the vectors.
2. Absence of immunogenicity.
3. Ability to target specific tissue or cell population.
4. Incorporation of elements that limit the expression of delivered genes to specific cell types.
5. Ability to modulate the levels of gene expression in response to exogenous and endogenous signals.
6. Stability and ease of productivity in large quantities at high concentrations.

RETROVIRAL VECTORS

Retroviruses are animal viruses which have the capacity to infect various cell types in animals. Retroviruses are the first type of viruses to be engineered to serve as a gene vector in gene therapy. The major reasons for considering retroviruses to be used in gene therapy are:

a. They can splice copies of their genes permanently into the chromosomes of the cells they invade (as they have integrase enzyme).
b. Retroviruses exhibit reverse transcription, unlike other RNA viruses which do not.

The life cycle of retroviruses includes the following stages of infection to the release of new generation of virus particles:

1. Infection of a target cell.
2. Production of a DNA copy of the RNA genome, using the reverse transcriptase that is carried within the virion particles.
3. Transport of viral DNA in to the nucleus.
4. Integration of the viral genomic DNA into the host genome at a number of different chromosomal sites.
5. Transcription of mRNA from the viral DNA that is driven by a strong promoter within the 5′-LTR (left terminal repeat) sequence.
6. Translation in the cytoplasm of the various viral (**Gag, Pol and Env**) proteins.
7. Formation and filling of the capsid with two RNA strands (as wild type of retroviruses consists of two identical ssRNA strands) and reverse transcriptase molecules.
8. Shedding of single membrane enveloped virions to the exterior of the cell.

After an entire sequence of retroviral DNA is cloned into a plasmid, it can be constructed into a vector. A variety of forms of retroviral vectors have been devised using this basic scheme. Regardless of the format, the upper limit of the amount of cloned DNA that can be carried by a retroviral vector is 8 kb.

Genome nature of retroviruses

The natural wild type of retrovirus particles consists of two identical single stranded RNA (ss RNA) genomes. The each strand of RNA is organized into six regions and they are:

The regions from 5′ end of RNA, the long terminal repeat called **5′ LTR** sequence,

A non-coding sequence that is required for encapsidation of the RNA strand during the virion particles formation and is designated as **psi + (ψ^+)**,

A *gag* gene that encodes for the capsid proteins,

The *pol* gene which codes for the reverse transcriptase and integrase enzymes,

A *env* gene which codes for the envelope proteins,

The **3′-LTR**, the 3' left terminal repeat sequence.

Production of packaged retroviral vector RNA

The cell line was genetically engineered to prepare a packaging cell line. The one genome of viral genome is genetically engineered to carry at one chromosomal site within its genome a **5′-LTR-*gag*-3′LTR** fragment and at other chromosomal site a **5′-LTR-*pol*-*env*-3′LTR.** Both these fragments lack the encapsidation sequence, **psi + (ψ^+)** that is present in the infectious union particles (Fig. 1.1).

In both the chromosomal integrated fragments or regions, the transcription is driven by promoter sequence within 5′LTR.

The packaging cell line thus produced synthesis of viral proteins, but due to lack of encapsidation sequence within either of the retroviral mRNA's the empty viral capsids are produced.

After transfection of a packaging cell line with a retroviral vector, the viral proteins continue to be synthesized, while the full length of RNA from the retrovirus vector that contains full length of genome including encapsidation region, **psi + (ψ^+)**, are transcribed and package the genetically engineered genome into viral capsids.

The released viral particles are replication defective in nature and carry the required remedial or corrective gene (gene X of interest) and a selective marker (Neomycin +, Neo^+) for selection of transgenic.

The therapeutic gene replaces the retroviral *gag*, *pol* and *env* genes. Expression of the therapeutic gene is driven off the viral long terminal repeat (LTR) containing the retroviral packaging signal sequence. Modified viral genome contains the therapeutic gene in place of *gag*, *pol* and *env* viral genes.

Table 1.4: Some of the major genetic diseases that can be treated with *ex-vivo* gene therapy

1. ADA-deficient SCID	14. Lung cancer
2. Adenoleukodystrophy	15. Macular degeneration
3. Chediak-Higachi syndrome	16. Malignant melanoma
4. Chronic granulomatous disease	17. Maroteaux-Lamy syndrome
5. Cystic fibrosis	18. Metachromatic leukodystrophy
6. Fanconi anemia	19. Neuroblastoma
7. Gaucher disease	20. Osteoporosis
8. The gpl-115 deficiency	21. Parkinson's disease
9. Granulocyte actin deficiency	22. Renal cell carcinoma
10. Hunter disease	23. Sickle-cell anemia
12. Hurler syndrome	24. Thalassemia
13. Infantile agranulocytosis	

Adenovirus

Adenovirus is a double-stranded DNA virus that can carry more than 30 kb of insert DNA and has a low pathogenicity in humans. To test this delivering system, muscle cells of mice were injected with an adenovirus vector that carried a reporter gene under the control of muscle specific regulating sequences. The gene product was synthesized for at least 75 days after the initial injection.

Some of the attempts to improve the adenovirus as a therapeutic gene vector include:

1. Reduction of immunogenicity by further mutation of genome to reduce viral protein production.
2. Increasing potency of the virus to reduce the immune challenge.
3. Co-administration of immunosuppressants with the vector therapy.

IN-VIVO GENE THERAPY

In-vivo gene therapy encompasses the direct delivery of a corrected version or remedial gene into the cells of a particular tissue or organs of a suffering patient (Fig. 1.2).

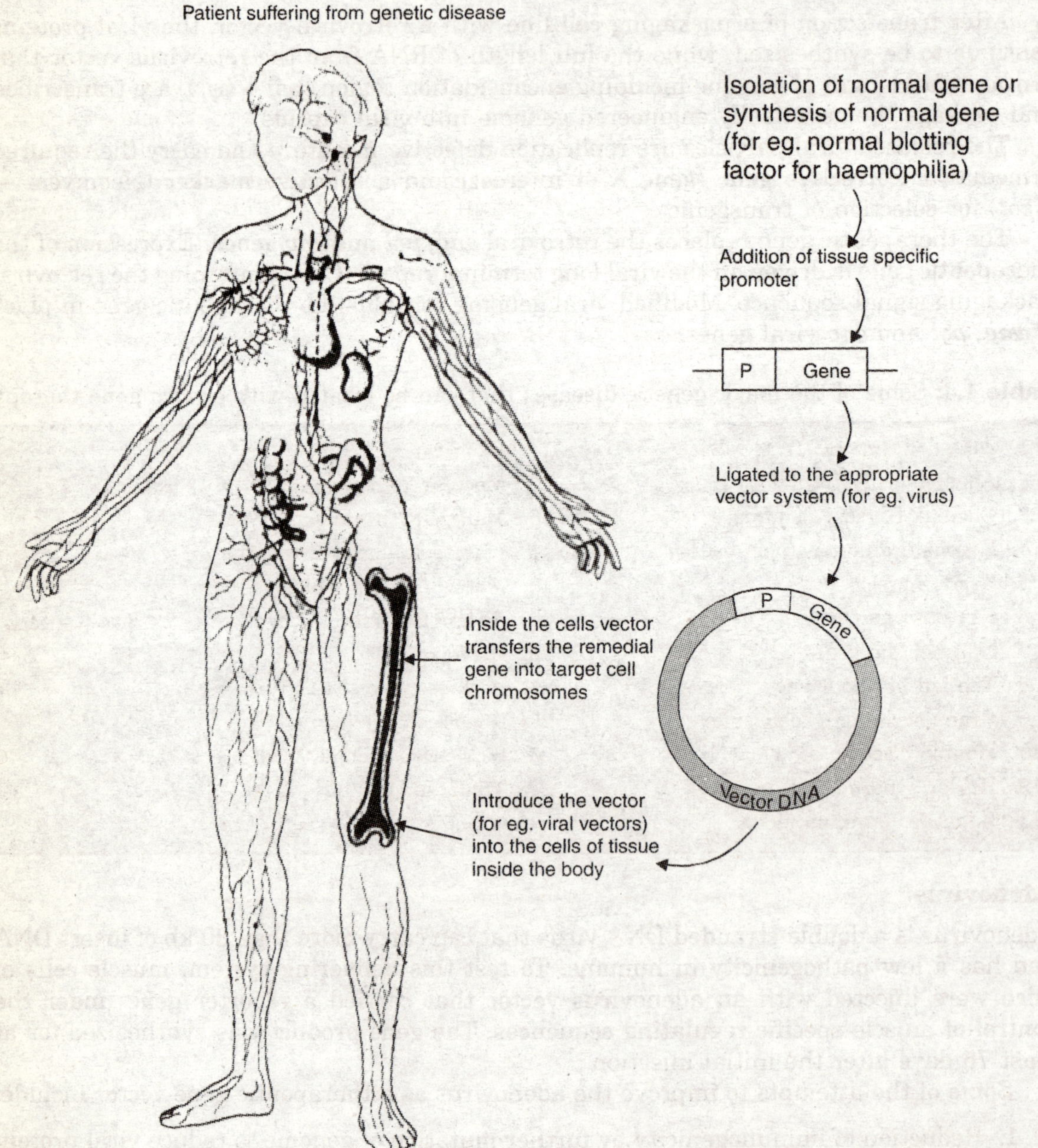

Fig. 1.2. Diagram showing procedure involved in the *in-vivo* gene therapy in patients suffering from genetic diseases.

This therapy includes various viruses to deliver the remedial gene and determining whether the administration of pure DNA is sufficient for uptake into the target cells, is just beginning to be examined. Despite the successful transfer of eukaryotic genes to animal cells using retroviral vectors, plasmid DNA constructs are less time consuming to prepare and easier to collect in a large concentrations. In addition, the use of retroviral vectors requires that the target cells must be dividing in order to be infected. Whereas in many of the tissues that are treated using gene therapy, majority of cells are in the quiescent state. Consequently, alternative strategies to *ex-vivo* gene therapy, which include using various benign viruses to deliver the remedial gene and efforts are made to study the possibility of injection of pure DNA is enough for uptake by the target cells. The lot of procedures and techniques have to be standardized for this method of therapy processes.

As to avoid large number of technical difficulties, now-a-days *in-vivo* method of gene therapy is also gaining much impetus. The large number of diseases such as cystic fibrosis, haemophilia are considered under this category of the gene therapy.

Table 1.5: Some of the major genetic diseases that can be treated with *in-vivo* gene therapy

1. Cystic fibrosis	4. Dystrophin muscular dystropy
2. Cardiovascular disease	5. Familial hypercholesterolemia
3. Cancer	6. Haemophilia

The various vector system that are used for gene therapy has been show in the following Table 1.6:

Table 1.6: Comparison of the properties of various vector system

	Features	Retro Viral	Adeno Viral	AAV	Herpes
1.	Maximum insert size	7–7.5 kb	30 kb	3.5–4 kb	150 kb
2.	Concentrations (Viral particles/ml)	> 10	> 10	> 10	> 10
3.	Route of gene delivery	*ex-vivo*	*ex/in-vivo*	*ex/in-vivo*	*ex/in-vivo*
4.	Integration	yes	no	yes/no	no
5.	Duration of expression *in-vivo*	short	short	long	long
6.	Stability	good	good	good	good
7.	Ease of preparation	pilot scale up to 20-50 litres	easy to scale up	difficult to purify to scale up	–
8.	Immunologic problems	few	extensive	not known	–
9.	Pre-existing host immunity	unlikely	yes	yes	–
10.	Safety problems	insertional mutagenesis	inflammatory response toxicity (IRT)	(IRT)	

Some of the specific barriers to the *in vivo* delivery of DNA have been identified:

1. Rapid degradation of DNA within tissues or blood by nucleases.
2. Limited dispersion of DNA from the site of interstitial administration.
3. The inability of DNA to cross intact basement membrane of the endothelium or epithelium effectively.
4. The rapid clearance of DNA from the vascular compartment by cells of the reticulo endothelial system.
5. The need for effective interaction with the surface of the target cell to induce internalization.
6. Destruction of DNA in the endosomal or lysosomal compartment by nuclease, acid and reducing agents.

ANTISENSE GENE THERAPY

In contrast to the basic types of *ex-vivo* and *in-vivo* gene therapies, antisense gene therapy is designed to block or at least lower the expression of a specific gene that is responsible for disease. As result this blockage results in complete suppression of the diseased gene product in the affected persons or patients.

Antisense gene therapy follows the introduction of a nucleic acid sequences that is complementary to all or a portion of a specific mRNA into a desired target cell. Later cell synthesizes the mRNA from the introduced nucleic acid sequences and self gene sequences. After the two RNA sequences, the endogenous mRNA and exogenous antisense RNA are produced by the cells in normal way of transcription process.

Later both mRNA and antisense mRNA hybridize to form a double stranded RNA molecule. The amount of mRNA available for translation is drastically reduced in the target cell. Consequently a smaller amount of the target protein is synthesized in the cells or tissues. It is also possible that antisense sequence will hybridize to the complementary gene sequence, thereby blocking the transcription of the target mRNA and thus gene expression. In either of the case, the formation of a specific protein (*i.e.* diseased gene product) is significantly reduced within a particular cell type or tissues.

In some types of human genetic diseases and cancers, genes can be either over expressed or unregulated with these consequences, that either high concentrations of a gene product (proteins) or its continued presence in the cell changes the normal functioning of the cell. In such cases the disease state is not resulted because of the abnormal gene product. Thus, the protocols that are developed and designed to provide the normal functional gene would be completely ineffective approach. However, defective regulation might be overcome by controlling the over expression of the protein from the over producing gene.

Antisense technology has been used more often to discover the function of specific gene products by blocking their production. This can be done in two ways:

1. Antisense oligonucleotide strategy
2. Antisense gene strategy

1. Antisense oligonucleotide strategy

In this method an oligonucleotide that has an antisense sequence to a region of a target mRNA is introduced into a cell. Inside the cell introduced antisense oligonucleotide forms the base pairing with the target mRNA due to its complementarities. This completely prevents the target mRNA's translation into its product. Thus, the target product will not be able to show its effect in the cell.

2. Antisense gene strategy

In antisense gene strategy a highly expressible gene is cloned in its reverse orientation and then introduced into a cell. The RNA is transcribed from this construct in a regular transcription process inside the cell. But, the RNA transcribed from this one is not a RNA that carries message for the protein product. Instead it forms an antisense RNA sequence that is complementary to the target normal mRNA. Thus, the translation of the target mRNA is prevented. Though it blocks the translation, antisense RNA does not contain signals for the initiation of translation. Hence, the complete translation process cannot take place in the cell and thus will not produce the target protein.

STRATEGIES FOR GENE THERAPY

In attempting to carry out gene therapy on somatic cells there are several requirements that must be met in order to have even a remote chance of success. The gene for the particular disorder should be isolated along with its regulatory sequences. There must also be an efficient way of targeting the gene to a particular set of cells either *in-vivo* or *ex-vivo*.

Genes may be also being thought of either house keeping genes or tissue specific genes. The house keeping genes are not tightly regulated and are generally expressed in all tissues at all levels of development. Since regulation of gene expression is one of the difficult areas of gene therapy, the house keeping genes would be a simplified approach for gene therapy since regulation is not an important feature of the gene function.

The tissue specific genes are generally tightly regulated and abnormal expression of these genes could lead to major disease problems. The thalassemias are good example of this type of gene in that poorly regulated expression of subunit of hemoglobin results in disruption in the structure of hemoglobin with the ultimate problem of poor oxygen binding to the molecule.

DIAGNOSIS

Before any attempts can be made to correct a genetic disorder, one must first determine which gene responsible for this disorder. In many cases nothing is known about the gene responsible for this disorder. In many cases nothing is known about the gene or the gene product and one must then begin by determining the chromosomal location of the gene. The approximate position of the gene can be determined through positional cloning. Positional cloning, once called reverse genetics, consists of first establishing a linkage between the gene and a genetic marker in the human genome then using linkage to locate the position of the gene. Markers are genes, genome location of which is known and linkage analysis follows the inheritance of both the marker and the gene under study.

If the gene is linked to a particular marker and can be located within approximately two mega bases of the marker, then the region can be searched for coding sequences. A technique called chromosome walking uses DNA probes to find fragments of the region being searched and these fragments then produce new probes that are again used to find new fragments further along the DNA sequence. If a coding sequence is found, the entire gene can be sequenced, the amino acid sequence of the gene product deduced, the function postulated and finally the relationship between the gene and the abnormality estimated by comparing the mutant with a wild type gene. Variations in the DNA sequence that occurs in 1% or more of the population are called polymorphism. Occasionally polymorphism may be found in an enzyme restriction site that is unique to a particular gene of interest in a genetic disorder and the polymorphism is used to determine the presence of the gene by digesting the gene into fragments and analyzing the fragment sizes by southern blotting. These types of polymorphisms are called restriction fragment length polymorphisms (RFLPs) and are very useful because they may be used as markers of the genes in which they arise. Once a gene is related to a particular inherited disorder it is sometimes possible to use the RFLPs to identify the presence of that particular allele in an individual.

The use of RFLPs in a carrier or prenatal diagnosis entails three steps:

1. An appropriate RFLP marker is chosen that is either within or closely linked to the disease lows and for which an individual at risk of carrying the disease is heterozygous.
2. Determine which of the marker alleles is on the chromosome carrying the disease allele.
3. Using the markers the DNA sample is examined to see whether the individual has inherited the chromosome carrying the abnormal phenotype.

A number of genetic diagnostic tests are in the process of being developed and commercial diagnostic kits are becoming available for some tests. With the first phase of the Human Genome Project complete, more genes related to disease are being sequenced by positional cloning and many more tests will become available for carrier and prenatal detection of monogenetic disorders and subject to complications by environmental effects. However the new DNA chips are available or capable of detecting all 6200 genes in the yeast genomes reading frame through the use of 260,00025 bp of oligonucleotide probes. DNA chips should soon have the capacity of analyzing all 30,000–40,000 genes in the human genome.

Genetic polymorphism in the CYP450 drug metabolizing enzymes is also being exploited using the above technologies to tailor drug therapies to genetic makeup. Knowledge of CYP450 polymorphism is becoming important in the prescribing of drug therapy since knowledge of a particular CYP450 polymorphism in a particular patient could prevent serious side effects or wasted, expensive, drug therapy.

GENE DELIVERY

A major problem in gene therapy is similar to the problem encountered in all forms of drug therapy that being the assurance of drug efficacy through efficient delivery of the therapeutic agent to its biologic targets in fully functional form.

Gene delivery is unique in the sense that it is the product of gene function, the protein, not the gene itself that is the therapeutic agent. Hence, we must not only deliver the gene to its proper target and it will produce the therapeutic agent in such a form that it too will be assured of reaching its specific target.

What is the most critical in gene delivery is the effective modulation of the production of the therapeutic protein.

Constitutive expression of some proteins may occur without adversely affecting the therapeutic efficacy. However constitutive expression of most genes is not desirable and can even be life threatening.

The steps involved in the delivery of the therapeutic agent include:

1. *Administration:* Introduction of the nucleic acid (DNA) into the body, which usually involves either *in-vivo* or *in-vitro* techniques.
2. *Delivery:* Delivery of the nucleic acid from the site of administration to the nucleus, which includes directing the bioavailability uptake of the nucleic acid from the cytosol to the nucleus.
3. *Expression:* Expression of the nucleic acid product including the normal steps in gene expression—transcription, translation and post translational modification. Expression is vital to the therapeutic efficacy of the gene in the sense that controlled expression of genes is important to proper biological function. With few exceptions delivery technologies result in continuous gene expression.

BARRIERS TO GENE DELIVERY

The blood brain barrier, connective tissue and epithelial cell linings are particular examples of cellular structure that may impede delivery of DNA to the target cells. Humoral and cellular immune response also pose serious barriers to successful delivery of DNA to target cell.

There are many genetic diseases that passed the Phase-I trials for somatic cell gene therapy and have been listed in the Table 1.7. They have been approved in mid-1990s.

Table 1.7: The somatic cell gene therapies that have been approved in the Phase I trials.

Gene Product	Disease	Target Cell Type
1. Adenosine deaminase	SCID	T cell, Stem cell
2. Tumour necrosis factor	Malignant melanoma	Tumour infiltrating lymphocytes
3. Interleukin-2	Advanced cancer	Tumour cells
4. Factor 1X	Hemophilia B	Fibroblasts
5. LDL receptor	Hypercholesterolemia	Hepatocytes
6. HLA-B7	Malignant melanoma	Melanoma cells

At present some of the diseases are generally considered for the gene therapies and their current therapy protocol are listed in the Table 1.8. These genetic diseases are cured using this therapy method and can get a complete solution to the problem in case of the suffering individuals.

Table 1.8: Current therapies for single gene human genetic disease.

Gene Product	Disease Symptom	Frequency	Current Therapy	Prognosis
1. Adenosine deaminase	SCID Loss of T and B cells	1:1,000,000	Bone marrow transplants, adenosine deaminase replacement	Without therapy fatal by 2 yrs, with therapy clinical improvement
2. Blood clotting factor VIII	Hemophilia altered plasma causes defective blood clotting, chronic internal bleeding into joints, excessive bleeding after wounding	1:10,000	Concentrate of factor VIII by transfusion	Extended life expectancy, requires continual treatment, risk of viral infection from transfusion
3. Phenylalanine hydroxylase	Phenylketonuria excess phenylalanine in the bloodstream of newborns causes mental retardation	1:10,000 males	Restriction of dietary phenylalanine	Good in many cases, if treatment is started early and maintained
4. Cystic fibrosis transmembrane regulator	Cystic fibrosis multisystem diseases, pancreatic insufficiency, in some cases, intestinal blockage, blocked airways of the lungs	1:2,500	Antibiotics physical clearing of the lungs upgrading diet	Fatal by late 20s
5. Dystrophin	Duchenne muscular dystrophy, progressive muscle wasting	1:75,000 males	Only supportive treatments; good nutrition aid in respiratory function, confinement to a wheel chair	Fatal by early 20s

GERM LINE GENE THERAPY

In genetic engineering, changes made only in the germ line can pass on to the next generation. The germ line starts as a fertilized egg or zygote in an individual. This cell divides into a cluster of physiologically identical blastomere which forms a hollow ball called blastocyst. Several days later the dividing cells become differentially committed to form a particular part of the embryo. One group of cells is set aside very early from the embryonic yolk sac to continue the germ line, migrating to the gonads (testes and ovaries) and later form the gametes. The gametes, sperm and eggs pass their genes to the next generations.

After fertilization with sperm the egg divides to form zygote. The zygote divides to form number of blastomeres and on continued cleavage it forms the blastocyst. Both the egg and the blastomeres have attractive features for genetic engineering.

The egg is very large cell, relatively easy to manipulate and inject with DNA in to it. Remarkably, if DNA is injected in to an egg, it will often form stably integrated into one of the chromosomes. Thus, it can be incorporated into all subsequent cells of body. In rare cases it can result in the complications. That is sometimes a segment of DNA introduced into a fertilized egg will not integrate into a chromosome until egg cell divides. In such cases only some of the cells of the embryo will contain the copy of the foreign DNA. The resulting animal will be of mosaic of the modified and unmodified cell types. And also the several copies of the introduced DNA can become integrated into different chromosomes. In this case different genotypes will segregate during subsequent generations.

The blastomere stage is especially convenient for genetic engineering because cells can be grown and manipulated in the *in vitro* conditions. This group of cells at this stage of development is known as embryonic stem (ES) cells. These embryonic stem cells can be propagated indefinitely in the cell culture *in vitro.* One or several embryonic stem cells can be injected in to blastocyst that is obtained by culturing a fertilized egg for several division cycles in Petri dish. This blastocyst is then implanted into a surrogate mother. Both these types of cells can contribute patches of cells to the final adult animal.

The resulting animal is therefore mosaic or chimera—an individual composed of a mixture of genetically engineered cells and normal cells thus genetically they are different cells. Thus, the individual sperm or eggs may be derived from either the embryonic stem cells or those of the fertilized egg.

Germ line gene therapy is still in its infancy stage. This has been important for producing a variety of genetically altered animals for special characters, but for germ line gene therapy has to be standardized for practical purposes.

Genetic technology has evolved over the past 40 yrs to the stage where we are on the verge of determining the entire sequence of the 3 billion base pairs in the human genome. The Human Genome Project (HGP) is developing scientific technology for the manipulation and analysis of human genome. Already relatively new science of Bio-informatics is developing methods for storing and retrieving the vast amounts of information that HGP is generating all over the world.

One can see that the HGP is not only creating a revolution in the medical science but the social, legal, ethical and economic sciences are also being challenged to reevaluate their vision of the social structure and its application. However, gene therapy has considerable amount of promise for treating the approximately 5000 known genetic diseases. Gene therapy approaches are particularly valuable and ultimate for those diseases that are not having alternative treatment methods.

SUMMARY

Gene therapy is a process of introduction of a normal gene into individuals who are suffering from genetic disease because of the abnormal or mutant gene. The normal gene is introduced through the vectors into target cells or tissues. Gene therapy was first performed in 1990 at National Institute of Health, U.S.A. for treating a rare genetic disease known as severe combined immunodeficiency disease caused by absence of adenosine deaminase gene.

Gene therapy can be done by gene addition, gene replacement, gene switching and gene regulation or alteration of specific genes involved in causing the disease in the affected individuals. By all the four types the normal gene will be introduced. Based on the types of cells used for the introduction of normal genes, gene therapy can be of two types viz., germ line gene therapy and somatic gene therapy. In germ line gene therapy, germ cells are used whereas in somatic gene therapy, somatic or vegetative cells are used for gene introduction. The introduced genes will be passed on to their offsprings in case of germ line gene therapy but in case of somatic gene therapy modification is restricted to the affected individuals only.

Somatic gene therapy methods, based on therapy processes, can be categorized into *ex-vivo* therapy, *in-vivo* therapy and antisense gene therapy. Various types of vectors are used in these types of gene therapy. The retroviral vectors, adenoviral vectors, adeno associated viral vectors and herpes simplex viral vectors are commonly employed. Genes can also be introduced through non-viral vectors such as naked DNA injection, gene gun, microinjection, electroporation, calcium phosphate or liposome mediated and *in vitro* fertilization methods. Over 500 known human genetic diseases can be considered for the gene therapy and out of this at present nearly 210 gene therapy protocols are available for use.

EXERCISE

1. Define gene therapy.
2. Explain the disadvantages of the conventional therapy used for the genetic diseases.
3. Discuss in detail types of gene therapy based on types of cells used.
4. Name different types of genetic diseases in humans and add a note on SCID and its gene therapy.
5. Distinguish between germ line gene therapy and somatic gene therapy.
6. Write notes on different types of vectors used in gene therapy.
7. What do you mean by *in-vivo* gene therapy and explain its usage in gene therapy protocol.
8. Give an account on *ex-vivo* gene therapy.
9. Elucidate the procedure involved in antisense gene therapy method.
10. Write an account on future of gene therapy in medicinal field.

2

Stem Cells and Their Applications

2.1 INTRODUCTION

A new era in animal cell biotechnology, stem cell biology began in 1998 with the derivation of cells from human blastocysts and fetal tissue with the unique ability of differentiating into cells of all tissues in the body, i.e., the cells are pluripotent in nature. Since then, several research teams from all over the world have characterized many of the molecular characteristics of these stem cells and improved the methods for culturing them. Even in the time of extraordinary advances made in the prevention, diagnosis, and treatment of human diseases, devastating illnesses such as heart disease, diabetes, cancer, and diseases of the nervous system, such as Parkinson's Disease and Alzheimer's Disease, continue to deprive people of health, independence, and well-being. These genetic diseases affect people throughout the world irrespective of social status, environmental condition and economic status.

Research and understanding of processes in human developmental biology has led to the discovery of human stem cells (precursor cells that can give rise to multiple tissue types), including embryonic stem (ES) cells, embryonic germ (EG) cells, and adult stem cells by the researchers throughout the world. Stem cells differ from other kinds of cells in the body. Recently, techniques have been developed for the *in vitro* culture of stem cells, providing unprecedented opportunities for studying and understanding human embryology and factors that affect the human development. As a result, scientists can now carry out experiments aimed at determining the mechanisms underlying the conversion of a single, undifferentiated cell, the fertilized egg, into the different cells comprising the organs and tissues of the human body. Although it is impossible to predict the outcomes, scientists and the public will gain immense new knowledge in the biology of human development that will likely hold remarkable potential for large numbers of therapies and cures.

The stem cell is a special kind of cell that has a unique capacity to renew itself and to give rise to specialized cell types of the body (more than 220 different known cell types of the human body). It has a multifarious capacity to form different cells depending upon its signal, environment or induction. A stem cell is a relatively unspecialized cell that, when it divides, can do two things: make another cell similar to itself, or make any of a number of cells with more specialized functions. For example, just one kind of stem cell in our blood can make new red blood cells, or white blood cells, or other kinds—depending on what the body needs. These cells are like the stem of a plant that spreads out in different directions as it grows and forms various organs of the plant.

The mature adult human body has 75 trillion number of cells. In an individual most cells of the body, such as blood cells, bone cells, pancreatic cells, heart cells or skin cells, that are committed to conduct a specific function in the body that is, it is said to be of differentiated. But stem cells uncommitted and remain uncommitted throughout, until they receive a specific signal for development into a specialized cell. The *stem* cells are unspecialized and undifferentiated in nature. One of the fundamental properties of a stem cell is that it does not have any tissue-specific structures that allow it to perform specialized functions. A stem cell cannot work with its neighbors to pump blood through the body when it is present in heart (like a heart muscle cell); it cannot carry molecules of oxygen through the bloodstream in the blood (like a red blood cell); and it cannot pass electrochemical signals to other cells that allow the body to move or speak (like a nerve cell). However, unspecialized stem cells can give rise to specialized cells, including heart muscle cells, blood cells, or nerve cells or majority of the cell types of the human body.

Stem cells are capable of dividing and renewing themselves for long periods. Unlike muscle cells, blood cells, or nerve cells that do not normally replicate themselves, but stem cells may replicate many times. When cells replicate themselves many times it is known as **proliferation.** A starting population of stem cells that proliferates for many months in the laboratory can yield millions of cells. If the resulting cells continue to be unspecialized, like the parent stem cells, the cells are said to be capable of long-term self-renewal.

This extraordinary capacity of division along combined with the ability to form specialized cells type makes stem cells unique in the biology. Medicinal surgeons and biologists have tried to replace damaged or diseased cells or tissues which has lost its essential functions in the body by using the stem cells from many years. The damaged or diseased condition makes an individual to suffer from the damage or disease life-long without any cure. But during the last decades development, stem cells have received much focus and attention. In breakthrough investigations researchers in 1998 successfully isolated this class of pluripotent (having capacity to form different cell types) stem cell from early human embryos and established cell lines in culture system. Within few years of their discovery it was evident from various research works that these stem cells are considered to have multifarious applications. These cells have considerable amount of capacity to form almost all of the specialized mature cells of the body. Hence, they have greater potential to form replacement cells for a wide variety of damaged and/or dead tissues and organs, such as the heart, the pancreas, the muscle, and the nervous system. As a result these human stem cells have the great promise in repair or replacements of cells or tissues that are damaged or destroyed or diseased or malfunctioned by many of our most devastating diseases and disabilities. Through this type of repair or replacements the individuals that would have been suffered from these life-long diseases can be cured. The cure is not only permanent but also in the most effective manner.

Thus, large number of researchers started attracted towards stem cells biology. Meanwhile researcher and biologists were beginning to explore human pluripotent stem cells from different sources such as embryos and fetal tissue. During such exhaustive work an entirely new information was emerging about a class of stem cells that have been in clinical use for years and they are generally called adult stem cells. An adult stem cell is an undifferentiated cell that is found in a completely differentiated (specialized) tissue in the adult individual, such as blood, liver. These stem cells can yield the specialized cell types of the tissue from which it originated. In the adult body condition also it has the capacity to renew itself, when it receives the signal.

Research made during the past decade, scientists discovered adult stem cells in tissues that were previously not thought to contain them, such as the brain. In the recent time, they reported that adult stem cells from one tissue appear to be capable of developing into cell types that are characteristic of other tissues. For example, although adult hematopoietic stem cells from bone marrow have long been recognized as capable of developing into blood and immune cells, recently scientists reported that under certain conditions, the same stem cells could also develop into cells that have many of the characteristics of neurons. So, a new concept and a new term emerged—adult stem cell plasticity. This also helps in adopting the better therapy process in highly effective way.

The human adult and embryonic stem cells equivalent are having potential for generating replacement cells and tissues. Current science knowledge indicates that, although both of these cell types hold enormous promise, adult and embryonic stem cells differ in important ways. What is not known is the extent to which these different cell types will be useful for the development of cell-based therapies to treat disease and other accidental alignments.

Derivation of ES cells from early human embryos, and embryonic germ cells and fetal stem cells from aborted, fetal tissues raise ethical, legal, religious, and policy questions. Further, the potential uses of stem cells for generating human tissues and, perhaps, organs, is a subject of ongoing public debate. The large number of debate is underway in many developed and developing countries throughout the world.

Definitions of commonly used terms in stem cells biology

Stem cell. A stem cell is a cell from the embryo, fetus, or adult that has, under certain conditions, the ability to reproduce itself for indefinite periods. But in the case of adult stem cells it can do so for throughout the life of the organism in natural condition. Stem cells can also give rise to specialized cells that make up the tissues and organs of the body *i.e.*, nearly 220 cell types that make up the entire human body.

These cells are also known as magic seeds. Information regarding the basic understanding about embryonic stem cells has come from animal research. Stem cells in the laboratory condition can proliferate indefinitely, a characteristic that is not observed in the adult stem cells.

Pluripotent stem cell. A single pluripotent (pluri means many) stem cell has the ability to give rise to types of cells that develop from the three germ layers (mesoderm, endoderm, and ectoderm) from which all the cells of the body arise. The only known sources of human pluripotent stem cells are those isolated and cultured from early human embryos and from fetal tissue that was destined to be part of the gonads.

Embryonic stem cell. An embryonic stem cell is derived from a group of cells called the inner cell mass, which is part of the early (4-to 5-day) embryo called the blastocyst. Once removed from the blastocyst, the cells of the inner cell mass can be cultured into embryonic stem cells. These embryonic stem cells are not themselves embryos. In fact, evidence is emerging that these cells do not behave in the laboratory as they would in the developing embryo—that is, the conditions in which these cells develop in culture are likely to differ from those in the developing embryo.

Embryonic Germ Cell. An embryonic germ cell is derived from fetal tissue. Specifically, they are isolated from the primordial germ cells of the gonadal ridge of the 5- to 10-week fetus. Later in development, the gonadal ridge develops into the testes or ovaries and the primordial germ cells give rise to eggs or sperm. Embryonic stem cells and embryonic germ cells are pluripotent, but they are not identical in their properties and characteristics.

Differentiation. Differentiation is the process by which an unspecialized cell (such as a stem cell) becomes specialized into one of the many cells that make up the body. During differentiation, certain genes become activated and other genes become inactivated in an intricately regulated fashion. As a result, a differentiated cell develops specific structures and performs certain functions.

For example, a mature, differentiated nerve cell has thin, fiber-like projections that send and receive the electrochemical signals that permit the nerve cell to communicate with other nerve cells. In the laboratory, a stem cell can be manipulated to become specialized or partially specialized cell types (*e.g.*, heart muscle, nerve, or pancreatic cells) and this is known as directed differentiation.

Adult stem cell. An adult stem cell is an undifferentiated (unspecialized) cell that occurs in a differentiated (specialized) tissue, renews itself, and becomes specialized to yield all of the specialized cell types of the tissue from which it originated. Adult stem cells are capable of making identical copies of themselves for the lifetime of the organism. This property is referred to as "self-renewal". Adult stem cells usually divide to generate progenitor or precursor cells, which then differentiate or develop into "mature" cell types that have characteristic shapes and specialized functions, *e.g.*, muscle cell contraction or nerve cell signaling.

Sources of adult stem cells include bone marrow, blood, the cornea and the retina of the eye, brain, skeletal muscle, dental pulp, liver, skin, the lining of the gastrointestinal tract, and pancreas. The most abundant information about adult human stem cells comes from studies of hematopoietic (blood-forming) stem cells isolated from the bone marrow and blood. These adult stem cells have been extensively studied and applied therapeutically for various diseases treatment. At this point, there is no isolated population of adult stem cells that is capable of forming all the kinds of cells of the body. Adult stem cells are rare. Often they are difficult to identify, isolate, and purify. There are insufficient numbers of cells available for transplantation and adult stem cells do not replicate indefinitely in culture. This makes its wide applications in different fields.

Plasticity. Plasticity is the ability of an adult stem cell from one tissue to generate the specialized cell type(s) of another tissue. A recently reported example of plasticity is that, under specific experimental conditions, adult stem cells from bone marrow generated cells that resemble neurons and other cell types that are commonly found in the brain. The concept of adult stem cell plasticity is new, and the phenomenon is not thoroughly understood. Evidence suggests that, given the right environment, some adult stem cells are capable of being "genetically reprogrammed" to generate specialized cells that are characteristic of different tissues.

Clonality or clonally derived stem cell. A cell is said to be clonally derived or to exhibit clonality if it was generated by the division of a single cell and is genetically identical to that cell. In stem cell research, the concept of clonality is important for several reasons. For researchers to fully understand and harness the ability of stem cells to generate replacement cells and tissues, the exact identity of those cells' genetic capabilities and functional qualities must be known. Human pluripotent stem cells from embryos and fetal tissue are by their nature clonally derived. However, very few studies have shown clonal properties of the cells that are developed from adult stem cells. It is crucial to know whether a single cell is capable of developing an array of cell types, or whether multiple stem cell types, that when grown together, are capable of forming multiple cell types.

For instance, recent research has shown that a mixture of cells removed from fat tissue or umbilical cord blood are capable of developing into blood cells, bone cells, and perhaps others. Researchers have not shown that a single cell is responsible for giving rise to other cell types or, if so, what kind of cell it is. These results may well be attributable to multiple types of precursor cells in the starting tissue; such results from fat cells may, in fact, be due to the presence of hematopoietic stem cells in the fat tissue. The importance of showing that one cell type can reproducibly become another and self-replicate cannot be overemphasized.

Progenitor or precursor cell. A progenitor or precursor cell occurs in fetal or adult tissues and is partially specialized; it divides and gives rise to differentiated cells. Researchers often distinguish precursor/progenitor cells from adult stem cells in the following way: when a stem cell divides, one of the two new cells is often a stem cell capable of replicating itself again. In contrast, when a progenitor/precursor cell divides, it can form more progenitor/precursor cells or it can form two specialized cells, neither of which is capable of replicating itself. Progenitor/precursor cells can replace cells that are damaged or dead, thus maintaining the integrity and functions of a tissue such as liver or brain. Progenitor/precursor cells give rise to related types of cells—lymphocytes such as T cells, B cells, and natural killer cells, for example—but in their normal state do not generate a wide variety of cell types.

Totipotent stem cells. The stem cells that can give rise to a fully functional mature organism as well as to every cell type of the body. In animals this type of development is not completely understood.

Pluripotent stem cells. The stem cells which are capable of giving rise to virtually any tissue type or many tissue types, but not to a functioning mature complete organism.

Multipotent stem cells. The stem cells that are more differentiated cells (that is, their possible lineages are less plastic/more determined) and thus can give rise only to a limited number of tissue types.

Human Stem Cell Research birth and evolution

Most cells within an animal or human being are committed to fulfilling a single function within the body. In contrast, stem cells are a unique and important set of cells that are not specialized. Stem cells retain the ability to become many or all of the different cell types in the body and thereby play a critical role in repairing organs and body tissues throughout life. Although the term stem cells is often used in reference to these repair cells within an adult organism, a more fundamental variety of stem cells is found in the early stage embryc. Embryonic stem cells may have a greater ability to become different types of body cells than adult stem cells. It is important to understand some of the difficulties that researchers have faced in isolating various types of stem cells, working with the cells in the laboratory, and proving experimentally that the cells are true stem cells. Most of the basic research discoveries on embryonic and adult stem cells come from research using animal models, particularly mice.

During early part of 1981, researchers reported methods for growing mouse embryonic stem cells in the laboratory, and it took nearly about 20 years before similar achievements could be made with human embryonic stem cells. Animal embryos were the only source for research on embryonic stem cells until November 1998. Much of the knowledge obtained in embryonic stem cells has emerged from the efforts made in the two fields of applied research: applied reproductive biology, *i.e.*, *in vitro* fertilization technologies, and basic research on mouse embryology.

There have been many technical challenges that have been overcome in adult stem cell research as well. Some of the barriers include: the rare occurrence of adult stem cells among other, differentiated cells, difficulties in isolating and identifying the cells (researchers often use molecular "markers" to identify adult stem cells), and in many cases, difficulties in growing adult stem cells in tissue culture. Much of the research demonstrating the plasticity of adult stem cells comes from studies of animal models in which a mixture of adult stem cells from a donor animal is injected into another animal, and the development of new, specialized cells is traced.

The two independent groups of U.S. scientists announced the successful isolation of human embryonic stem cells. One group, at the University of Wisconsin, derived stem cells from one-week-old embryos produced via *in vitro* fertilization (IVF). The work is controversial, in the opinion of some individuals, because the stem cells are located within the embryo and the process of removing them destroys the embryo. The second group, at Johns Hopkins University, derived cells with very similar properties from five- to nine week old embryos or fetuses obtained through elective abortion.

Professor James Thomson in 1998, at the University of Wisconsin-Madison isolated cells from the inner cell mass of the early embryo, called the blastocyst, and developed the first human embryonic stem cell lines. During the same period Professor **John Gearhart** at Johns Hopkins University reported the first derivation of human embryonic germ cells from an isolated population of cells in fetal gonadal tissue, known as the primordial germ cells, which are destined to become the eggs and sperm. From both of these sources, the researchers developed pluripotent stem cell "lines", which are capable of renewing themselves for long periods and giving rise to many types of human cells or tissues. Human embryonic stem cells and embryonic germ cells differ in some characteristics, however, and do not appear to be equivalent.

Human Embryonic Stem Cell Research origin and its path

Even though stem cells were identified in 1998 but its exact research orientation started in as early as 1878 when first mammalian egg was used for the fertilization purpose. This showed that cells potency determination is possible in the animals.

- **1878:** First reported attempts to fertilize mammalian eggs outside the body.
- **1959:** First report of animals (rabbits) produced through IVF in the United States.
- **1960s:** Studies of teratocarcinomas in the testes of several inbred strains of mice indicates they originated from embryonic germ cells. The work establishes embryonal carcinoma (EC) cells as a kind of stem cell. For a more detailed discussion of human embryonal carcinoma cells.
- **1968:** Edwards and Bavister fertilize the first human egg *in vitro.*
- **1970s:** EC cells injected into mouse blastocysts produce chimeric mice. Cultured SC cells are explored as models of embryonic development, although their complement of chromosomes is abnormal.
- **1978:** Louise Brown, the first IVF baby, is born in England.
- **1980:** Australia's first IVF baby, Candace Reed, is born in Melbourne.
- **1981:** Evans and Kaufman, and Martin derive mouse embryonic stem (ES) cells from the inner cell mass of blastocysts. They establish culture conditions for growing pluripotent mouse ES cells *in vitro.* The ES cells yield cell lines with normal, diploid karyotypes and generate derivatives of all three primary germ layers as well as primordial germ cells. Injecting the ES cells into mice induces the formation of teratomas. The first IVF baby, Elizabeth Carr, is born in the United States.
- **1984-88:** Andrews et al., develop pluripotent, genetically identical (clonal) cells called embryonal carcinoma (EC) cells from Tera-2, a cell line of human testicular teratocarcinoma. Cloned human teratoma cells exposed to retinoic acid differentiate into neuron-like cells and other cell types.
- **1989:** Pera et al., derive a clonal line of human embryonal carcinoma cells, which yields tissues from all three primary germ layers. The cells are aneuploid (fewer or greater than the normal number of chromosomes in the cell) and their potential to differentiate spontaneously *in vitro* is typically limited. The behavior of human EC cell clones differs from that of mouse ES or EC cells.
- **1994:** Human blastocysts created for reproductive purposes using IVF and donated by patients for research, are generated from the 2-pronuclear stage. The inner cell mass of the blastocyst is maintained in culture and generates aggregates with trophoblast-like cells at the periphery and ES-like cells in the center. The cells retain a complete set of chromosomes (normal karyotype); most cultures retain a stem cell-like morphology, although some inner cell mass clumps differentiate into fibroblasts. The cultures are maintained for two passages.
- **1995-96:** Non-human primate ES cells are derived and maintained *in vitro,* first from the inner cell mass of rhesus monkeys, and then from marmosets. The primate ES cells are diploid and have normal karyotypes. They are pluripotent and differentiate into cells types derived from all three primary germ layers. The primate ES cells resemble human EC cells and indicate that it should be possible to derive and maintain human ES cells *in vitro.*
- **1998:** Thomson et al., derive human ES cells from the inner cell mass of normal human blastocysts donated by couples undergoing treatment for infertility. The cells are cultured through many passages, retain their normal karyotypes, main-

tain high levels of telomerase activity, and express a panel of markers typical of human EC cells non-human primate ES cells. Several (non-clonal) cell lines are established that form teratomas when injected into immune-deficient mice. The teratomas include cell types derived from all three primary germ layers, demonstrating the pluripotency of human ES cells. Gearhart and colleagues derive human embryonic germ (EG) cells from the gonadal ridge and mesenchyma of 5- to 9-week fetal tissue that resulted from elective abortions. They grow EG cells *in vitro* for approximately 20 passages, and the cells maintain normal karyotypes. The cells spontaneously form aggregates that differentiate spontaneously, and ultimately contain derivatives of all three primary germ layers. Other indications of their pluripotency include the expression of a panel of markers typical of mouse ES and EG cells. The EG cells do not form teratomas when injected into immune-deficient mice.

- **2000:** Scientists in Singapore and Australia led by Pera, Trounson, and Bongso derive human ES cells from the inner cell mass of blastocysts donated by couples undergoing treatment for infertility. The ES cells proliferate for extended periods *in vitro*, maintain normal karyotypes, differentiate spontaneously into somatic cell lineages derived from all three primary germ layers, and form teratomas when injected into immune-deficient mice.
- **2001:** As human ES cell lines are shared and new lines are derived, more research groups report methods to direct the differentiation of the cells *in vitro*. Many of the methods are aimed at generating human tissues for transplantation purposes, including pancreatic islet cells, neurons that release dopamine, and cardiac muscle cells.

Sources of Stem Cells

The major possible sources of stem cells are

- embryos created via IVF (for infertility treatment or for research purposes)
- embryos or fetuses obtained through elective abortion
- embryos created via SCNT (somatic cell nuclear transfer, or cloning)
- adult tissues (bone marrow, umbilical cord blood)

Bone Marrow

The classic source of hematopoietic stem cells (HSCs) is bone marrow. For more than 40 years, doctors performed bone marrow transplants by anesthetizing the stem cell donor, puncturing a bone—typically a hipbone—and drawing out the bone marrow cells with a syringe. About 1 in every 100,000 cells in the marrow is a long-term, blood-forming stem cell; other cells present include stromal cells, stromal stem cells, blood progenitor cells, and mature and maturing white and red blood cells.

Peripheral Blood

As a source of HSCs for medical treatments, bone marrow retrieval directly from bone is quickly fading into history. For clinical transplantation of human HSCs, doctors now prefer to harvest donor cells from peripheral, circulating blood. It has been known for decades that a small number of stem and progenitor cells circulate in the bloodstream, but in the past 10 years, researchers have found that they can coax the cells to migrate

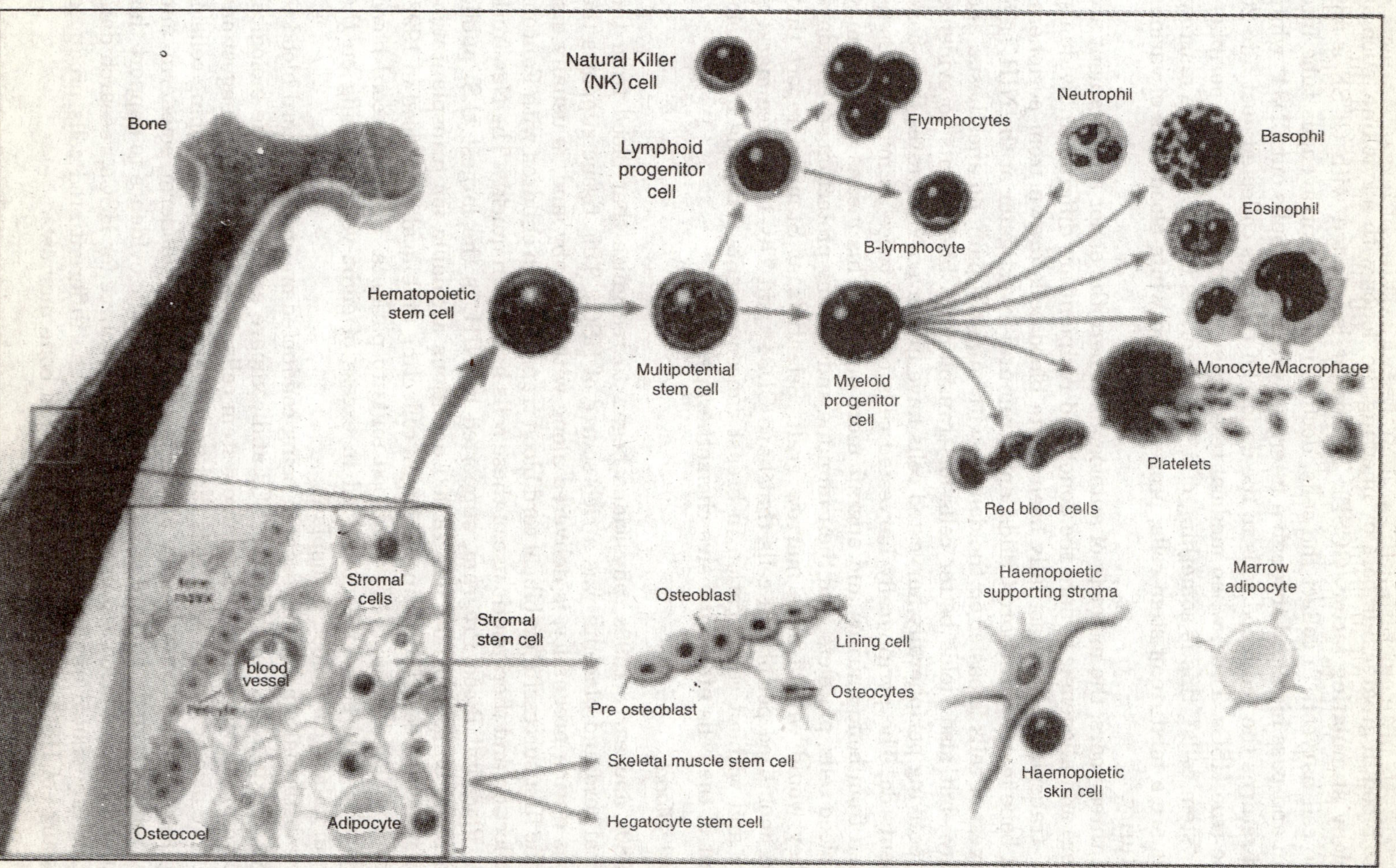

Fig. 2.1. Hematopoietic and stromal stem cell differentiation.

from marrow to blood in greater numbers by injecting the donor with a cytokine, such as granulocyte-colony stimulating factor (GCSF). The donor is injected with GCSF a few days before the cell harvest. To collect the cells, doctors insert an intravenous tube into the donor's vein and pass his blood through a filtering system that pulls out $CD34^+$ white blood cells and returns the red blood cells to the donor. Of the cells collected, just 5 to 20 per cent will be true HSCs. Thus, when medical researchers commonly refer to peripherally harvested "stem cells", this is something of a misnomer. As is true for bone marrow, the $CD34^+$ cells are a mixture of stem cells, progenitors, and white blood cells of various degrees of maturity.

In the past three years, the majority of autologous (where the donor and recipient are the same person) and allogeneic (where the donor and recipient are different individuals) "bone marrow" transplants have actually been white blood cells drawn from peripheral circulation, not bone marrow. Richard Childs, an intramural investigator at the NIH, says peripheral harvest of cells is easier on the donor—with minimal pain, no anesthesia, and no hospital stay—but also yields better cells for transplants. Childs points to evidence that patients receiving peripherally harvested cells have higher survival rates than bone marrow recipients do. The peripherally harvested cells contain twice as many HSCs as stem cells taken from bone marrow and engraft more quickly. This means patients may recover white blood cells, platelets, and their immune and clotting protection several days faster than they would with a bone marrow graft. Scientists at Stanford report that highly purified, mobilized peripheral cells that have $CD34^+$ and $Thy\text{-}1^+$ surface markers engraft swiftly and without complication in breast cancer patients receiving an autologous transplant of the cells after intensive chemotherapy.

Umbilical Cord Blood

In the late 1980s and early 1990s, physicians began to recognize that blood from the human umbilical cord and placenta was a rich source of HSCs. This tissue supports the developing fetus during pregnancy, is delivered along with the baby, and, is usually discarded. Since the first successful umbilical cord blood transplants in children with Fanconi anemia, the collection and therapeutic use of these cells has grown quickly. The New York Blood Center's Placental Blood Program, supported by NIH, is the largest U.S. public umbilical cord blood bank and now has 13,000 donations available for transplantation into small patients who need HSCs. Since it began collecting umbilical cord blood in 1992, the center has provided thousands of cord blood units to patients. Umbilical cord blood recipients—typically children—have now lived in excess of eight years, relying on the HSCs from an umbilical cord blood transplant.

There is a substantial amount of research being conducted on umbilical cord blood to search for ways to expand the number of HSCs and compare and contrast the biological properties of cord blood with adult bone marrow stem cells. There have been suggestions that umbilical cord blood contains stem cells that have the capability of developing cells of multiple germ layers (multipotent) or even all germ layers, *e.g.*, endoderm, ectoderm, and mesoderm (pluripotent). To date, there is no published scientific evidence to support this claim. While umbilical cord blood represents a valuable resource for HSCs, research data have not conclusively shown qualitative differences in the differentiated cells produced between this source of HSCs and peripheral blood and bone marrow.

Fetal Hematopoietic System

An important source of HSCs in research, but not in clinical use, is the developing blood-producing tissues of fetal animals. Hematopoietic cells appear early in the development of all vertebrates. Most extensively studied in the mouse, HSC production sweeps through the developing embryo and fetus in waves. Beginning at about day 7 in the life of the mouse embryo, the earliest hematopoietic activity is indicated by the appearance of blood islands in the yolk sac. The point is disputed, but some scientists contend that yolk sac blood production is transient and will generate some blood cells for the embryo, but probably not the bulk of the HSCs for the adult animal. According to this proposed scenario, most stem cells that will be found in the adult bone marrow and circulation are derived from cells that appear slightly later and in a different location. This other wave of hematopoietic stem cell production occurs in the AGM—the region where the aorta, gonads, and fetal kidney (mesonephros) begin to develop. The cells that give rise to the HSCs in the AGM may also give rise to endothelial cells that line blood vessels. These HSCs arise at around 10 to 11 days in the mouse embryo (weeks 4 to 6 in human gestation), divide, and within a couple of days, migrate to the liver. The HSCs in the liver continue to divide and migrate, spreading to the spleen, thymus, and—near the time of birth—to the bone marrow.

Whereas an increasing body of fetal HSC research is emerging from mice and other animals, there is much less information about human fetal and embryonic HSCs. Scientists in Europe, including Coulombel, Peault, and colleagues, first described hematopoietic precursors in human embryos only a few years ago. Most recently, Gallacher and others reported finding HSCs circulating in the blood of 12- to 18-week aborted human fetuses that was rich in HSCs. These circulating cells had different markers than did cells from fetal liver, fetal bone marrow, or umbilical cord blood.

The Stem Cell Database

Ihor Lemischka and colleagues at Princeton University and the Computational Biology and Informatics Laboratory at the University of Pennsylvania are collaborating to record all the findings about hematopoietic stem cell (HSC) genes and markers in the Stem Cell Database.

The collaborators started the database five years ago. Its goal is listing and annotating all the genes that are differentially expressed in mouse liver HSCs and their cellular progeny. The database is growing to include human HSCs from different blood sources, and a related database, constructed in collaboration with Kateri A. Moore, also at Princeton University, will document all genes active in stromal cells, which provide the microenvironment in which stem cells are maintained. The combined power of the two databases, along with new tools and methods for studying molecular biology, will help researchers put together a complete portrait of the hematopoietic stem cell and how it works. The databases will continue to grow and take advantage of other efforts, such as those to complete the gene sequences of mammals. Data will be publicly available to researchers around the world.

The factors which involve in making of stem cells and the long-term proliferation ability and pluripotency of embryonic stem cells and embryonic germ cells are very important. First, for basic research purposes, it is important to understand the genetic and molecular basis by which these cells continue to make many copies of themselves over long periods of time. Second, if the cells are to be manipulated and used for transplantation, it is important to have sufficient quantities of cells that can be directed to differentiate into the desired cell type(s) and used to treat many patients suffering from a particular disease.

In recent months, other investigators have been successful in using somewhat different approaches to deriving human pluripotent stem cells. At least 5 other laboratories have been successful in deriving pluripotent stem cells from human embryos and one additional laboratory has created cell lines from fetal tissue. In each case, the methods for deriving pluripotent stem cells from human embryos and embryonic germ cells from fetal tissue are similar, yet they differ in the isolation and culture conditions as initially described by Thomson and Gearhart, respectively. It is not known to what extent U.S.-based researchers are using these additional sources of embryonic stem and germ cells.

At present, there have been multiple human adult stem cell lines that have been created through a combination of public and private resources (*e.g.*, hematopoietic stem cells). Substantial adult stem cell research has been underway for many years, and in recent years this has included basic studies on the "plasticity" of such cells.

Human Pluripotent Stem Cells and their Characters

Since 1998, research teams have refined the techniques for growing human pluripotent cells in culture systems. Collectively, the studies indicate that it is now possible to grow these cells for up to two years in a chemically defined medium.

The cell lines have been shown to have a normal number of chromosomes and they generate cell types that originate from all three primary germ layers.

Cultures of human pluripotent stem cells have active telomerase, which is an enzyme that maintains the length of telomeres and is important for cells to maintain their capacity to replicate. Human pluripotent stem cells appear to maintain relatively long telomeres, indicating that they have the ability to replicate for many, many generations.

Evidence of structural, genetic, and functional cells characteristic of specialized cells developed from cultured human and mouse embryonic stem cells has been shown for:

1. pancreatic islet-cell-like cells that secrete insulin (mouse and human);
2. cardiac muscle cells with contractile activity (mouse and human);
3. blood cells (human and mouse);
4. nerve cells that produce certain brain chemicals (mouse).

Up to date, there are approximately more than 40 cell lines of human pluripotent stem cells that have been derived from human blastocysts or fetal tissue.

Overall, it appears human embryonic cells and embryonic germ cells are not equivalent in their potential to proliferate or differentiate.

Stem Cells and its Mysteries

- What are the mechanisms that allow human embryonic stem cells and embryonic germ cells to proliferate *in vitro* without differentiating?
- What are the intrinsic controls that keep stem cells from differentiating?
- Is there a universal stem cell? That is, could a kind of stem cell exist (possibly circulating in the blood) that can generate the cells of any organ or tissue?

- Do adult stem cells exhibit plasticity as a normal event in the body or is it an artifact of the culture conditions? If plasticity occurs normally, is it a characteristic of all adult stem cells? What are the signals that regulate the proliferation and differentiation of stem cells that demonstrate plasticity?
- What are the factors responsible for stem cells to "home" to sites of injury or damage?
- What are the intrinsic controls that direct stem cells along a particular differentiation pathway to form one specialized cell over another? How are such intrinsic regulators, in turn, influenced by the microenvironment, or niche, where stem cells normally reside?
- Will the knowledge about the genetic mechanisms regulating the specialization of embryonic cells into cells from all embryonic germ layers during development enable the scientists to engineer adult stem cells to do the same?
- What are the sources of adult stem cells in the body? Are they "leftover" embryonic stem cells, or do they arise in some other way? And if the latter is true—which seems to be the case—exactly how do adult stem cells arise, and why do they remain in an undifferentiated state when all the cells around them have differentiated?
- How many kinds of adult stem cells exist, and in which tissues do they exist?
- Is it possible to manipulate adult stem cells to increase their ability to proliferate in a culture dish so that adult stem cells can be used as a sufficient source of tissue for transplants?
- Does the genetic programming status of stem cells play a significant role in maintaining the cells, directing their differentiation, or determining their suitability for transplant?
- Are the human embryonic stem and germ cells that appear to be homogeneous and undifferentiated in culture, in fact, homogeneous and undifferentiated? Or are they heterogeneous and/or "partially" differentiated?
- What are the cellular and molecular signals that are important in activating a human pluripotent stem cell to begin differentiating into a specialized cell type?
- Will analysis of genes from human pluripotent stem cells reveal a common mechanism that maintains cells in an undifferentiated state?
- Do all pluripotent stem cells pass through a progenitor/precursor cell stage while becoming specialized? If so, can a precursor or progenitor cell stage be maintained as optimal cells for therapeutic transplantation?
- What stage of differentiation of stem cells will be best for transplantation? Would the same stage be optimal for all transplantation applications, or will it differ on a case-by-case basis?
- What differentiation stages of stem cells would be best for screening drugs or toxins, or for delivering potentially therapeutic drugs?

Adult Stem Cells and their Characters

- To date, published scientific papers indicate that adult stem cells have been identified in brain, bone marrow, peripheral blood, blood vessels, skeletal muscle, epithelia of the skin and digestive system, cornea, dental pulp of the tooth, retina, liver, and pancreas. Thus, adult stem cells have been found in tissues that develop from all three embryonic germ layers.

- There is no evidence of an adult stem cell that is pluripotent. It has not been demonstrated that one adult stem cell can be directed to develop into any cell type of the body. That is, no adult stem cell has been shown to be capable of developing into cells from all three embryonic germ layers.
- In the body, adult stem cells can proliferate without differentiating for a long period (the characteristic referred to as long-term self-renewal), and they can give rise to mature cell types that have characteristic shapes and specialized functions of a particular tissue.
- Adult stem cells are rare. Often they are difficult to identify, isolate, and purify.
- One important, limiting factor for the use of adult stem cells in future cell-replacement strategies is that there are insufficient numbers of cells available for transplantation. This is because most adult stem cell lines when grown in a culture dish are unable to proliferate in an unspecialized state for long periods of time. In cases where they can be grown under these conditions, researchers have not been able to direct them to become specialized as functionally useful cells.
- Stem cells from the bone marrow are the most-studied type of adult stem cells. Currently, they are used clinically to restore various blood and immune components to the bone marrow via transplantation. There are two major types of stem cells found in bone: hematopoietic stem cells which form blood and immune cells, and stromal (mesenchymal) stem cells that normally form bone, cartilage, and fat. The restricted capacity of hematopoietic stem cells to grow in large numbers and remain undifferentiated in the culture dish is a major limitation to their broader use for research and transplantation studies. Researchers have reported that at least two other populations of adult stem cells occur in bone marrow and blood, but these cells are not well characterized.
- Evidence to date indicates that umbilical cord blood is an abundant source of hematopoietic stem cells. There do not appear to be any qualitative differences between the stem cells obtained from umbilical cord blood and those obtained from bone marrow or peripheral blood.
- Several populations of adult stem cells have been identified in the brain, particularly in a region important in memory, known as the hippocampus. Their function in the brain is unknown. When the cells are removed from the brain of mice and grown in tissue culture, their proliferation and differentiation can be influenced by various growth factors.
- Current methods for characterizing adult stem cells depend on determining cell-surface markers and making observations about their differentiation patterns in culture dishes.
- Some adult stem cells appear to have the capability to differentiate into tissues other than the ones from which they originated; this is referred to as plasticity. Reports of human or mouse adult stem cells that demonstrate plasticity and the cells they differentiate or specialize into include: (1) blood and bone marrow (unpurified hematopoietic) stem cells differentiate into the 3 major types of brain cells (neurons, oligodendrocytes, and astrocytes), skeletal muscle cells, cardiac muscle cells, and liver cells; (2) bone marrow (stromal) cells differentiate into cardiac

muscle cells, skeletal muscle cells, fat, bone, and cartilage; and (3) brain stem cells differentiate into blood cells and skeletal muscle cells.

- Very few published research reports on the plasticity of adult stem cells have shown that a single, identified adult stem cell can give rise to a differentiated cell type of another tissue. That is, there is limited evidence that a single adult stem cell or genetically identical line of adult stem cells demonstrates plasticity. Researchers believe that it is most likely that a variety of populations of stem cells may be responsible for the phenomena of developing multiple cell types.
- A few experiments have shown plasticity of adult stem cells by demonstrating the development of mature, fully functional cells in tissues other than which they were derived and the restoration of lost or diminished function in an animal model.

Embryonic Stem Cells and Embryonic Germ Cells

In 1985, it was shown that it is possible to obtain precursors to many different blood cells from mouse embryonic stem cells. Perkins was able to obtain all the major lineages of progenitor cells from mouse embryoid bodies, even without adding hematopoietic growth factors.

Mouse embryonic stem cells in culture, given the right growth factors, can generate most, if not all, the different blood cell types, but no one has yet achieved the "gold standard" of proof that they can produce long-term HSCs from these sources—namely by obtaining cells that can be transplanted into lethally irradiated mice to reconstitute long-term hematopoiesis.

The picture for human embryonic stem and germ cells is even less clear. Scientists from James Thomson's laboratory reported in 1999 that they were able to direct human embryonic stem cells—which can now be cultured in the lab—to produce blood progenitor cells. Israeli scientists reported that they had induced human ES cells to produce hematopoietic cells, as evidenced by their production of a blood protein, gamma-globin. Cell lines derived from human embryonic germ cells (cultured cells derived originally from cells in the embryo that would ultimately give rise to eggs or sperm) that are cultured under certain conditions will produce $CD34^+$ cells. The blood-producing cells derived from human ES and embryonic germ (EG) cells have not been rigorously tested for long-term self-renewal or the ability to give rise to all the different blood cells.

Comparisons between adult Stem Cells and Embryonic Stem Cells

Biomedical research on stem cells is at an early stage, but is advancing rapidly. After many years of isolating and characterizing these cells, researchers are just now beginning to employ stem cells as discovery tools and a basis for potential therapies. This new era of research affords an opportunity to use what has already been learned to explore the similarities and differences of adult and embryonic stem cells. (In this discussion, comments about embryonic stem cells derived from human embryos, and embryonic germ cells derived from fetal tissue, will be referred to equally as embryonic stem cells, unless otherwise distinguished.)

Similarities of Adult and Embryonic Stem Cells

By definition, stem cells have in common the ability to self-replicate and to give rise to specialized cells and tissues (such as cells of the heart, brain, bone, etc.) that have specific functions. In most cases, stem cells can be isolated and maintained in an unspecialized state. Scientists use similar techniques (*i.e.*, cell-surface markers and monitoring the expression of certain genes) to identify or characterize stem cells as being unspecialized. Scientists then use different genetic or molecular markers to determine that the cells have differentiated—a process that might be compared to distinguishing a particular cell type by reading its cellular barcode.

Stem cells from both adult and embryonic sources can proliferate and specialize when transplanted into an animal with a compromised immune system. (Immune-deficient animals are less likely to reject the transplanted tissue). Scientists also have evidence that differentiated cells generated from either stem cell type, when injected or transplanted into an animal model of disease or injury, undergo "homing", a process whereby the transplanted cells are attracted by and travel to the injured site. Similarly, researchers are finding that the cellular and non-cellular "environment" into which stem cell-derived tissues are placed (*e.g.*, whether they are grown in a culture dish or transplanted into an animal) prominently influences how the cells differentiate.

Another important area that requires substantially more research concerns the immunologic characteristics of human adult and embryonic stem cells. If any of these stem cells are to be used as the basis for therapy, it is critical to understand how the body's immune system will respond to the transplantation of tissue derived from these cells. At this time, there is no clear advantage of one stem cell source over the other in this regard.

Differences between Adult and Embryonic Stem Cells

Perhaps the most distinguishing feature of embryonic stem cells and adult stem cells is their source. Most scientists now agree that adult stem cells exist in many tissues of the human body (*in vivo*), although the cells are quite rare. In contrast, it is less certain that embryonic stem cells exist as such in the embryo. Instead, embryonic stem cells and embryonic germ cells develop in tissue culture after they are derived from the inner cell mass of the early embryo or from the gonadal ridge tissue of the fetus, respectively.

Depending on the culture conditions, embryonic stem cells may form clumps of cells that can differentiate spontaneously to generate many cell types. This property has not been observed in cultures of adult stem cells. Also, if undifferentiated embryonic stem cells are removed from the culture dish and injected into a mouse with a compromised immune system, a benign tumor called a teratoma can develop. A teratoma typically contains a mixture of partially differentiated cell types. For this reason, scientists do not anticipate that undifferentiated embryonic stem cells will be used for transplants or other therapeutic applications. It is not known whether similar results are observed with adult stem cells.

Stem cells in adult tissues do not appear to have the same capacity to differentiate as do embryonic stem cells or embryonic germ cells. Embryonic stem and germ cells are clearly pluripotent; they can differentiate into any tissues derived from all three germ layers of the embryo (ectoderm, mesoderm, and endoderm). But are adult stem cells also pluripotent? When they reside in their normal tissue compartments—the brain, the bone marrow, the epithelial lining of the gut, etc.,—they produce the cells that are specific to that kind of tissue and they have been found in tissues derived from all three embryonic layers. But can adult stem cells be taken out of their normal environment and be manipulated or otherwise induced to have the same differentiation potential as embryonic stem and germ cells? To date, there are no definitive answers to these questions, and the answers that do exist are sometimes conflicting.

These sources of stem cells do not seem to have the same ability to proliferate in culture and at the same time retain the capacity to differentiate into functionally useful cells. Human embryonic stem cells can be generated in abundant quantities in the laboratory and can be grown (allowed to proliferate) in their undifferentiated (or unspecialized) state for many, many generations. From a practical perspective in basic research or eventual clinical application, it is significant that millions of cells can be generated from one embryonic stem cell in the laboratory. In many cases, however, researchers have had difficulty finding laboratory conditions under which some adult stem cells can proliferate without becoming specialized. This problem is most pronounced with hematopoietic stem cells isolated from blood or bone marrow. These cells when cultured in the laboratory either fail to proliferate or do so to a limited extent, although they do proliferate if transplanted into an animal or human. This technical barrier to proliferation has limited the ability of researchers to explore the capacity of certain types of adult stem cells to generate sufficient numbers of specialized cells for transplantation purposes.

These differences in culturing conditions contribute to the contrasts in the experimental systems used to evaluate the ability to become specialized under particular laboratory conditions. Much of the information on the directed differentiation of embryonic stem cells into cells with specialized function comes from studying mouse or human embryonic cell lines grown in laboratory culture dishes. In contrast, most knowledge about the differentiation of adult stem cells are from observations of cells and tissues in animal models in which mixtures of cells have been implanted.

Stem cells also differ in their capacity to specialize into various cell and tissue types. Current evidence indicates that the capability of adult stem cells to give rise to many different specialized cell types is more limited than that of embryonic stem cells. A single embryonic stem cell has been shown to give rise to specialized cells from all three embryonic layers. However, it has not yet been shown that a single adult stem cell can give rise to specialized cells derived from all three embryonic germ cell layers. Therefore, a single adult stem cell has not been shown to have the same degree of pluripotency as embryonic stem cells.

Derivation of Human Embryonic Stem Cells

The first documentation of the isolation of embryonic stem cells from human blastocysts was in 1994. Since then, techniques for deriving and culturing human ES cells have been refined. The ability to isolate human ES cells from blastocysts and grow them in culture seems to depend in large part on the integrity and condition of the blastocyst from which the cells are derived. In general, blastocysts with a large and distinct inner cell mass tend to yield ES cultures most efficiently (see Fig. 2.2. Human Blastocyst Showing Inner Cell Mass and Trophectoderm).

Blastocyst *in vitro*

After a human oocyte is fertilized *in vitro* by a sperm cell, the following events occur according to a fairly predictable timeline. At 18 to 24 hours after *in vitro* fertilization of the oocyte is considered day 1. By day 2 (24 to 25 hours), the zygote (fertilized egg) undergoes the first cleavage to produce a 2-cell embryo. By day 3 (72 hours), the embryo reaches the 8-cell stage called a morula. It is at this stage that the genome of the embryo begins to control its own development. This means that any maternal influences—due to the presence of mRNA and proteins in the oocyte cytoplasm—are significantly reduced. By day 4, the cells of the embryo adhere tightly to each other in a process known as compaction and by day 5, the cavity of the blastocyst is completed. The inner cell mass begins to separate from the outer cells, which become the trophectoderm that surrounds the blastocyst. This represents the first observable sign of cell differentiation in the embryo.

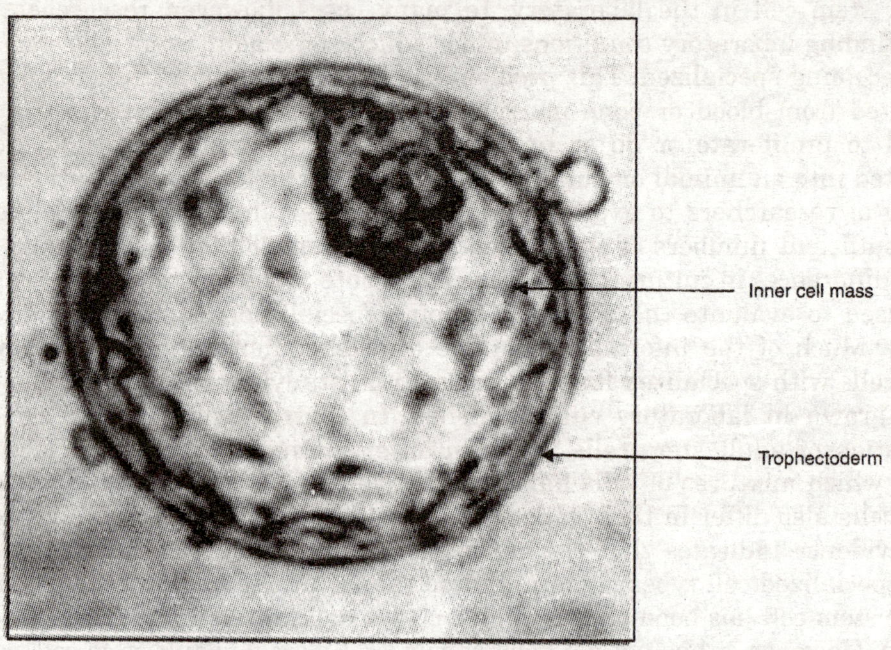

Fig. 2.2. Human blastocyst showing inner cell mass and trophectoderm.

Many IVF clinics now transfer day-5 embryos to the uterus for optimal implantation, a stage of development that more closely parallels the stage at which a blastocyst would implant in the wall of the uterus *in vivo*. This represents a change—and a greatly improved implantation rate—from earlier IVF procedures in which a 2-cell embryo was used for implantation.

Day-5 blastocysts are used to derive ES cell cultures. A normal day-5 human embryo *in vitro* consists of 200 to 250 cells. Most of the cells comprise the trophectoderm. For deriving ES cell cultures, the trophectoderm is removed, either by microsurgery or immunosurgery (in which antibodies against the trophectoderm help break it down, thus freeing the inner cell mass). At this stage, the inner cell mass is composed of only 30 to 34 cells.

The *in vitro* conditions for growing a human embryo to the blastocyst stage vary among IVF clinics and are reviewed elsewhere. However, once the inner cell mass is obtained from either mouse or human blastocysts, the techniques for growing ES cells are similar.

Applications of Stem Cells and its Research

From the new discoveries of various stem cell types and their properties, it is found to be having wide applications from simple understanding the biology to the treatment to the longlasting cancer and other dreaded diseases. Importantly, these cells are research tools and they open many avenues of opportunity for biomedical research and gene therapy. This can provide the longterm complete recovery tools to the therapeutic methods used in varieties of diseases.

Recovery of Vital Body Functions through Transplantation

Stem cells may hold the key to replacing cells lost in many crippling and devastating diseases. There is little doubt that this potential benefit underpins the vast interest about stem cell research. Some of the major diseases that are effecting human beings are Parkinson's disease, diabetes, chronic heart disease, end-stage kidney disease, liver failure, and cancer. These are just a few of the diseases for which stem cells have therapeutic potential in the regular course.

There are number of diseases that shorten lives, there are no effective treatments but the goal is to find a way to replace what natural processes have taken away. For example, today, science has brought us to a point where the immune response can be subdued, so that organs from one person can be used to replace the diseased organs and tissues of another. But, despite recent advances in transplantation sciences, there is a shortage of donor organs that makes it unlikely that the growing demand for lifesaving organ replacements will be fully met through organ donation strategies. The use of stem cells to generate replacement tissues for treating neurological diseases is a major focus of research. Spinal cord injury, multiple sclerosis, Parkinson's disease, and Alzheimer's disease are among those diseases for which the concept of replacing destroyed or non-functional cells in the brain or spinal cord is a practical goal. Many research report features several recent advances that demonstrate the regenerative properties of adult and embryonic stem cells.

Another major discovery frontier for research on adult and embryonic stem cells is the development of transplantable pancreatic tissues that can be used to treat diabetes. Scientists in academic and industrial research are vigorously pursuing all possible avenues of research, including ways to direct the specialization of adult and embryonic stem cells to become pancreatic islet-like cells that produce insulin and can be used to control blood glucose levels. Researchers have recently shown that human embryonic stem cells to be directly differentiated into cells that produce insulin.

There are common misconceptions about both adult and human embryonic stem cells. First, the lines of unaltered human embryonic stem cells that exist will not be suitable for direct use in patients. These cells will need to be differentiated or otherwise modified before they can be used clinically. Current challenges are to direct the differentiation of embryonic stem cells into specialized cell populations, and also to devise ways to control their development or proliferation once placed in patients.

A second misconception is that adult stem cells are ready to use as therapies. With the exception of the clinical application of hematopoietic stem cells to restore the blood and immune system, this is not the case. The therapeutic use of this mixture of cells has proven safe because the mixture is place back into the environment from which it was taken, *e.g.*, the bone marrow. In fact, many of the adult stem cell preparations currently being developed in the laboratory represent multiple cell types that are not fully characterized. In order to safely use stem cells or cells differentiated from them in tissues other than the tissue from which they were isolated, researchers will need purified populations (clonal lines) of adult stem cells.

In addition, the potential for the recipient of a stem cell transplant to reject these tissues as foreign is very high. Modifications to the cells, to the immune system, or both will be a major requirement for their use. In sum, with the exception of the current practice of hematopoietic stem cell transplantation, much basic research lies ahead before direct patient application of stem cell therapies is realized.

Applications in Basic Research

Embryonic stem cells will undoubtedly be key research tools for understanding fundamental events in embryonic development that one day may explain the causes of birth defects and approaches to correct or prevent them. Another important area of research that links developmental biology and stem cell biology is understanding the genes and molecules, such as growth factors and nutrients, that function during the development of the embryo so that they can be used to grow stem cells in the laboratory and direct their development into specialized cell types.

Delivery Systems in Therapeutic Use

Stem cells are already being explored as a vehicle for delivering genes to specific tissues in the body. Stem cell-based therapies are a major area of investigation in cancer research. For many years, restoration of blood and immune system function has been used as a component in the care of cancer patients who have been treated with chemotherapeutic agents. Now, researchers are trying to devise more ways to use specialized cells derived from stem cells to target specific cancerous cells and directly deliver treatments that will destroy or modify them.

Many Other Applications of Stem Cells

Future uses of human pluripotent cell lines might include the exploration of the effects of chromosomal abnormalities in early development. This might include the ability to monitor the development of early childhood tumors, many of which are embryonic in origin. Another future use of human stem cells and their derivatives include the testing of candidate therapeutic drugs. Although animal model testing is a mainstay of pharmaceutical research, it cannot always predict the effects that developmental drug may have on human cells. Stem cells will likely be used to develop specialized liver cells to evaluate drug detoxifying capabilities and represents a new type of early warning system to prevent adverse reactions in patients. The coupling of stem cells with the information learned from the human genome project will also likely have many unanticipated benefits in the future.

Most of the evidence that stem cells can be directed to differentiate into specific types of cells suitable for transplantation—for example, neurons, heart muscle cells, or pancreatic islet cells—comes from experiments with stem cells from mice. And although more is known about mouse stem cells, not all of that information can be translated to the understanding of human stem cells. Mouse and human cells differ in significant ways, such as the laboratory conditions that favor the growth and specialization of specific cell types.

Another important aspect of developing therapies based on stem cells will be devising ways to prevent the immune system of recipients from rejecting the donated cells and tissues that are derived from human pluripotent stem cells. Modifying or evading the immune rejection of cells or tissues developed from embryonic stem cells will not be able to be done exclusively using mouse models and human adult stem cells.

As with any new research tool, it will also be important to compare the techniques and approaches that various laboratories are using to differentiate and use human embryonic stem cells. Such research will provide a more complete understanding of the cells' characteristics. One key finding about the directed differentiation of pluripotent stem cells learned thus far is that relatively subtle changes in culture conditions can have dramatic influences on the types of cells that develop.

Potential Uses of Human Embryonic Stem Cells

Many uses have been proposed for human embryonic stem cells. The most-often discussed is their potential use in transplant therapy—*i.e.*, to replace or restore tissue that has been damaged by disease or injury.

Using Human Embryonic Stem Cells for Therapeutic Transplants

Diseases that might be treated by transplanting human ES-derived cells include Parkinson's disease, diabetes, traumatic spinal cord injury, Purkinje cell degeneration, Duchenne's muscular dystrophy, heart failure, and osteogenesis imperfecta. However, treatments for any of these diseases require that human ES cells be directed to differentiate into specific cell types prior to transplant. The research is occurring in several laboratories, but is limited because so few laboratories have access to human ES cells. Thus, at this stage, any therapies based on the use of human ES cells are still hypothetical and highly experimental (see Fig. 2.3. Major Goals in the Development of Transplantation Therapies from Human ES Cell Lines).

One of the current advantages of using ES cells as compared to adult stem cells is that ES cells have an unlimited ability to proliferate *in vitro*, and are more likely to be able to generate a broad range of cell types through directed differentiation. Ultimately, it will also be necessary to both identify the optimal stage(s) of differentiation for transplant, and demonstrate that the transplanted ES-derived cells can survive, integrate, and function in the recipient. The potential disadvantages of the use of human ES cells for transplant therapy include the propensity of undifferentiated ES cells to induce the formation of tumors (teratomas), which are typically benign. Because it is the undifferentiated cells—rather than their differentiated progeny—that have been shown to induce teratomas, tumor formation might be avoided by devising methods for removing any undifferentiated ES cells prior to transplant. Also, it should be possible to devise a failsafe mechanism—*i.e.*, to insert into transplanted ES-derived cells suicide genes that can trigger the death of the cells should they become tumorigenic.

Human ES derived cells would also be advantageous for transplantation purposes if they did not trigger immune rejection. The immunological status of human ES cells has not been studied in detail, and it is not known how immunogenic ES-derived cells might be. In general, the immunogenicity of a cell depends on its expression of Class I major histocompatability antigens (MHC), which allow the body to distinguish its own cells from

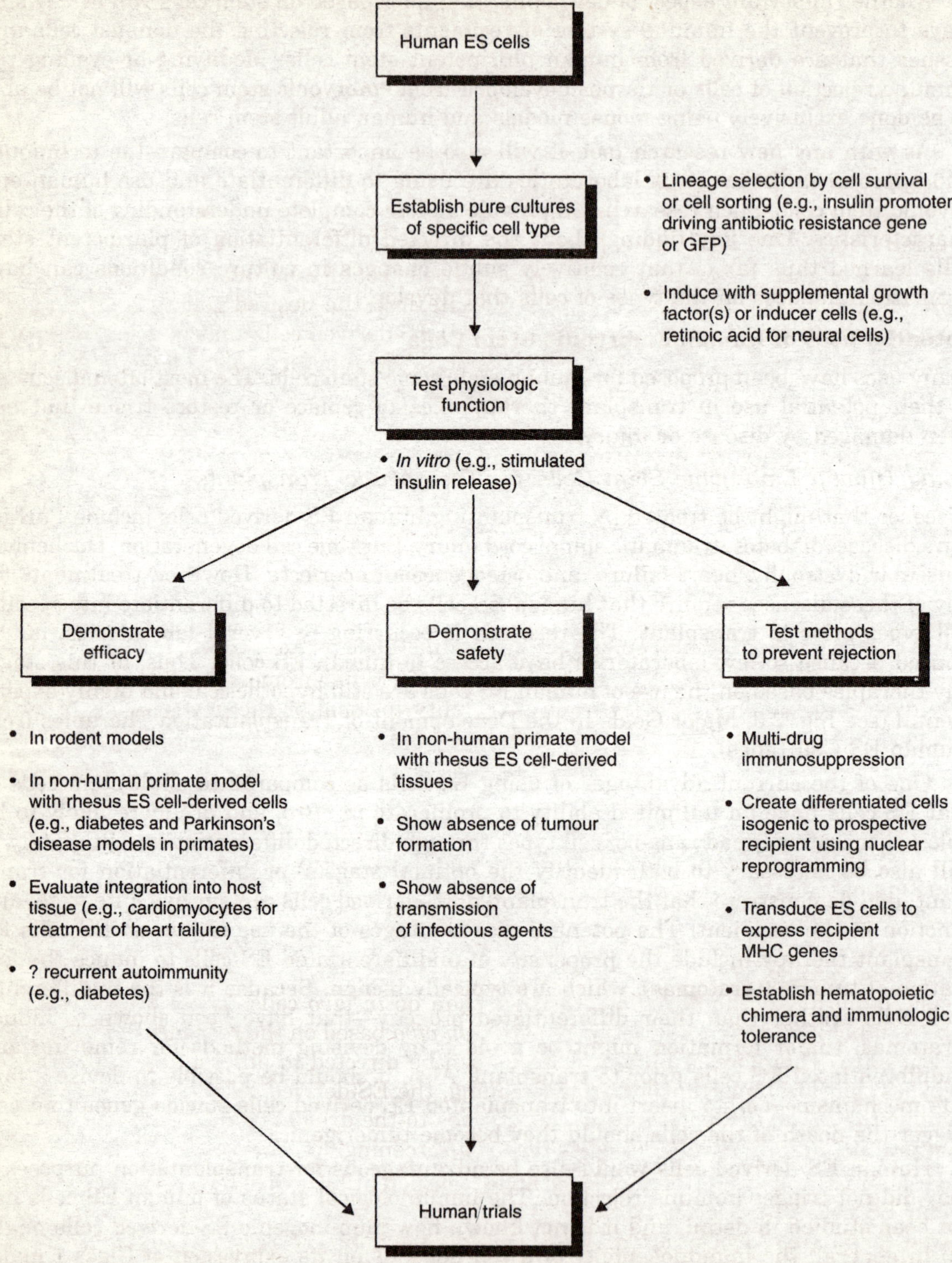

Fig. 2.3. Major goals in the development of transplantation therapies from human es cell lines.

foreign tissue, and on the presence of cells that can bind to foreign antigens and "present" them to the immune system.

The potential immunological rejection of human ES-derived cells might be avoided by genetically engineering the ES cells to express the MHC antigens of the transplant recipient, or by using nuclear transfer technology to generate ES cells that are genetically identical to the person who receives the transplant. It has been suggested that this could be accomplished by using somatic cell nuclear transfer technology (so-called therapeutic cloning) in which the nucleus is removed from one of the transplant patient's cells, such as a skin cell, and injecting the nucleus into an oocyte. The oocyte, thus "fertilized", could be cultured *in vitro* to the blastocyst stage. ES cells could subsequently be derived from its inner cell mass, and directed to differentiate into the desired cell type. The result would be differentiated (or partly differentiated) ES-derived cells that match exactly the immunological profile of the person who donated the somatic cell nucleus, and who is also the intended recipient of the transplant—a labor intensive, but truly customized therapy.

Other Potential Uses of Human Embryonic Stem Cells

Many potential uses of human ES cells have been proposed that do not involve transplantation. For example, human ES cells could be used to study early events in human development. Still-unexplained events in early human development can result in congenital birth defects and placental abnormalities that lead to spontaneous abortion. By studying human ES cells *in vitro,* it may be possible to identify the genetic, molecular, and cellular events that lead to these problems and identify methods for preventing them.

Such cells could also be used to explore the effects of chromosomal abnormalities in early development. This might include the ability to monitor the development of early childhood tumors, many of which are embryonic in origin.

Human ES cells could also be used to test candidate therapeutic drugs. Currently, before candidate drugs are tested in human volunteers, they are subjected to a barrage of preclinical tests. These include drug screening in animal models—*in vitro* tests using cells derived from mice or rats, for example, or *in vivo* tests that involve giving the drug to an animal to assess its safety. Although animal model testing is a mainstay of pharmaceutical research, it cannot always predict the effects that a candidate drug may have on human cells. For this reason, cultures of human cells are often employed in preclinical tests. These human cell lines have usually been maintained *in vitro* for long periods and as such often have different characteristics than do *in vivo* cells. These differences can make it difficult to predict the action of a drug *in vivo* based on the response of human cell lines *in vitro*. Therefore, if human ES cells can be directed to differentiate into specific cell types that are important for drug screening, the ES-derived cells may be more likely to mimic the *in vivo* response of the cells/tissues to the drug(s) being tested and so offer safer, and potentially cheaper, models for drug screening.

Human ES cells could be employed to screen potential toxins. The reasons for using human ES cells to screen potential toxins closely resemble those for using human ES-derived cells to test drugs (above). Toxins often have different effects on different animal species, which makes it critical to have the best possible *in vitro* models for evaluating their effects on human cells.

Finally, human ES cells could be used to develop new methods for genetic engineering (see Fig. 2.4. Genetic Manipulation of Human Embryonic Stem Cells). Currently, the genetic complement of mouse ES cells *in vitro* can be modified easily by techniques such as homologous recombination. This is a method for replacing or adding genes, which requires that a DNA molecule be artificially introduced into the genome and then expressed. Using this method, genes to direct differentiation to a specific cell type or genes that express a desired protein product might be introduced into the ES cell line. Ultimately, if such techniques could be developed using human ES cells, it may be possible to devise better methods for gene therapy.

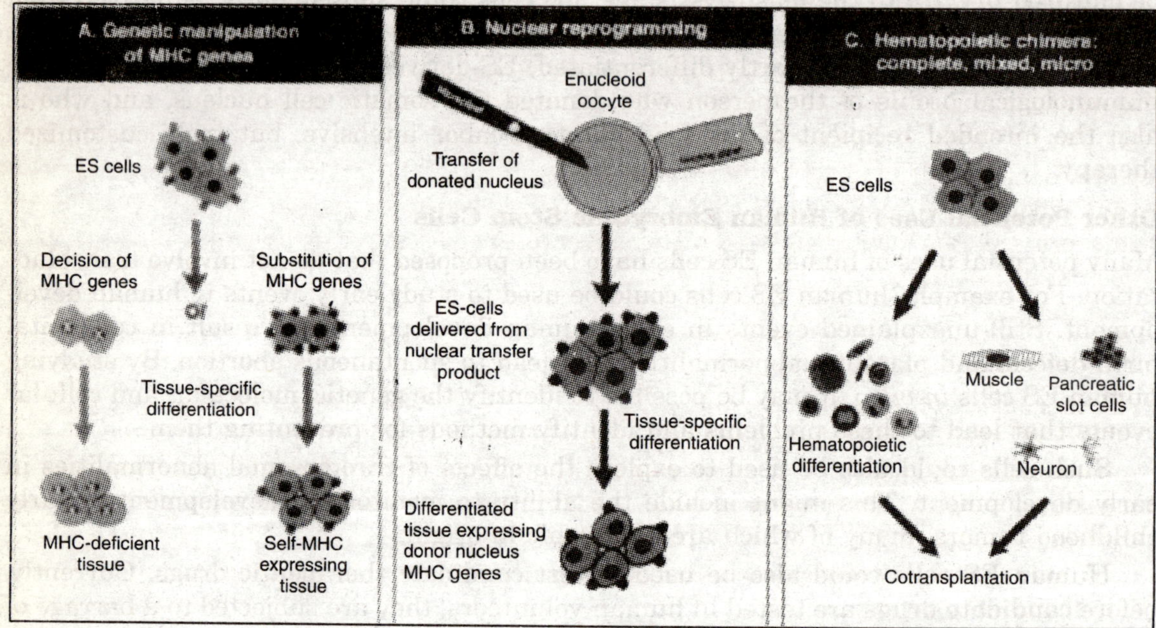

Fig. 2.4. Genetic manipulation of human embryonic stem cells.

The major Clinical Uses of Hematopoietic Stem Cells

Leukemia and Lymphoma

Among the first clinical uses of HSCs were the treatment of cancers of the blood—leukemia and lymphoma, which result from the uncontrolled proliferation of white blood cells. In these applications, the patient's own cancerous hematopoietic cells were destroyed via radiation or chemotherapy, then replaced with a bone marrow transplant, or, as is done now, with a transplant of HSCs collected from the peripheral circulation of a matched donor. A matched donor is typically a sister or brother of the patient who has inherited similar human leukocyte antigens (HLAs) on the surface of their cells. Cancers of the blood include acute lymphoblastic leukemia, acute myeloblastic leukemia, chronic myelogenous leukemia (CML), Hodgkin's disease, multiple myeloma, and non-Hodgkin's lymphoma.

Thomas and Clift describe the history of treatment for chronic myeloid leukemia as it moved from largely ineffective chemotherapy to modestly successful use of a cytokine, interferon, to bone marrow transplants—first in identical twins, then in HLA-matched siblings. Although there was significant risk of patient death soon after the transplant either from infection or from graft-versus-host disease (GVHD), for the first time, many patients survived this immediate challenge and had survival times measured in years or even decades, rather than months. The authors write, "In the space of 20 years, marrow transplantation has contributed to the transformation of [chronic myelogenous leukemia] CML from a fatal disease to one that is frequently curable. At the same time, experience acquired in this setting has improved our understanding of many transplant-related problems. It is now clear that morbidity and mortality are not inevitable consequences of allogeneic transplantation, [and] that an allogeneic effect can add to the anti-leukemic power of conditioning regimens..."

In a recent development, CML researchers have taken their knowledge of hematopoietic regulation one step farther. On May 10, 2001, the Food and Drug Administration approved Gleevec™ (imatinib mesylate), a new, rationally designed oral drug for treatment of CML. The new drug specifically targets a mutant protein, produced in CML cancer cells, that sabotages the cell signals controlling orderly division of progenitor cells. By silencing this protein, the new drug turns off cancerous overproduction of white blood cells, so doctors do not have to resort to bone marrow transplantation. At this time, it is unknown whether the new drug will provide sustained remission or will prolong life for CML patients.

Inherited blood disorders

Another use of allogeneic bone marrow transplants is in the treatment of hereditary blood disorders, such as different types of inherited anemia (failure to produce blood cells), and inborn errors of metabolism (genetic disorders characterized by defects in key enzymes needed to produce essential body components or degrade chemical byproducts). The blood disorders include aplastic anemia, beta-thalassemia, Blackfan-Diamond syndrome, globoid cell leukodystrophy, sickle-cell anemia, severe combined immunodeficiency, X-linked lymphoproliferative syndrome, and Wiskott-Aldrich syndrome. Inborn errors of metabolism that are treated with bone marrow transplants include: Hunter's syndrome, Hurler's syndrome, Lesch Nyhan syndrome, and osteopetrosis. Because bone marrow transplantation has carried a significant risk of death, this is usually a treatment of last resort for otherwise fatal diseases.

Hematopoietic stem cell rescue in cancer chemotherapy

Chemotherapy aimed at rapidly dividing cancer cells inevitably hits another target—rapidly dividing hematopoietic cells. Doctors may give cancer patients an autologous stem cell transplant to replace the cells destroyed by chemotherapy. They do this by mobilizing HSCs and collecting them from peripheral blood. The cells are stored while the patient undergoes intensive chemotherapy or radiotherapy to destroy the cancer cells. Once the drugs have washed out of a patient's body, the patient receives a transfusion of his or her stored HSCs. Because patients get their own cells back, there is no chance of immune mismatch or graft-versus-host disease. One problem with the use of autologous HSC transplants in cancer therapy has been that cancer cells are sometimes inadvertently collected and reinfused back into the patient along with the stem cells. One team of investigators finds that they can prevent reintroducing cancer cells by purifying the cells and preserving only the cells that are $CD34^+$, $Thy\text{-}1^+$.

The National Marrow Donor Program

Launched in 1987, the National Marrow Donor Program (NMDP) was created to connect patients who need blood-forming stem cells or bone marrow with potential nonrelated donors. About 70 per cent of patients who need a life-saving HSC transplant cannot find a match in their own family.

The NMDP is made up of an international network of centers and banks that collect cord blood, bone marrow, and peripherally harvested stem cells and that recruit potential donors. As of February 28,2001, the NMDP listed 4,291,434 potential donors. Since its start, the Minneapolis-based group has facilitated almost 12,000 transplants—75 per cent of them for leukemia. Major recruiting efforts have led to substantial increases in the number of donations from minorities, but the chance that African Americans, Native Americans, Asian/Pacific Islanders, or Hispanics will find a match is still lower than it is for Caucasians.

Graft-versus-tumor treatment of cancer

One of the most exciting new uses of HSC transplantation puts the cells to work attacking otherwise untreatable tumors. A group of researchers in NIH's intramural research program recently described this approach to treating metastatic kidney cancer. Just under half of the 38 patients treated so far have had their tumors reduced. The research protocol is now expanding to treatment of other solid tumors that resist standard therapy, including cancer of the lung, prostate, ovary, colon, esophagus, liver, and pancreas.

This experimental treatment relies on an allogeneic stem cell transplant from an HLA-matched sibling whose HSCs are collected peripherally. The patient's own immune system is suppressed, but not totally destroyed. The donor's cells are transfused into the patient, and for the next three months, doctors closely monitor the patient's immune cells, using DNA fingerprinting to follow the engraftment of the donor's cells and regrowth of the patient's immune cells, using DNA fingerprinting to follow the engraftment of the donor's cells and regrowth of the patient's own blood cells. They must also judiciously suppress the patient's immune system as needed to deter his/her T cells from attacking the graft and to reduce graft-versus-host disease.

A study by Joshi *et al.*, shows that umbilical cord blood and peripherally harvested human HSCs show antitumor activity in the test tube against leukemia cells and breast cancer cells. Grafted into a mouse model that tolerates human cells, HSCs attack human leukemia and breast cancer cells. Although untreated cord blood lacks natural killer (NK) lymphocytes capable of killing tumor cells, researchers have found that at least in the test tube and in mice, they can greatly enhance the activity and numbers of these cells with cytokines IL-15.

Ethical Issues

Stem cell research is controversial not because of its goals, but rather because of the means of obtaining some of the cells. Research involving most types of stem cells, such as those derived from adult tissues and umbilical cord blood, is uncontroversial, except when its effectiveness as an alternative to embryonic stem cells is debated. The crux of the debate centers around embryonic stem cells, which enable research that may facilitate the development of medical treatments and cures, but which require the destruction of an embryo to derive. In addition, because cloning is one method of producing embryos for research, the ethical issues surrounding cloning are also relevant. As previously mentioned, the Bush Administration, a group of Representatives, a group of Senators, and a group of Nobel Laureates have each presented their respective positions on embryonic stem cell research. In addition, various other organisations, individuals, and councils have issued opinions and reports on the topic. Some groups, such as the Christian Legal Society, Focus on the Family, and the Christian Coalition, support the 2001 Bush policy. Others, such as the National Academies, the Coalition for the Advancement of Medical Research.

The *President's Council* was created by President Bush in Nov. 2001 to "advise the President on bioethical issues that may emerge as a consequence of advances in biomedical science and technology". The President's Council on Bioethics, *Monitoring Stem Cell Research* is crucial in deciding the future of stem cell research around the world. The former First Lady Nancy Reagan and former Presidents Gerald Ford, Jimmy Carter, Bill Clinton favor more embryonic stem cell research than the Bush policy allows. Still others, such as the National Right to Life Committee and the United States Conference of Catholic Bishops, oppose all embryonic stem cell research. Two presidential bioethics advisory panels have considered the issues involved in embryonic stem cell research. The President's Council on Bioethics published one report directly on the topic, *Monitoring Stem Cell Research,* 75 in which it sought to characterize the issues. While the Council made no recommendations there, in two other reports it has recommended that "Congress should ... prohibit the use of human embryos in research beyond a designated stage in their development (between 10 and 14 days after fertilization)", and unanimously recommended "a ban on cloning-to-produce-children."

A predecessor to the President's Council, the National Bioethics Advisory Committee (NBAC), recommended federal funding for stem cell research using "embryos remaining after infertility treatments," but not for the "derivation or use of embryos ... made for research purposes." Detailed review of the assorted reports and statements reveals that, while positions on embryonic stem cell research may be broadly categorized as *for* or *against,* there is an array of finer distinctions present. These finer distinctions in turn reveal the variation in ethical and moral as well as factual beliefs. The following discussion breaks down the arguments about embryonic stem cell research according to these finer distinctions, demonstrating both the complexity of the issues and the points of resonance among the groups.

Embryo Destruction and Relief of Human Suffering. Most positions on embryonic stem cell research rest at least in part on the relative moral weight accorded to embryos and that accorded to the prospect of saving, prolonging, or improving others' lives. For some, the inquiry begins and ends with this question. For instance, one opponent of the research, the American Life League, posits that "human life begins at conception/ fertilization and that there is never an acceptable reason for intentionally taking an innocent human life." Similarly, the United States Conference of Catholic Bishops states that the research is immoral because it 'relies on the destruction of some defenseless human beings for the possible benefit to others'. Some groups explore the moral standing of human embryos, and also consider the "duty to relieve the pain and suffering of others." Others take the position that embryos do not have the same moral status as persons. They acknowledge that embryos are genetically human, but hold that they do not have the same moral relevance because they lack specific capacities, including consciousness, reasoning and sentience. They conclude that performing research to benefit persons justifies the destruction of embryos. Acceptance of the notion that the destruction of embryos can be justified in some circumstances forms the basis of pro-stem cell research opinions, and is usually modified with some combination of the distinctions and limitations that follow.

SUMMARY

Stem cells are pluripotent cells having the capacity to grow indefinitely in the culture and can be differentiated into any cell type of the body. Stem cells may be embryonic stem cells, embryonic germ cells and adult stem cells. The embryonic stem cells are derived from inner cell mass of the blastocyst and embryonic germ cells are obtained from 5 to 10 week old foetus. The adult stem cells are derived from various tissues of the adult of the individuals such as bone marrow, blood, skeletal muscle, cornea and retina of eye, brain, dental pulp, liver, skin, gastrointestinal tract lining and pancreas. Even with wide range of studies on stem cells, still lot of characters are not thoroughly understood and much more mysteries should be unrevealed for full potential applications.

Stem cells are better tools for treating a large number of major crippling and devastating diseases such as Parkinson's disease, diabetes, chronic heart disease, end-stage kidney disease, liver failure, and cancer for which there is effective and permanent or alternative therapies available.

With the immense applications and wide characteristics, stem cells research attracts the scientists throughout the world in all the fields of biology from the simple cell culture to the complex gene therapy and transplantation. Grounded in half a century of research, the study of stem cells is one of the most exciting and rapidly advancing disciplines in biomedicine today. Breakthrough discoveries in both the laboratory and clinic have sharply expanded the use and supply of life-saving stem cells. Yet even more promising applications are on the horizon and scientists' current inability to grow hematopoietic stem cells outside the body could delay or thwart progress with these new therapies. New treatments include graft-versus-tumor therapy for currently incurable cancers, autologous transplants for autoimmune diseases, and gene therapy and tissue repair for a host of other problems. The techniques, cells, and knowledge that researchers have now are inadequate to realize the full promise of stem cells-based therapy.

Key issues for tapping the potential of hematopoietic stem cells will be finding ways to safely and efficiently expand the numbers of transplantable human hematopoietic stem cells *in vitro* or *in vivo*. It will also be important to gain a better understanding of the fundamentals of how immune cells work—in fighting infections, in causing transplant rejection, and in graft-versus-host disease as well as master the basics of stem cells differentiation. Concomitant advances in gene therapy techniques and the understanding of cellular plasticity could make stem cells one of the most powerful tools for treating many of the diseases for which no other alternative therapies are available.

EXERCISE

1. Define stem cells and mention its application in medicine.
2. Give a comparison between embryonic stem cells and adult stem cells.

3. What do you mean by pluripotent stem cells and totipotent stem cells?
4. Explain the sources of stem cells.
5. Write a note on the major sources of hemopoietic stem cells.
6. Elucidate the potential uses of the human embryonic stem cells.
7. List out the major historical events of stem cells research.
8. Give an account on applications of hemopoietic stem cells in medicine.
9. Blastocyst and its uses.
10. How adult stem cells are better than embryonic stem cells in treatment processes?
11. What are the similarities of adult stem cells and embryonic stem cells.
12. Name some of the mysteries that are not still understood in the stem cells biology.
13. What are the ethical issues associated with stem cells research?
14. How hematopoietic stem cells can be used to treat leukemia and lymphoma?
15. Mention importance of establishing National Marrow Donor Program.
16. List out the major inherited blood disorders.

3

Animal Cell Culture and Metabolism

3.1 INTRODUCTION

During the last fifty years, the prerequisites for the maintenance and propagation of animal cells in culture have been worked out systematically. The present state of development is characterized by the fact that the cultivation of animal cells has been established in many laboratories and clinics in order to deal with biochemical, physiological and morphological questions.

The development of animal cell biotechnology owed much to the needs of two major branches of medical research; the production of antiviral vaccines and the understanding of cancer biology. The standardization of conditions and cell lines for the production and assay of viruses undoubtedly provided much impetus to the development of modern animal cell biotechnology, particularly the production of large numbers of cells suitable for biochemical analysis. This and other technical improvements made possible by the commercial supply of reliable media, sera and by the greater control of contamination with antibiotics and clean-air equipment, have made animal biotechnology accessible to a wide range of interests.

In addition to cancer biology and virology, other areas of research have come to depend heavily on animal cell culture systems. The introduction of cell fusion, genetic manipulation, animal cloning, gene expression analysis, monoclonal antibodies production, cell interactions, intracellular control mechanisms in cell differentiation, nerve function analysis, hormone industries and transplantation studies are dependent upon the animal culture systems for understanding and production technologies.

The cultivation of animal cells on a large technical scale started with BHK (baby hamster kidney) cells which were adapted to growth in suspension in 1962 and have been used industrially since 1967 in the United Kingdom, Italy, Germany, Spain, Denmark and France, particularly for the production of foot and mouth disease vaccines. Girard (1977) reported the construction of a factory in which every year 500,000 liters of cell suspension are produced in 3000 liter fermenters. More advanced processes are already based on fermenters with a capacity up to 10,000 liters.

As majority of animal cells are suitable for *in vitro* cultivation, the present annual demand of around 300 millions of experimental animals worldwide will be reduced as further developments become available.

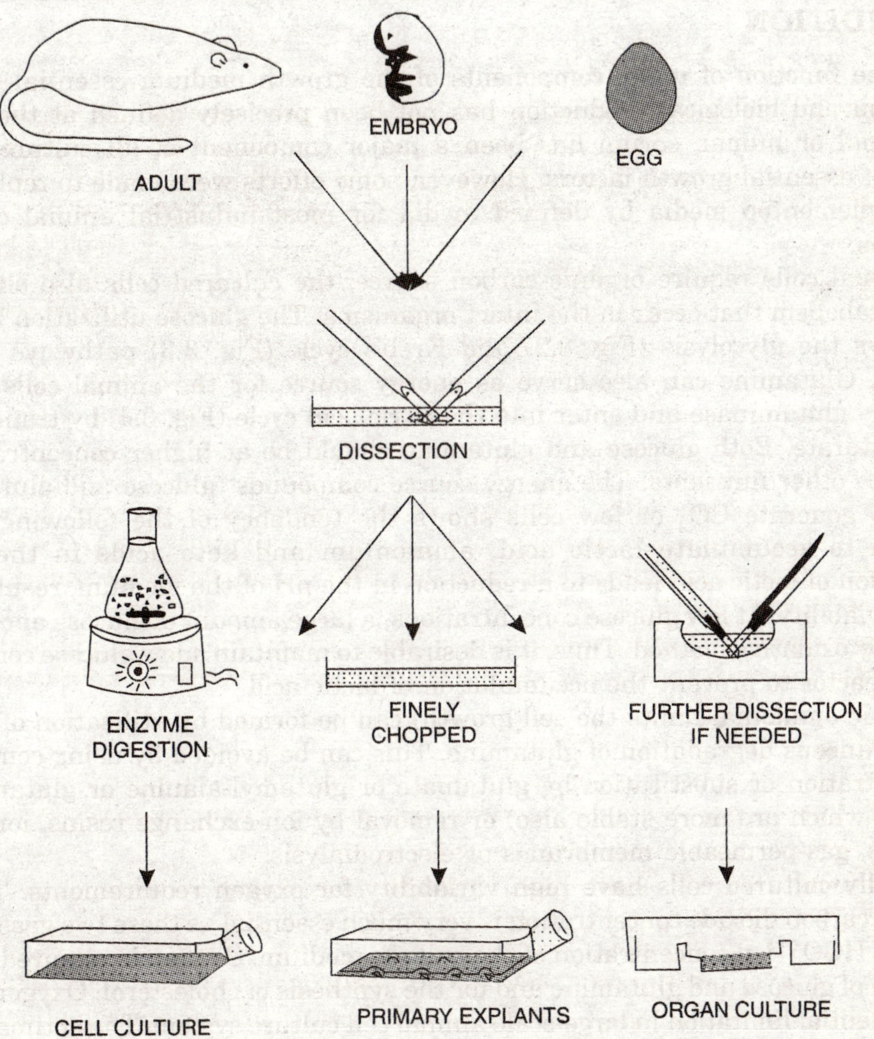

Fig. 3.1. Different types of animal tissue culture.

Animal cell culture deals with culturing the parts of organs, tissues or individual cell *in vitro*. The starting material for a culture initiation is known as **explant**. As long as the material used maintains its intact structure is called **tissue** or **organ culture** (Fig. 3.1). If the organization of a tissue is destroyed by mechanical, chemical or enzymatic action that forms individual cells is known as **cell culture**. Cells or tissues taken from an organism for culture form the **primary culture**. The generations of cells obtained after the first subcultivation and all subsequent ones are referred as **cell lines**.

At present in animal biotechnology varieties of normal and transformed cell types and cell lines are established and maintained for utilization in different purposes from simple study of cell culture characteristics to large scale production of vaccines and other therapeutic and commercial proteins (Table 3.1).

3.2 GENERAL METABOLISM OF ANIMAL CELLS IN CULTURE CONDITION

To date, the function of many components of the growth medium essential for cellular proliferation and biological production has not been precisely defined at the molecular level. Animal or human serum has been a major component of all culture media for provision of essential growth factors. However, some efforts were made to replacement of serum-supplemented media by defined media for most industrial animal cell culture applications.

As animal cells require organic carbon source, the cultured cells also show similar type of metabolism that occur in the intact organisms. The glucose utilization by cultured cells follows the glycolysis (Fig. 3.2) and Kreb's cycle (Fig. 3.3) pathways for energy production. Glutamine can also serve as energy source for the animal cells in culture condition by glutaminase and enter into the citric acid cycle (Fig. 3.4) by transamination to 2-oxoglutarate. Both glucose and glutamine should be at higher concentration when compared to other nutrients. The energy source compounds (glucose and glutamine) are oxidized to generate CO_2 or few cells shows the tendency of the following anaerobic conversion to accumulate lactic acid, ammonium and keto acids in the medium. Accumulation of lactic acid leads to a reduction in the pH of the medium, resulting in the loss of cell viability. At low glucose concentrations, a large amount of glucose and glutamine is utilized by oxidative method. Thus, it is desirable to maintain a low glucose concentration in the bioreactor to prevent the accumulation of lactic acid.

The toxic ammonia (limits the cell growth) can be formed by utilization of glutamine or by spontaneous degradation of glutamine. This can be avoided by using comparatively low concentration or substitution by glutamate or glutamyl-alanine or glutamyl-glycine (dipeptides which are more stable also) or removal by ion-exchange resins, ion-exchange membranes, gas-permeable membranes or electrodialysis.

Generally cultured cells have high variability for oxygen requirements. The proper oxygen and carbon dioxide concentration is very much essential as these two gases influence the pH and HCO_3^- ion concentration of the culture medium. Oxygen is required for energy metabolism of glucose and glutamine and for the synthesis of cholesterol. Oxygen transport can be a potential limitation in large-scale animal cell culture system. The optimal dissolved oxygen (DO) is not generally the same for cell growth and product formation. Typically, the oxygen uptake rate (OUR) increases with an increase in DO up to about 10 to 30% saturation with air. Increase in DO beyond 30% does not lead to any further increase in OUR. Higher DO concentrations are toxic and lead to oxidative damage, including DNA degradation, lipid peroxidation, polysaccharide depolymerization and metabolism of nutrients in culture medium at rates greater than that required for consumption.

The range of optimal concentrations of amino acids in the medium for cell growth is wide. Amino acid consumption is often cell line dependent and affects cell growth but not protein synthesis. The balance of amino acids that is critical for the cell growth.

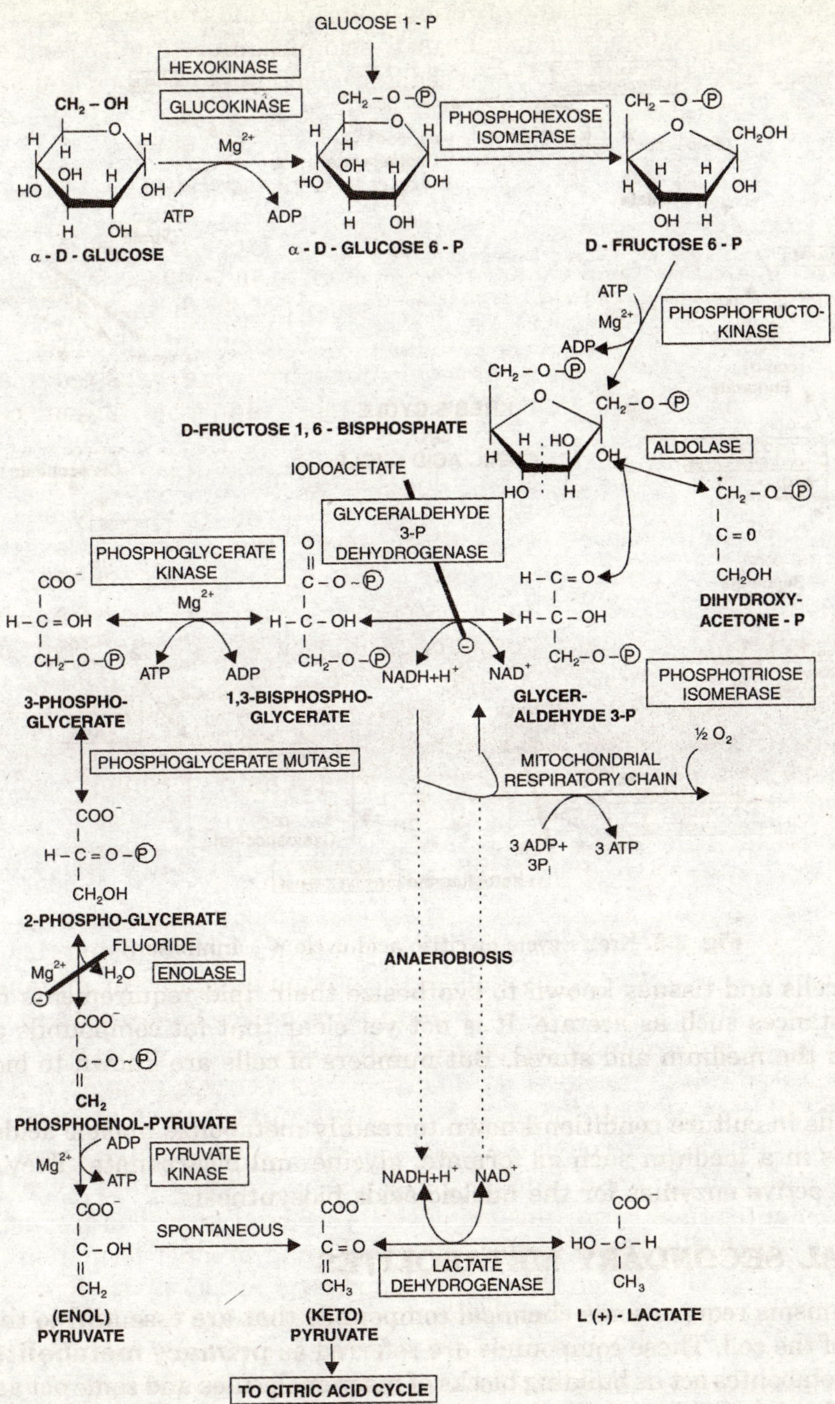

Fig. 3.2. Embden-Meyerhof pathway of glycolysis.
(Ⓟ, $-PO_3^{2-}$; P_i, $HOPO_3^{2-}$; ⊖, inhibition)

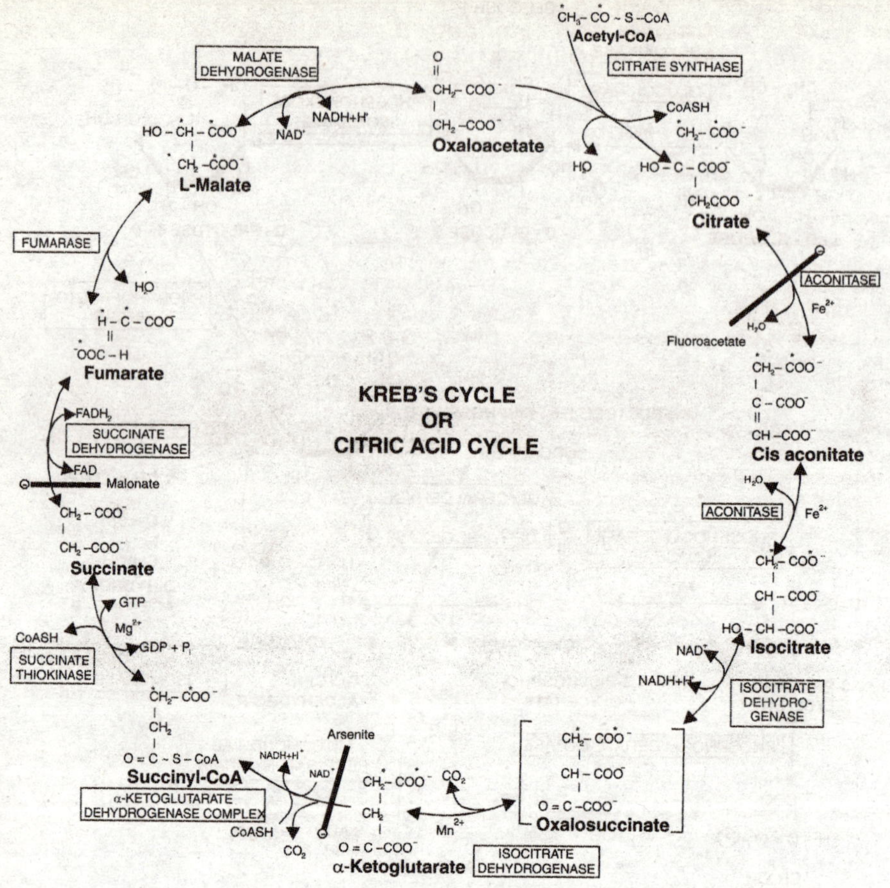

Fig. 3.3. Kreb's cycle or citric acid cycle (⊖ inhibition)

Cultured cells and tissues known to synthesize their lipid requirements from simple chemical substances such as acetate. It is not yet clear that fat compounds are directly taken up from the medium and stored. But numbers of cells are known to biosynthesize cholesterol.

Animal cells in culture condition known to readily metabolize nucleic acids from simple compounds in a medium such as formate, glycine and bicarbonate. They contain all the functional active enzymes for the nucleic acids biosynthesis.

3.3 SPECIAL SECONDARY METABOLITES

All living organisms require some chemical compounds that are essential to the structure and function of the cell. These compounds are referred as **primary metabolites**. Some of the primary metabolites act as building blocks of macromolecules and some act as coenzyme. The example for primary metabolites include carbohydrates, amino acids, organic acids, purines, pyramidines and vitamins. Metabolism of these compounds is known as primary metabolism. The **secondary metabolites** are compounds which are not essential for the structure and function, but necessary for maintaining the proper health and state of the cell or body. The best known example of secondary metabolites are antibiotics from fungi and bacteria. In case of animals also secondary metabolites are produced which are much essential to maintain proper health and metabolic activity. Hence they are considered special secondary metabolites. These include hormones, interferons, polypeptide growth factors and proteins of therapeutic value (Table 3.1).

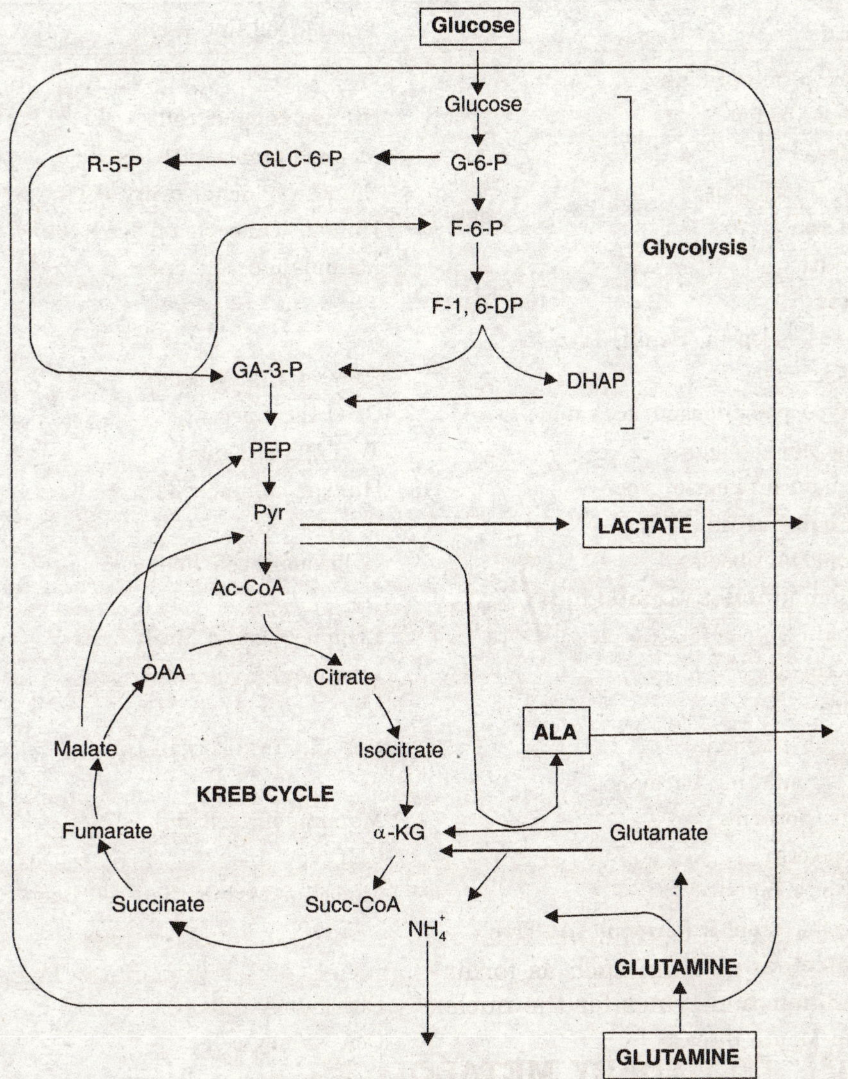

Fig. 3.4. Model of glucose and glutamine catabolism in mammalian cells in culture condition. Compounds in bold letters are consumed from or excreted into the medium. Abbreviations used: G-6-P, glucose-6-phosphate: GLC-6-P, 6-phosphategluconate; R-5-P, ribose-5-phosphate; DHAP, dihydroxyacetons phosphate; PEP, phosphoenolpyruvate; Pyr, pyruvate; Lac, lactate; Ac-CoA, acetyl-CoA; α-KG, α-CoA; OAA, oxaloacetate; Ala, alanine; Succ-CoA, Succinyl CoA.

Human cell contains more than 29,000 genes, most of them code essential proteins. Sometime a gene is defective due to mutation resulted in production of defective protein.

In some cases protein is not produced at all. A deficiency of the protein may not be noticed, but sometimes the deficiency results in disease. The deficiency is generally inherited and thus the disease is genetically linked.

Table 3.1: Human therapeutic proteins produced from cultured animal cells

Therapeutic Protein	Type of cell line used
Cellular biochemicals	
Angiogenic factor	Human tumour cells
β-Interferon	Mouse fibroblast MG 63
β- and γ-Interferons	Chinese hamster ovary (CHO) cell lines
α-Interferon	Human leukocytes (CRL 1432)
Factor VIII	Mammalian cell lines
Enzymes	
Collagenase, Pepsin, Renin, Trypsin	–
Urokinase	Human kidney cells
Tissue-type plasminogen activator	CHO cell lines
Plasminogen activator	TCL 598 cell lines
Single chain urokinase type	Human kidney cells
Immunoglobulins	
Monoclonal antibodies	Hybridoma cell lines
Immunomodulator Compounds	
Interleukin-2	Lymphoblastoid T-cell lines
Lymphokines	Lymphocyte cell lines
Hormones	
Luteinizing hormone	Specific tissue/organ cells
Follicle stimulating hormone	
Chorionic hormone	Mammalian cell line
Erythropoietin	CHO cell line
Growth hormone	Mouse or Mammalian cell lines
Adrenocorticotrophic hormone (ACTH)	
Viral Vaccines	
Canine distemper	Dog kidney cells
Foot and Mouth disease	Cow kidney cells
Hepatitis-B	Human plasma-desired
Influenza	Chick embryo
Measles	Chick embryo
Mumps	Chick embryo
Polio	Primate kidney
Rabies	Chick embryo/human diploid cells

Now it is possible to treat certain genetically associated diseases provided the protein deficiency is identified. There are many human therapeutic proteins which are produced by the recombinant DNA technology in large scale to meet the demand of the needy people.

3.4 INSULIN: PRODUCTION OF HUMAN INSULIN

An example of genetically-linked disease is diabetes mellitus. It is the most common and widespread disease. The beta cells of the pancreas are responsible for the production of hormone Insulin. If these cells fail to produce the required quantities of the hormone causes diabetes mellitus. In 1961, Sir Edard Sharpy-Schafer discovered insulin, while studying the Islets of Langherans of the pancreas. Frederick Banting and Howard Best in Canada reported antidiabetic role of insulin in 1921. It actually facilitates the cellular uptake of glucose. Without insulin, glucose remains in the bloodstream resulting in increase of blood sugar as well as lowering the production of energy rich ATP. The patient with diabetes mellitus is very weak and tired because the cells are very low in ATP. The kidney removes excessive glucose from the bloodstream and excretes it in the urine (mellitus= sweet). To keep the glucose diluted, the kidney also puts out an abnormally large amount of water (diabetes= siphon=large volume of water siphoned from the body).

Diabetics are prone to eye damage and blindness, as well as kidney disorders, nerve damage, and circulatory diseases, including gangrene and stroke. They have to be careful to avoid the diabetic coma, a loss of consciousness caused by pH imbalance arising from the body's breakdown of fats. Approximately 150 million people in the world are believed to be diabetic and it is estimated that an increase to 300 million in the year 2025. In India at present there are 35 million diabetics and is expected to rise 3 times in another 20 years. They must take regular injections of insulin to avoid the serious effects of the disease. According to a recent estimate, by year 2025, the global cost of treating diabetes and its complications could reach US $1 trillion (1000 billion) annually.

Insulin was for the first time applied to a diabetic nine-year-old boy in 1922 with success. Eli Lilly immediately set it on for production in 1923 and animal insulin (from Pancreas) was marketed. Frederick Sanger and his coworkers (1957) at Cambridge University finally succeeded in achieving the first complete description of the structure of protein molecule insulin. Sanger's work was not only to discover the overall chain configuration of the insulin molecule but also to learn the sequence of all the amino acids in the chain. Insulin is a protein made up of 51 amino acids in two inter-connected chains, an A chain of 21 amino acids and a B chain of 30 amino acids. Insulin has been purified from the pancreas of pigs and cows obtained at slaughter houses. Animal insulin is similar to human insulin. However, there is a difference of one (pig) or three (cow) amino acids. Porcine (pig) insulin is similar to human insulin (Humulin) in that the C-terminal threonine on the A chain is replaced by alanine, while bovine (cow) is different at three positions (Table 3.2). This difference in chemical structure causes allergic responses in diabetic patients after taking the animal insulin. All patients treated with injections of bovine insulin develop circulating anti-insulin antibodies that are responsible for subcutaneous fat atrophy and also neutralize insulin in the circulation. Consequently the patient needs a higher dose in due course. Moreover, some diabetics are allergic to animal hormones. Besides bovine insulin, shows a greater immunogenicity as compared to porcine. Therefore there is a requirement of human insulin. Another important factor is that to treat one diabetic patient for one year, 7-10 pounds of pancreas tissues (70 pigs) are needed. About 800-1000 kilograms of pancreas produce 100 gm of insulin. Since then it has been produced commercially by several pharmaceutical concerns.

In 1970s biochemists used the recombinant DNA technology for the production of insulin. They separated the DNA segment (gene) coding for insulin. Then they attached the structural gene for insulin to the bacterial plasmid of *E. coli*. They also inserted promoter sequence of lac operon before insulin gene to produce insulin. In July 1980, for the first time 17 patients received injections of genetically engineered insulin at Guy' Hospital in London, England and this insulin worked as the animal insulin. Later on they inserted a short segment of DNA between the promoter site and the structural genes. The new segment directs the production of 24 amino acids peptide chain that is continued with the insulin. This encourages the bacterium to secrete the synthesized insulin into the surrounding media (outside the cell) which in turn saved the trouble of retrieving insulin

Table 3.2: Amino acid sequence of human, porcine and bovine insulins

Insulin	A8	A10	B30
Human	Threonine	Isoleucine	Threonine
Porcine	Threonine	Isoleucine	Alanine
Bovine	Alanine	Valine	Alanine

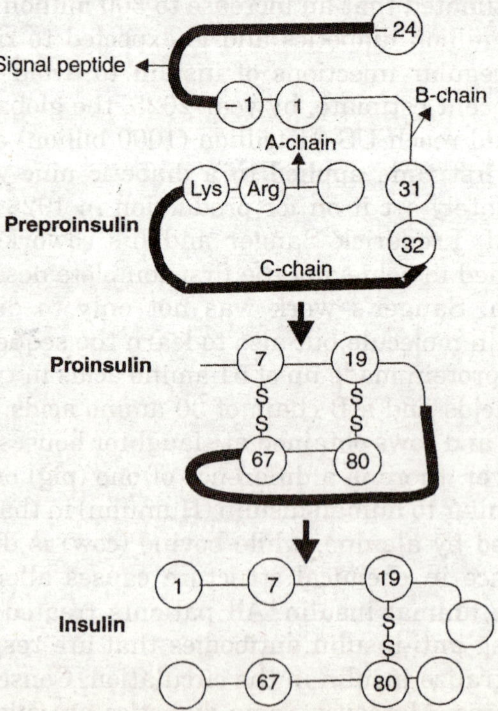

Fig. 3.5. Processing of preproinsulin to produce insulin.

from the cell body. Further, the bacterium continues its metabolism as the food is supplied from one end of the fermentation tank and the insulin is removed from the other end.

In the body pancreatic cells first synthesize a large chemical molecule called preproinsulin, which is 105 amino acids in length. This molecule includes a signal sequence (24 amino acids) and three sections of amino acids (A, B and C chains). During natural insulin production the signal sequence is removed to form proinsulin and later a section of 33 amino acids is cut free at the Golgi apparatus and the remaining two chains (A and B) are joined to form a molecule of insulin (Fig. 3.5).

Synthesizing of preproinsulin is too difficult in bacteria. However, the biotechnologists used two separate bacteria for synthesis of A and B chain of insulin molecule. These chains are extracted after the synthesis from the media and purified. Later they were joined chemically by disulfide linkages resulted in synthetic insulin. In 1986, the Eli Lilly company received approval to market Humulin, the first genetically engineered insulin for general use. In addition to this, biotechnologists discovered a way to alter the porcine insulin to insulin, which is identical to human insulin. Thus today there are two biotechnological methods to manufacture the insulin (Fig. 3.5).

3.5 PRODUCTION OF HUMAN GROWTH HORMONE (HGH)

HGH is a protein that stimulates overall body growth by increasing the absorption of amino acids by the cells which in turn helps in protein synthesis. It also promotes the use of fat as body fuel for energy production. This hormone is produced by pea-shaped gland present at the base of the pitutary gland. This pitutary human growth hormone is also known as somatotropin. Under production of this protein leads to pitutary dwarfism, which is another metabolic disorder. Giving the same protein controls this disease. When the pitutary gland produces insufficient amount of HGH in childhood, the growth is retarded. The child develops a chubby face; abundant body fat at the waist, height of about 4 feet and intelligence is normal. However, the adult has unusual body proportions and is a dwarf. If dwarfism is diagnosed at an early age and while bone growth is still occurring, injections of HGH can be given. This is very expensive and pitutary glands from nearly 80 cadavers must be used to obtain enough HGH for a year's therapy (8-10 year individual). Further, it is dangerous to give the HGH of infected brain (e.g. CJ syndrome).

DNA technologists can now produce HGH exclusively through the mechanisms of genetic engineering. The process is similar to that for insulin. Natural HGH comprises 191 amino acids and this structural detail was given by Genetech group consisting of Goeddel and coworkers. During production of HGH in the body the additional signal peptide of 26 amino acids is cut free to produce mature HGH during secretion. However, the complementary DNA molecules that biotechnologists isolate also contain this signal sequence. This is first removed by using restriction enzyme EcoRI. Then the gene is inserted to the plasmid (Fig. 3.6).

The synthetic HGH when injected to the undersized children, it stimulates body growth and helps a child to reach his or her normal height and size. The hormone is species specific and its main source has been human cadavers. It has been demonstrated that an intramuscular dose of 6-10 mg per week can increase the stature by more than 6 cm during the first year of treatment. However, the hormone has to be used while bone formation is actively occurring which means that adults would not benefit from HGH therapy.

Genetically engineered HGH is used after 1985 and marketed as Protropin by the Genentech Company and Humatrope by Eli Lilly & Company. The synthetic hormone has tripled the growth rate of hormone deficient babies in the first year and doubled the growth rate in the second and third years of therapy. Currently it is prescribed for children described as very short. To reach normal heights, daily intramuscular injections of the drug may be required for many years at a cost of about 20,000 dollars per year. As of 1998, 30,000 short children were taking hormone therapy each year in the US.

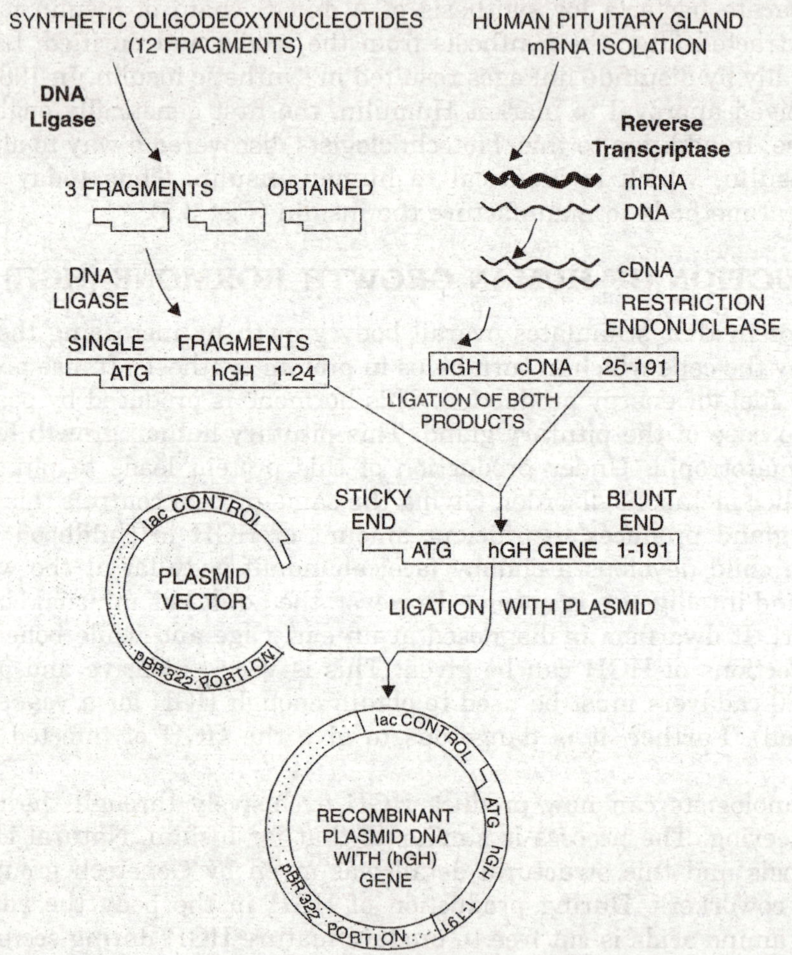

Fig. 3.6. Synthesis of recombinant molecule of human growth hormone.

Recently Somatotropin (HGH) gene was synthetically produced by a combination of chemical synthesis and isolation of natural molecule. This is a polypeptide of 191 amino acids. The segment of gene that codes for the first 24 amino acids part of the peptide chain was constructed chemically from different blocks of nucleotide. The recombinant human growth hormone plasmid was constructed by usual genetic engineering techniques. Reverse transcriptase enzyme was used to copy the gene for the hormone from mRNA obtained from human pituitary tissues. Restriction endonucleases which cut out the needed part of the DNA. Then DNA ligase was then used to join the natural and synthetic fragments. The complete gene was inserted into a modified version of plasmid vector pBR 322 with the lac operon. The synthetic part of the growth-hormone gene had been constructed with its own initiation codon (ATG) required for the protein synthesis. The hormone could, therefore, be produced independently in prokaryotic bacterial cells, without the need for attachment to a bacterial protein. Thus purification becomes very easy. This growth hormone may have many clinical uses, but the extremely limited availability of the human growth hormones is a serious limitation. Microbiological production may not only increase the commercial availability of the drug but also make it possible to investigate in detail its potential applications. Now rat tumor cell lines and mammalian cell lines are used for production of human growth hormone. Currently several industries including Genentech, Pasteur Institute, Kabi Vitrum AB involved in large scale production of somatotropin. Biosynthesis of somatotropin might reduce the cost of treatment.

3.6 FACTOR VIII: PRODUCTION OF FACTOR VIII

It is a blood protein that contributes to the formation of a blood clot and in doing so prevents hemophilia. In US the most common inherited blood clotting disorder is hemophilia A. This disorder affects about 1 in 10,000 males in the USA and is caused by a genetic defect for normal blood clotting. The defect is traced to be the inability of the body to produce an important factor responsible for blood clotting process called Factor VIII.

Previously this was obtained from units of whole blood (8000 pints of blood are needed to yield enough clotting factor for a single patient for a single year). Unfortunately contamination by unknown agents is possible and this possibility was observed during 1980s by the HIV infection. More than 10,000s of patients have got AIDS due to contaminated blood. Afterwards, the DNA technologists worked to produce synthetic factor VIII using genetic engineering. They noticed that the gene for factor VIII is located on the X-chromosome and it comprises 186, 000 bp with in 26 exons. They isolated m RNA that contains 9000 bases and encodes the final factor VIII, a protein consists of 2332 amino acids. Scientists synthesized a cDNA molecule using this RNA as a template. Then inserted it into hamster kidney cells where it encoded the factor VIII protein. Mammalian cell lines were used for the production because factor VIII is very large and complex, with at least 25 sites where carbohydrate molecules are attached.

By 1992, two types of factor VIII were synthesized and FDA (Federal Food and Drug Administration) was offered licence within a year. Recombinate, a product of Genetics Institute in Cambridge, Massachusettes and Kogenate was produced by Miles Lab.

Symptoms of immunological intolerance have been observed in some recipients.

3.7 INTERFERONS: PRODUCTION OF INTERFERON

Alick Isaacs and Jean Lindemann (1957) reported the production of some antiviral substances in cells of many vertebrates after viral infections. Isaacs (1963) called such substances as interferons because these play an important role in virus interference. Similar such proteins were also observed in plants such as tomatoes and tobacco (Wallace, 1944). Thus the virus-infected cells produce a protein that protects uninfected cells against the infection of virus or inhibits the productive multiplication cycle of the virus called interferons.

Interferon is not a single substance but a group of more than twenty substances with molecular weights between 20,000-30,000 Daltons, and are broadly classified into three types on the basis of physicochemical and antigenic properties: alpha (IFN-α), beta (IFN-β) and gamma (IFN-γ) interferons. All interferons are protein molecules and many are also glycoproteins. They have carbohydrate molecules attached to the amino acids at various points along the chain. The interferons trigger a nonspecific reaction that protects cells against numerous different viruses, including the ones that stimulated their production. Interferons are produced by cells after a virus releases its DNA or RNA into the cell's cytoplasm. The interferons secreted by the cells then bind to receptor sites on nearby cells. Here they trigger the production of protective proteins. These proteins inhibit viral replication by methods, which are not fully understood. However, recently it has been shown that one protein binds to mRNA encoded by virus. In early days it was very difficult to get one gram of interferon to work with (90,000 pints of blood were required to produce a single gram of interferon). However, after deciphering the genetic code of interferon by Swiss and Japanese biotechnologists (1980), it was possible to synthesize the interferon using genetic engineering techniques. It was found that the genetically engineered (bacteria derived) interferons reduced the symptoms of hepatitis, diminished the spread of herpes zoster (Shingles) and shrank certain tumors.

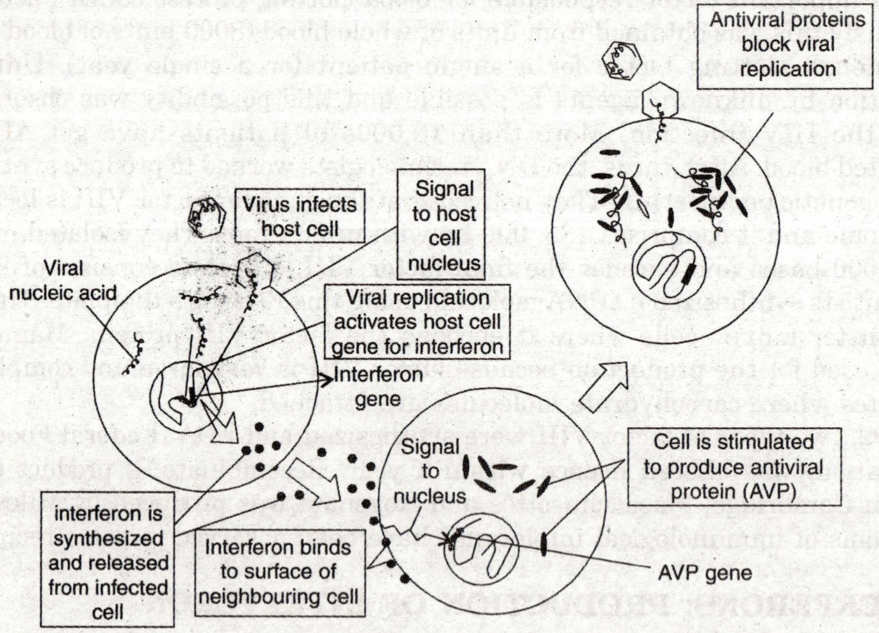

Fig. 3.7. The mechanism of interferons α and β action on viral replication.

In 1984, Swiss biotechnology firm started marketing alpha interferon as 'Intron' and in 1986, FDA approved the use of this against leukemia; in 1988 approved the use of this against genital warts, sarcoma, malignant melanoma, multiple myeloma and certain kidney cancers. Interferon beta licensed in 1993 for use against multiple sclerosis. Interferons can be used to treat common cold and broad spectrum of viral diseases. It is produced by separating WBCs from human blood and exposing them to Sendai Virus. The viral capsid stimulates a human cell chromosome (presumably number 5) to produce interferon. However, these interferons are unable to affect viral replication in the cell in which they were produced. These interferons are absorbed by adjacent cells where they induce chromosome number 21 to produce an unknown protein that binds to mRNA molecules specified by the virus (Fig 3.7). Interferon is finally released in the culture medium and is purified using monoclonal antibodies by employing affinity chromatography technique. Negative aspect of this is that the interferons production is costly because bacteria cannot synthesize glycoproteins such as interferons as they lack such modification mechanisms.

Throughout the world a large number of firms are manufacturing leukocyte interferon in the United States including Genetech, Interferon Sciences, Schering Plough as well as Roche Institute of Molecular Biology. In Japan also, Yamanouchi Pharmaceutical Company, Tokyo, has manufactured interferon. In Europe there are a large number of public and private concerns which are attempting/having produced interferon such as Searle Laboratories and Wellcome Laboratories in U.K., University of Louvain in Belgium, Snafu Company in France, Hoechst in Federal Republic of Germany and Kabi Vitrum AB in Sweden, Israel is also on the world market for successful production of Interferon. However, the major difficulty is how to reduce the cost of production to produce interferon to be utilized by an average customer.

When compared to earlier methods which used leukocytes in serum-free medium for α-interferon production, the recent methods employing mammalian cell lines. The lymphoblastoid cell line called Namalva (ATCC No. CRL 1432) obtained from tumor is used for α-interferon production. This cell line produced up to 20,000 units of α-interferons per 10^6 cells. The Wellcome Foundation (Beckenhan, U.K.) and Sumitomo Chemical Company (Japan) are major producers of α-interferon.

The β-interferon are produced by diploid fibroblasts, particularly cell line the FS-4. This cell line has a limited life span corresponding to about 30 passages. Now the MG63 (ATCC No. CRL 1427) obtained from osteosarcoma patient is considered as a superior cell line for β-interferon production. This cell line produces up to 10^8 units per litre of culture medium. The Bioferon (Germany) company and Rega Institute (University of Louvain, Belgium) are major producers of β-interferon in the market.

The γ-interferon was produced from B-lymphocytes or T-lymphocytes obtained from fresh blood collections. The mammalian cell lines such as monkey tumor cell lines are used for the large scale production of γ-inteferons. Mizrali in 1983 listed more than ten firms producing γ-interferons in the large scale bioreactors.

3.8 TISSUE PLASMINOGEN ACTIVATOR (TPA): PRODUCTION OF TPA

Biotechnology has also extended to disease therapy as researchers and physicians seek to develop new drugs and approaches to treating disease using DNA technology. The two examples are Tissue plasminogen activator and Interferon. Both drugs once in very short supply, are now produced in large quantities for therapeutic use. In addition entirely new class of compounds called antisense molecules has been developed to stimulate the specific molecules associated with disease.

The blockage of blood vessels by clots is the commonest cause of death in the developed world. Blood clots in the vessels, which supply the heart (coronary heart disease), brain (stroke) or lungs (pulmonary embolism) account for more than half of all deaths in the West. The processes involved in the formation and subsequent destruction of a blood clot (thrombus) are complex and finely balanced. An array of enzymes and other substances ensure that clots normally form only when and where they are required to prevent blood loss, and that they are broken down once their task is completed. The basis of a blood clot is a network of fibres of the protein, fibrin. This network can be destroyed by the action of plasmin, a fibrinolytic enzyme, which is formed in the blood from an inactive precursor, plasminogen. The role of plasminogen activators (Pas) is to remove part of the plasminogen molecule to yield plasmin (Fig. 3.8). The Pas are found in blood, tears, saliva, urine, semen and cerebrospinal and other body fluids. They are also found in many tissues and cells, including cancer cells. In addition, *Streptomyces* bacteria manufacture an enzyme, streptokinase, which is also able to activate plasminogen.

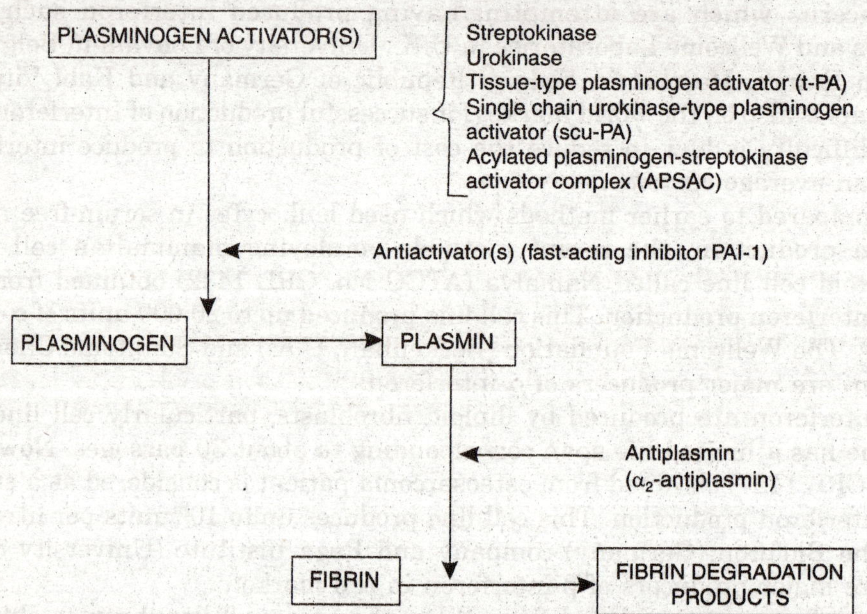

Fig. 3.8. A diagram of the fibrinolytic system stimulated by tissue plasminogen activator.

Since many years, coronary thrombosis has been treated with streptokinase and other enzymes that destroy blood clots in the blood vessels. A recent addition to the list of therapies is a protein called tissue plasminogen activator (TPA). It is a naturally occurring protease expressed in mammalian cells. TPA exhibits its protease activity by degrading an inactive precursor enzyme, namely, plasminogen, changing it into plasmin. Plasmin is also a protease that degrades fibrin, the protein of which blood clots are formed. TPA attaches itself to blood clots and stimulates other blood borne components to break down the clot without inhibiting the blood clotting power in other parts of the body (wounds). Whereas the another clot dissolving enzyme used in the therapy is urokinase decreases the clotting ability of the blood throughout the body resulted in risk of internal bleeding. TPA can be administrated intravenously whereas other clot dissolver must be administrated directly into the blocked vessel. TPA appears to work quickly than other dissolvers of clots. This is another advantage of TPA.

TPA was the first drug produced by a mammalian cell culture. The production of tissue plasminogen activator from a continuous line of procine kidney cells (LCC-PK_1; ATCC No. CL 101) has been patented by Lilly Research Laboratories (US patent No. 3904480). During 1980s the DNA technologists first synthesized complimentary DNA molecule for TPA. Then they biochemically attached it to a synthetic plasmid containing a TPA signal sequences, promoter and terminator sites. Then plasmid was introduced to mammalian cells, and cells secreting high levels of TPA were obtained from the mixture. Cultivated in large fermenters, the cells supplied sufficient TPA for pharmaceutical use.

In 1987, the biotechnology company Genentech received a licence from the FDA to market TPA under the trade name Activase. TPA is not used as a heart attack preventive but it is used as therapy for those having heart attacks. It is a protein product of DNA technology that attaches to blood clots and stimulates blood components to break down clot. The protein works locally rather than throughout the body where it could cause hemorrhaging.

3.9 VACCINES: PRODUCTION OF VACCINES

Vaccines are preparations of either dead or living attenuated (weakened) microorganisms. Vaccines may be administered to man or animals to stimulate their immunity to infection by the same or closely related organisms. Vaccine can also be defined as an antigenic material that produces specific immunity by stimulating an immune response. Edward Jenner in 1778 showed the cowpox pustule materials are known to protect the individuals from small pox (Vaccinia means cow pox). This lead to the process of vaccination and later Louis Pasteur advocated to the Rabies disease successfully. If an individual's immune system has responded to a particular antigen once, as state of resistance will remain for a long period of time. Vaccines are of three different kinds killed vaccines, Attenuated vaccines (avirulent) or toxoids.

Killed vaccines are suspensions of virulent microorganisms (bacteria/virus), which are killed by heat or antiseptic agents (phenol or formalin). But its antigenic characteristic should not denature during the heating or chemicals treatment. These vaccines are given twice or thrice at four weeks or longer time intervals. The major examples for killed vaccines are TAB vaccine, whooping cough vaccine, salk vaccine, rabies vaccine, cholera vaccine, influenza vaccine.

TAB vaccine

It is prepared by using three species of Salmonella. This suspension contain 1000 million bacilli of *Salmonella typhi,* 750 million bacilli each of *S. paratyphi* A and *S. paratyphi* B per ml. Then suspension was sterilized by heating at 60^0C for 30 minutes and preserved in 0.5% phenol. This vaccine is used against enteric fever and known to provide immunity for about two years. Booster dose of 0.5 ml is given after every two years to give a long lasting immunity.

Whooping cough vaccine

Bacterium *Bordetella pertusis* cells are killed by formalin or formalin merthiolate and suspension made at the rate of 200,000 million cells per ml. Intramuscularly three doses of one ml each was given at four week interval to all the children.

Poliomyelitis salk vaccine

Three polio viruses (type 1, type 2 and type 3) are killed by formalin and two doses of intramuscular injection containing 1 ml each was given at an interval of three weeks. Then two booster doses are required for the long term immunity.

Rabies vaccine

This was prepared from sheep brain infected with poliomeylitis virus. The 4% suspension of sheep brain in which the virus is killed by 0.5% phenol at 37°C is used. Now-a-days a new rabies vaccine is used from virus grown in human diploid cells.

Influenza vaccine

Influenza vaccine consists of formalin killed viral strains. Influenza, on the other hand may be prevented by the parenteral injection of vaccines of influenza viruses that have been grown in chick embryos and rendered noninfectious by formalin or ultraviolet irradiation. With such a dead vaccine, the duration of immunity rarely exceeds a year and since the causative agent (virus) of influenza is capable of mutating frequently, different vaccines usually must be prepared to combat each mutant strain.

The French company, Transgene, has been aiming at producing a vaccine against rabies; in 1982 the workers strived to synthesized surface epitope of the rabies virus. They isolated a clone of *E. coli* that produced a 58,000-dalton viral epitope of which the encoding DNA had been introduced into the bacterial host. Rabies is a zoonois that occurs endemically in animals and humans in several countries of Africa, South America and Asia and is increasing in European countries also. Preventive vaccination is impeded by the high cost of the currently used vaccines. This is primarily due to difficulties countered in virus propagation in human cells and viral purification. Generally human cell culture requires high cost. Therefore, the production of cheap vaccine from a viral protein with the required immunogenic properties would be useful from the industrial production and public health viewpoints.

Both antibiotics and vaccines are used in modern medicine to fight infectious diseases. Vaccines are used as prophylactic agents to increase immunity and to prevent subsequently infection with more virulent strains. Antibiotics are employed most often as therapeutic agents to control disease after its onset, although they may also be used prophylactically against some organisms.

Recently, vaccine for Rickettsial Rocky Mountain spotted fever and typhus fever are also available. However, the major problem is allergic reactions often arise from residual egg protein in which the rickettsiae have been cultured. This has not been achieved successfully for the causative microbe of syphilis (*Treponema pallidum*) nor has it been possible to make vaccines with conventional techniques relating to the extraction and purification of antigens. The DNA of this microorganism was isolated from infected rabbit testicles and cloned in *E. coli* host cells with the help of a coliphage used as a vector. This has enabled the development of more specific diagnostic screening tests for syphilis causative agent and the manufacture of an effective vaccine against disease. The need for such an effective vaccine is particularly important, as the strains of *Treponema* tend to become resistant to the antibiotics currently in use for the treatment of syphilis.

In veterinary medicine efforts are also being made to produce effective vaccine in genetically modified microbes, which could be effective against most of the animal diseases such as food and mouth disease and swine fever. Foot and mouth disease is highly contagious and threatening for cattle, sheep and swine farming. The virus causing this disease belongs to the Aphthovirus (single stranded RNA) genus of the family Picornaviridae. Recently German scientists have been successful in cloning double-stranded DNA copies of the viral RNA into plasmid pBR 322 and grown in *E. coli* host. It was found that more than 1,000 molecules of viral protein were synthesized per bacterial cell with a positive serological test.

The Aphtho virus is surrounded by a capsid made of sixty copies of four polypeptides, VP_1, VP_2, VP_3 and VP_4. Treatment of the virus particles with trypsin enzyme results in the cleavage of VP_1 and a considerable decrease in infectivity and a loss of immunizing activity. It has, therefore, been advantageous to synthesize the polypeptide precursor of VP_1 and to use it as the protein conferring long-lasting immunity. A lot of work is currently going on to improve the immunizing ability of VP_1 protein. A vaccine made up of VP_3 has also been synthesized by genetically engineered bacteria by workers from Plum Island Animal disease Center, New York. Intervet International, pharmaceutical company from the Netherlands has produced two vaccines, cloned in bacteria, to prevent infectious diarrhoea of swine and calves caused by pathogenic strains of *E. coli*.

The development of synthetic vaccines produced from genetic recombination and gene cloning, is likely to increase their medical utility and availability. This offers a safe, less cost and more potent vaccines than the conventional vaccines. The development of synthetic vaccines has long term prospects as far as replacement of current bacterial and viral vaccines is concerned. Producing vaccines using peptides with specific immunogenic properties is also being investigated. This includes polio, cholera, tuberculosis and others.

Now-a-days a new approach for producing vaccines is the cell culture technique. The new vaccines based on cell culture are purer and safer than some of the existing vaccines. In the existing vaccines all the antigens are obtained after treatment of animals with particular bacteria or virus causing the disease are used. This causes "overloading". In the new method, cells are cultured with the specific bacteria or virus and only those antigens or antibodies, that are specifically required against that disease are injected into the human body.

3.10 HYBRIDOMA: PRODUCTION OF MONOCLONAL ANTIBODIES

Monoclonal antibodies are ultrapure antibodies synthesized by hybridoma cells. Hybridoma cells are hybrid cells formed by fusion of B-lymphocytes (antibody producing cell) and malignant myeloma (cancer) cells. B-lymphocytes produces antibodies specific for a particular antigens. Myeloma cells proliferate rapidly and can be cultured indefinitely (immortal) from a single cell. They cannot, however, be induced to yield antibodies to a specific antigen. That difficulty was overcome in 1975 when Cesar Milstein, G. Kohler and their colleagues at the Medical Research Council laboratory of Molecular Biology in Cambridge, England conceived the idea of fusing mouse myeloma cells with B-lymphocytes from the spleen of a mouse immunized with a specific antigen. The resulting hybrid myeloma or hybridoma cells had the properties of both parent cells: immortality and the ability to secrete large amounts of a single specific type of antibody. This discovery was a masterpiece in the research career of Kohler who had been a Ph.D. student in Milstein's laboratory. In 1947, Kohler succeeded in producing a hybridoma by fusing P3 myeloma

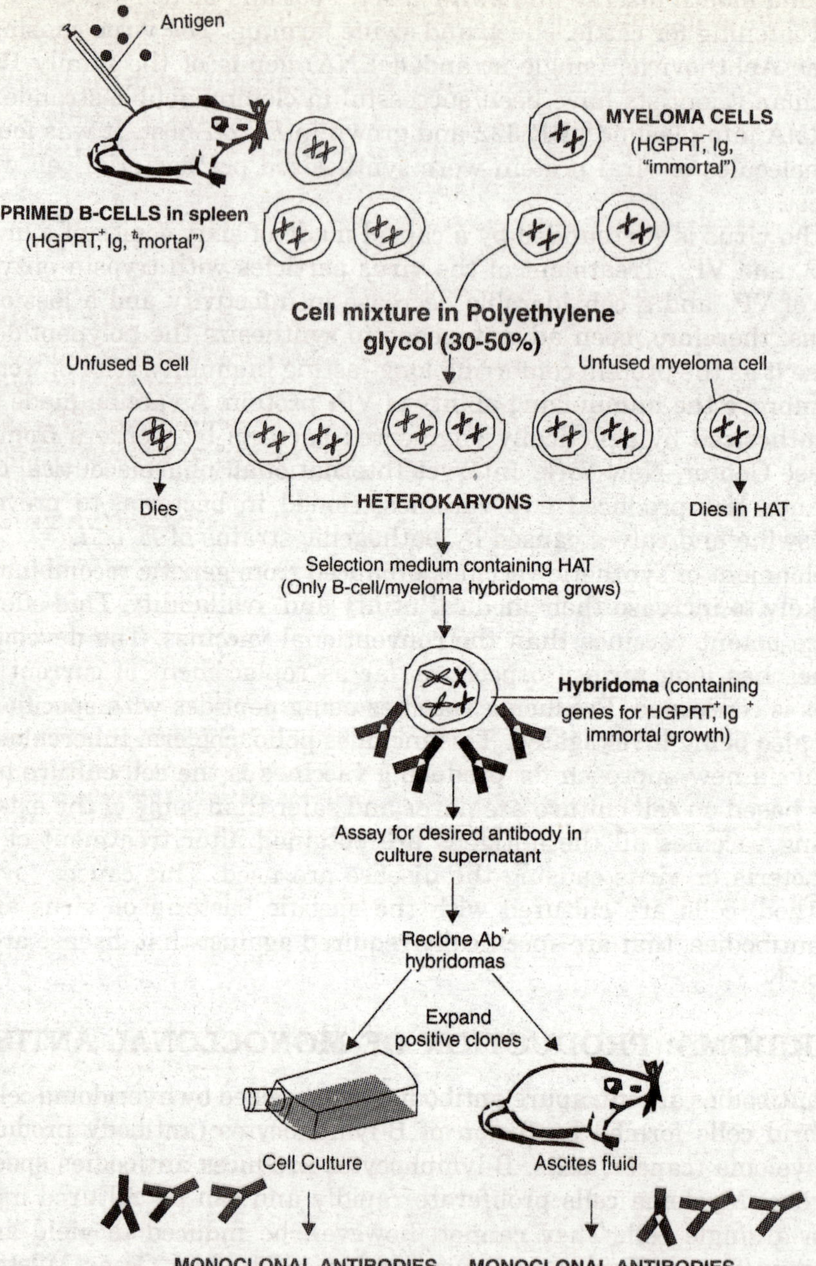

Fig. 3.9. The procedure followed for producing monoclonal antibodies specific for a given antigen developed by G. Kohler and C. Milstein. Spleen cells ($HGPRT^+$ and lg^+) from an antigen-immunized mouse are fused with mouse myeloma cells ($HGPRT^-$ and lg^-). The spleen cell provides the necessary enzymes for growth on HAT (Hypoxanthine, Aminopterin and Thymidine) medium, while the myeloma cell provides immortal-growth properties. Unfused myeloma cells or myeloma/myeloma fusions fail to grow because of their lack of HGPRT. Unfused spleen cells have limited growth capabilities *in vitro* and will die within a few days.

cells (resistant to azaguanine) and a lymphocyte from the spleen of a mouse immunized against red blood cells. Mice were first immunized against these R.B.C's; then P3 myeloma cells were mixed with spleen cells from immunized mice with the help of polyethylene glycol. The hybrid cells secreted immunoglobulins and some of these cells produced against the R.B.C.'s. Clones of these cells producing a single type of antibodies (Monoclonal antibody) were isolated and maintained in culture (Fig. 3.9).

The separated hybridoma clones can be maintained indefinitely at freezing temperature. These cells/clones could be injected into animals of same line (which produced the original cell required for fusion). These animals then develop tumors producing monoclonal antibodies, corresponding to those cells/clones, at a very high concentration (10-30 mg/ml). These antibodies could be produced against small haptens, proteins, enzymes, carbohydrates and glycolipid, viruses and even cell membrane components. Monoclonal antibodies, unlike the usual antiserum, can be reproduced many times. It is now possible to prepare human monoclonal antibodies and the motivation to produce these for therapeutic use is tremendous.

The technique for production of Monoclonal antibodies could be briefly summarized as follows:

1. Obtaining lymphotes secreting antibodies from immunized mouse.
2. Production of hybridomas by fusing lymphocytes (antibody secreting) with myeloma cells which would secrete specific antibodies.
3. Separating different hybridomas from each other by cloning.
4. Contınuous cultivation of hybridomas producing specific monoclonal antibody.

A suspension of immune lymphocytes are mixed with myeloma cells, the mixture is washed, concentrated and exposed under carefully controlled conditions to a fusing agent. Sendai virus was used earlier on lymphocyte hybridization. However, it is now being gradually replaced with polyethylene glycol (PEG) since this gives much better results, besides being inexpensive and easily available. The fusion is carried out by adding PEG to a concentrated suspension of lymphocyte-myeloma cells either in a tube or deposited on polycarbonate filter discs.

The PEG preparations with mean molecular weights between 1000-6000 are suitable for cell fusion; outside of this range they have an unacceptably high toxicity. The optimum concentration of PEG is normally between 30-50% and the time for which cells are exposed to PEG at this concentration is usually short (1 minute) to minimized the membrane damaging effects of the fusing agents. The cells are then gently resuspended, diluted and washed to free off PEG. The entire operations are conducted at 37°C. It is not necessary to purify B-lymphocytes before myeloma cells selectively fuse with these cells. The frequency of formation of viable hybrid cells is low. Most cells die at an early stage and this can be indirectly damaging to the surviving cells. Hybridoma cells are sensitive to alkaline culture conditions and are, therefore, maintained in an atmosphere containing at least 5% CO_2. The fusion conditions are such that the vast majority of myeloma cells are not fused to spleen B-lymphocytes and these need to be destroyed. This is done by growing the cells in the presence of HAT (Hypoxanthine, Aminopterin and Thymidine) medium containing aminopterin or amethopterin drugs which inhibit folic acid metabolism essential for purine synthesis (*de novo* pathway) and, therefore, kill unfused the myeloma cell deficient in HGPRT$^-$ (enzyme hypoxanthine guanine phosphoribosyl transferase).

These myeloma cells (HGPRT^{-}) are mutational variants of parent myeloma and produced for purposes of selecting the hybrid myeloma cells from that of HGPRT minus cells that cannot carry out salvage pathway of DNA synthesis. The unfused B-cells are able to survive in HAT medium, but these cells do not live for longer periods *in vitro* and thus die out within few days. Hybrid cell lines secreting specific monoclonal antibody are isolated and cloned by routine laboratory procedure. It is technically simpler to clone cells in fluid phase media.

Many monoclonal antibodies are now available for detecting pregnancy, diagnosing numerous pathogens, detecting tumor antigens, toxins, drugs in the blood and matching histocompatibility antigens. Several public and private enterprises have commercialized production of monoclonal antibodies all over the world since 1981. The price ranges between Rs. 1500-4000 per milli gram. Many pharmaceutical firms have been interested in large-scale production of ultrapure monoclonal antibodies because of enormous applications of these molecules.

Procedure followed for production of hybridoma and monoclonal antibodies

1. Inject an appropriate antigen into a mouse first by subcutaneously or in intraperitoneal cavity with successive doses and finally by intravenously.
2. Three days after the last intravenous dose animal was killed.
3. Spleen was asceptically removed and gently disrupted to release lymphocytes and red blood cells.
4. Lymphocytes are separated from RBC and spleenic fluid by density gradient centrifugation.
5. Lymphocytes are washed to remove all contaminating materials.
6. These lymphocytes are mixed with HGPRT – negative myeloma cell line.
7. Expose to Polyethylene glycol (PEG) at 30-50% concentration for one minute.
8. Cells are then washed to make free from PEG.
9. Washed cells comprise mixture of hybridomas and unfused myeloma cells and lymphocytes.
10. Hybridoma are selected on HAT medium. After 10-14 days the only surviving cells on HAT medium are the hybridoma cells.
11. Dilute hybridoma (different types of antibody producing cells) to a level so that aliquot contains single hybridoma cell (growth not easy on dilution).
12. Each clone is then examined (mass screening).
13. Store it in refrigerated conditions.
14. Culture whenever required on large scale.

APPLICATIONS OF MONOCLONAL ANTIBODIES

MCA has multifarious applications because of their specificities and thus gives a clear-cut results. A large number of hybridoma are produced and stored by cryopreservation technique. Hybridoma banks have been set up in a number of Institutes and Research centres.

(i) Monoclonal antibodies help in classification of blood groups into ABO types and Rh types.

(ii) By using monoclonal antibodies of an animal immunised against a particular medicine, dose determination of the medicine can be carried out. The reliability of such an immunodosage and the savings it entails are of substantial advantages.

(iii) Specific tools based on hybridoma technology could be produced to carry out hormone tests to diagnose viral diseases, to detect certain types of cancers, to monitor the presence or appearance of malignant cells after surgical or radio-therapeutic treatments and to detect allergies.

In 1982, Farrands *et al.*, succeeded in isolating an antibody reacting specifically to malignant tumors of the colon and rectum. Using same technique Mach detected cancer of thyroid and gastrointestinal tract and Tublana detected epithelial cancer.

(iv) Monoclonal antibodies could also be used for the purification of complex mixtures of substances which are biologically important (proteins, hormones, etc.)

(v) Monoclonal antibodies are also being used for the labelling and precise identification of specialized cells, such as neurons in order to gain better knowledge of the ways in which these cells associate and operate (localisation of chemical neutrotransmitters).

(vi) Cell membranes proteins are difficult to purify, since they are present in small amount and their biological activity is not easily measured. These difficulties are overcome by characterising membrane proteins by immunological method such as monoclonal antibody technique.

(vii) Monoclonal antibodies could be used as standard reagents to identify molecules present in cell membranes and also to distinguish between cell populations. Promising results are obtained with several blood forming and lymphoid cells.

(viii) The differential diagnosis of various leukaemias and related disorders is also possible by these technologies.

(ix) Serotherapy, a direct therapy, can be made more effective with the administration of a purified or monoclonal antibodies.

(x) Monoclonal antibodies could also be used in the preparation of very specific vaccines, particularly against certain viral strains and parasites.

Croce and his co-workers (1980) obtained a human hybridoma by fusing B lymphocytes from a patient suffering from multiple myeloma with lymphocytes from another patient with sub-acute panencephalitis. These hybridomas produced human M immunological molecules, specific for certain components of the measles virus. Such a result indicated the possibility of obtaining hybridomas from human B lymphocytes capable of secreting antibodies against pathogenic viruses.

(xi) Monoclonal antibodies could also neutralise the action of lymphocytes responsible for the rejection of grafts and destroy the auto-antibodies produced in auto-immune diseases (certain forms of diabetes, multiple sclerosis, rheumatic diseases).

(xii) Monoclonal antibodies could considerably increase the effectiveness of medicinal substances on the target cells while avoiding the serious side effects of cancer therapies.

3.11 NEED TO EXPRESS IN ANIMAL CELLS

Animal cell culture system has become an important aspect of biotechnology for the production of proteins, which are of medicinal importance. Now the major compounds that are in wide usage by animal cell culture system include varieties of vaccines, monoclonal antibodies, follicle stimulating hormone (FSH), IFN-α, IFN-β, IFN-γ, tissue plasminogen activating factors, factor VII, factor VIII, factor XI, factor X and erthryopoetins. Because efforts made to produce the required proteins (specifically glycoproteins) especially proteins which are of medicinal importance in the prokaryotic microorganisms and plants are known to having improper function, activity or antigenic characters.

Sometime the required proteins may not be obtained in large quantities or complex modification of final protein synthesized is not possible from the prokaryotic organisms. Generally prokaryotic organisms are not able to synthesize the foreign proteins in large quantities required for fulfilling the demand of the market. The complex modifications of protein include intricate tertiary structure formed by disulfide bond formation or "post-translation modifications" (changes in protein structure after the protein is made from m RNA). The post-translation modifications (Table 3.3) consist of addition of small chemical groups (e.g. an acetyl, methyl or phosphate group), glycosylation, lipid addition and gamma carboxylation.

Table 3.3: Post-translation modification of proteins

Modification	Amino acids that are modified
Addition of small chemical groups	
Acetylation	Lysine
Methylation	Lysine
Phosphorylation	Serine, Threonine, Tyrosine
Hydroxylation	Proline, Lysine
N-formylation	N-terminal glycine
Addition of sugar side chains	
O-linked glycosylation	Serine, Threonine
N-linked glycosylation	Asparagine
Addition of lipid side chains	
Acylation	Serine, Threonine, Cysteine
N-myristoylation	N-terminal glycine
Addition of biotin	
Biotinylation	Lysine

3.12 POST-TRANSLATION MODIFICATIONS

Addition of chemical groups was carried out in a highly specific manner, the same amino acids being modified in the same way in every copy of the protein. This chemical modification often plays an important role in determining the precise biochemical activity of the target protein.

Glycosylation of a protein means addition of carbohydrate moiety to the protein molecules to form glycoproteins. The above mentioned therapeutic proteins are good example for glycoproteins. This occurs in the endoplasmic reticulum and Golgi complex of eukaryotic cell and depends on the presence of specific enzymes; glycosyltransferases and glycosidases. Glycosylation are generally two types i.e. O-linked glycosylation; is the attachment of a sugar side chain through hydroxyl group of a serine or threonine amino acid and N-linked glycosylation involves attachment through the amino group on the side chain of asparagnine (Fig. 3.10).

Glycosylation can result in attachment to the protein α grand structures having networks of 10-12 sugar units of various types.

The prokaryotic organisms neither contain these organelles nor the enzymes and therefore not able to perform this post-translational modification. Eukaryotic yeast and filamentous fungi are able to glycosylate proteins but they do so differently from that of the animal cells.

The presence and conformation of the sugar moiety of a glycoprotein is very essential for a functional product in many cases. This functional properties usually prolongs the half-life in the blood stream, and diverts the glycoproteins to its specific location in the cell. Only the eukaryotic animal cells can do these changes appropriately. Also in many cases the production can be done by plants (especially secondary metabolites), but it involves several enzymatic functions most of which are not known. In some cases, such as immunoglobulins, progress has been made to assemble these molecules in bacterial cell or plant cell systems. However, these processes are not able to work efficiently at large scale. Hence these products can be obtained/expressed only from the cultivation of eukaryotic animal cells especially mammalian cell culture system. In recent years, the

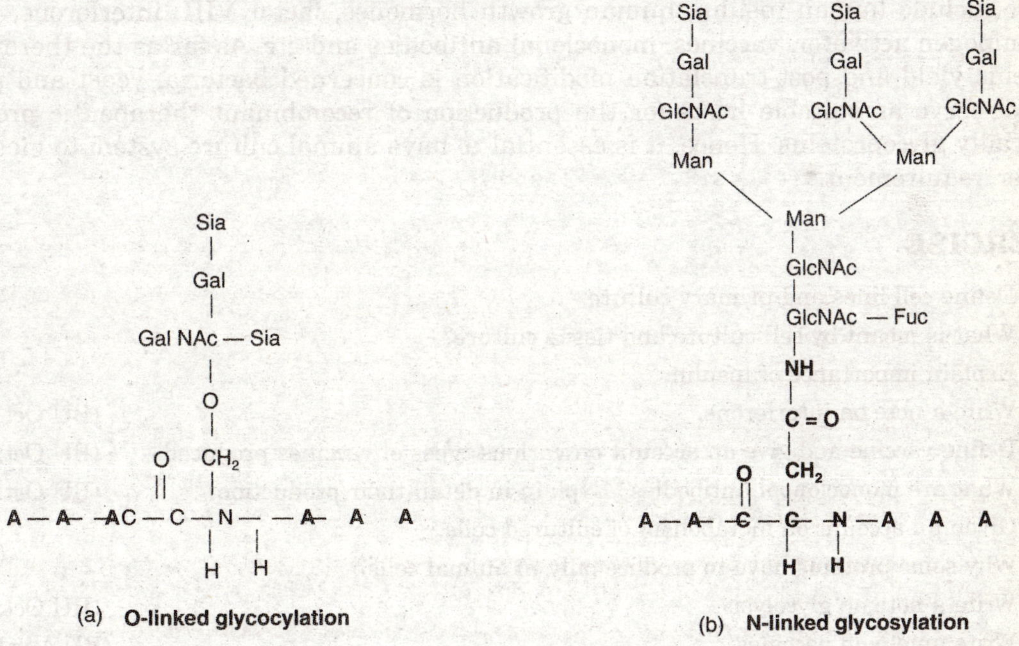

Fig. 3.10. Glycosylation of proteins (a) O-linked glycosylation. Sugar units added to serine amino acid (b) N-linked glycosylation. Larger sugar structure added to asparagine amino acid. Abbreviations used: A, different amino acids; Fuc, fucose; Gal, galactose; GalNAc, N-acetyl galactosamine; GlcNAc, N-acetyl glucosamine; Man, mannose; Sia; Sialic acid.

advance of insect cell culture system has allowed some of these molecules to be processed with post-translation modifications similar to higher eukaryotic cells. Pharmaceutical proteins that are not glycosylated or need not be glycosylated for proper function such as insulin, human growth hormone, human serum albumin and hemoglobin may be produced more cost-effectively with bacteria, yeasts and filamentous fungi.

SUMMARY

Growth of animal cells in a medium is known as cell culture. Currently the cultivation of animal cells has been established in many laboratories, clinics and research centres around the world. The importance of many components of the animal cell culture medium for cellular proliferation and biological production has been defined at molecular level. The cultured cells are also known to have same type of metabolism, similar to intact animal cells. Animal cells also need glucose as organic carbon source and utilized in glycolysis and Kreb's cycle for energy production. A cultured cell has capacity to utilize glutamine as energy source. Animal cells require varieties of amino acids for protein synthesis and they are capable of synthesizing lipid and nucleic acids from simple organic molecules used in the culture medium. Thus, in culture condition they could able to show the growth by synthesizing all the biomolecules required for cell multiplication.

Now-a-days it is possible to cure several human genetic disorders by using therapeutic proteins produced by recombinant DNA technology. There are many human therapeutic proteins being produced in large quantities to meet the demand of the affect population. These include human insulin, human growth hormones, factor VIII, interferons, tissue plasminogen activator, vaccines, monoclonal antibodies and etc. As far as the therapeutic proteins yield and post-translation modification is concerned bacteria, yeast and plants do not serve as suitable hosts for the production of recombinant therapeutic proteins, especially glycoproteins. Hence, it is essential to have animal culture system to meet this proper requirement.

EXERCISE

1. Define cell lines and primary culture.
2. What is meant by cell culture and tissue culture?
3. Explain importance of insulin.
4. Write a note on interferons. (BU Oct. 2003)
5. Define vaccine and give an account on various types of vaccines produced. (BU Oct. 2001)
6. What are monoclonoal antibodies? Explain in detail their production. (BU Oct. 2002)
7. Given an account on metabolism of cultured cells.
8. Why some proteins have to produce only in animal cells?
9. Write a note on glycolysis. (BU Oct. 2001)
10. Write a note on vaccines. (BU April 2001)
11. Give an account of glycosylation.
12. What is meant by post-translation modification? Add a note on different types of post-translation modifications.

13. Define secondary metabolites. Explain production of recombinant insulin and growth hormone. (BU April 2001)
14. Write a note on HAT medium. (BU April 2001, Oct. 2001)
15. Elucidate the applications of monoclonal antibodies. (BU Oct. 2002)
16. Give an account of types and properties of polio vaccines. (BU Oct. 2001)
17. Write a short note on monoclonal antibodies. (BU April 2003)
18. Discuss aminoacid catabolism with examples. (BU April 2003)
19. Explain the production of recombinant factor VIII. Add a note on interferons. (BU April 2003)
20. How tissue plasminogen activator is produced by animal cell biotechnology?

4

Growth Factors Promoting Proliferation of Animal Cell Culture

4.1 INTRODUCTION

In animal cell culture system it is now very clear that *in vitro* condition cultured cell undergo divisions only when blood serum is added to the medium. Serum means liquid portion of the blood (plasma) without fibrinogen and blood cells. First des Ligners in 1928 reported the successful cultivation of different mammalian tissues in dilute serum medium. Then in 1933 Parker showed cultivation of chick tissues in serum medium. Fischer *et al.* in 1948 gave a clue of low molecular weight growth factors present in serum and their importance.

Recently these serum proteins were identified and are used instead of whole serum in the culture medium (Table 4.1). They include epidermal growth factor (EGF), endothelial cell growth factor (ECGF), fibroblast growth factor (FGF), platelet derived growth factor (PDGF), Interleukin 1(IL-1), Interleukin 2 (IL-2), nerve growth factor (NGF), transforming growth factor (TGF), eye derived growth factor (EDGF), granulocyte-macrophage colony stimulating factor (GM-CSF), granulocyte colony stimulating factor (G-CSF), macrophage colony stimulating factor (M-CSF), stem cell factor (SCF) and erythropoietin (EPO). Thus serum provides some of the growth factors or stimuli necessary for the growth and maintenance.

4.2 COMMON FACTORS AFFECTING THE CELL CULTURE

Commonly in cultural condition addition of epidermal growth factor (EGF), platelet derived growth factor (PDGF) and transforming growth factor (TGF) can stimulate the division in most of the cell types. Whereas for specific cells such as nerve cells require **nerve growth factor** (NGF) and endothelial cells require endothelial cell growth factor **(ECGF)**. Now more specific type of growth factors for a particular cell type has been discovered. These factors are supplemented to the medium instead of whole serum. Thus it will decrease the non-defined medium usage with different lots, animals, strains and so on. Sometime the serum extracted may contain the infective viruses, which may be discovered or undiscovered type.

Table 4.1: Serum proteins that show growth factor activity on cellular proliferation and/or differentiation

Growth Factor	Main Source	Primary Activity on Cells	Characteristics
PDGF	platelets, endothelial cells, placenta	promotes proliferation of connective tissue, glial and smooth muscle cells	two different protein chains form 3 distinct dimer forms; AA, AB and BB
EGF	submaxillary gland, Brunner's gland	promotes proliferation of mesenchymal, glial and epithelial cells	
TGF-α	common in transformed cells	may be important for normal wound healing	related to EGF
FGF	wide range of cells; protein is associated with the ECM	promotes proliferation of many cells; inhibits some stem cells; induces mesoderm to form in early embryos	at least 19 family members, 4 distinct receptors
NGF		promotes neurite outgrowth and neural cell survival	several related proteins first identified as proto-oncogenes; trkA (*track A*), trkB, trkC
Erythropoietin	kidney	promotes proliferation and differentiation of erythrocytes	
TGF-β	activated TH_1 cells (T-helper) and natural killer (NK) cells	anti-inflammatory (suppresses cytokine production and class II MHC expression), promotes wound healing, inhibits macrophage and lymphocyte proliferation	at least 100 different family members
IGF-I	primarily liver	promotes proliferation of many cell types	related to IGF-II and proinsulin, also called Somatomedin C
IGF-II	variety of cells	promotes proliferation of many cell types primarily of fetal origin	related to IGF-I and proinsulin

Generally calf, bovine, porcine, horse or goat serum are used for the different cell culture. Culturing of human lymphocytes autologous AB serum was used. The serum from calf foetus is preferable because there was no gammaglobulins when compared to the adult animal serum, which are known to suppress the growth of many animal cell cultures. Serum factors generally act as mitogenic agents **(mitogens)**. Any factor which stimulates the cell division in culture conditions are referred to as mitogenic agents. These mitogenic agents are very much essential for the induction of division and maintenance of cells in culture condition.

4.3 EPIDERMAL GROWTH FACTOR (EGF)

Stanley Cohen in 1965 discovered this growth factor during the experiments of animal cell culture and EGF is a potent mitogen for epithelial and mesenchymal cells. This consists of 53 amino acid polypeptide chain and submaxillary gland and kidney produces this growth factor. The EGF known to stimulate the epidermal cell proliferation and nerve cell regeneration. EGF is having major action on stem cells, leukocytes, macrophages and tissue cells. More number of animal cells shows response to the epidermal growth factor in the medium. This is also required for the embryo development. It stimulates synthesis of DNA, collagen, hyaluronic acid and fibronectin.

4.4 ENDOTHELIAL CELL GROWTH FACTOR (ECGF)

This growth promoting factor is isolated from eye and brain. It is also produced in the developing embryo and having molecular weight of 16-17 kD. A large number of mesoderm and endoderm derived cells including vascular endothelial cells, glial cells, smooth muscle cells and fibroblasts are induced by this factor in the culture condition.

4.5 FIBROBLAST GROWTH FACTOR (FGF)

It is a 18 kD protein produced in many adult tissues and also in the embryo. There are two forms of FGF namely acidic FGF and basic FGF. It acts as a mitogenic substance, which induces cell proliferation in a number of cells derived from mesoderm and ectoderm such as fibroblasts, smooth muscle cells, vascular endothelial cells and glial cells. Heparin potentiate FGF activity by protecting it against denaturation and degradation. It also stimulates the production of collagenase and plasminogen activator.

4.6 PLATELET DERIVED GROWTH FACTOR (PDGF)

This is an endogenous growth promoting protein known to induce cell division in epidermal cells, leukocytes and T cells. The PDGF is the major nitogen found in serum. It stimulates the proliferation and chemotaxis of fibroblasts and smooth muscle cells. It also stimulates collagen, glycosaminoglycan and collagenase production by fibroblasts. There are two types of PDGF, namely, platelet derived growth factor-A, which is 125 amino acid polypeptide chain and platelet derived growth factor-B, which is a 109 amino acid polypeptide chain. These forms either homodimers (AA) or heterodimer (AB). PDGF is mainly involved in wound healing process.

4.7 INTERLEUKIN 1(IL-1)

This cytokine is earlier known as endogenous pyrogen, leucocyte activating factor (LAF), B cell activating factor (BAF), lymphocyte activating factor or catabolin which later named as interleukin-1 in 1979. This is known to produce by macrophages, monocytes, B cells, endothelial cells and fibroblasts but most abundantly by macrophages. In culture condition or *in vivo* they have effect on varieties of cell types. They help in proliferation of stem cells, lymphocytes, firbroblasts, endothelial cells, B cells, T_H cells and epithelial cells.

Interleukin 1 was known to induce protein synthesis (collagenase, plasminogen activator, fibrinogen, collagen), adhesion molecules production, extracellular matrix production, prostaglandin synthesis in various cell types and alters the permeability of cells. The receptor for IL-1 is present on all the cells of the animal body, hence they can act on all the cell types in culture system. IL-1 is a stable polypeptide retaining its activity upto 56°C and between pH 3-11. It occurs in two molecular forms IL-1 alpha (IL-1α) and IL-1 beta (IL-1β).

4.8 INTERLEUKIN 2 (IL-2)

This cytokine when discovered in 1976 is known as T cell growth factor (TCGF), later it was renamed as interleukin-2. It is 133 amino acid polypeptide chain. These are produced by T cells (majority of the time by CD4+ cells) and generally help in proliferation and differentiation of all types of T cells. It also supports long term growth/maintenance of the cultures. Sometimes they also show the activation and division of B cells.

4.9 NERVE GROWTH FACTOR (NGF)

This was the first "growth factor" to be isolated and characterised. Rita Levi-Montalcini and Hamburger discovered this in the 1951 and then purified by Cohen in 1960. This growth factor was also known as neurotrophins. It is 118 amino acid polypeptide chain with molecular weight of 26 kD and produced by submaxillary gland. The nerve growth factor regulates the development, differentiation and survival of neurons in cultural condition. The NGF is not a good mitogen (induces mitotic division) for the cells in culture.

4.10 TRANSFORMING GROWTH FACTOR (TGF)

It is produced in most adult tissues and many cell types in culture. The platelets, macrophages, lymphocytes and mast cells are major cells that secretes TGF. The TGF-β is a homodimeric 25 kD protein with 112 amino acids. There are two tyes of TGF namely TGF-α and TGF-β and also produced by embryo cells. This factor generally makes normal cells anchorage independent. The anchorage independent character is a feature of transformed animal cells (*e.g.* tumor cells). Hence this factor is known as transforming growth factor. This elicits multiple cellular responses such as stimulation of proliferation, reversible growth inhibitory activity against epithelial, endothelial, fibroblast, neuronal, lymphoid and hematopoietic cell types.

4.11 ERYTHROPOIETIN (EPO)

It is a glycoprotein and stimulates the division and differentiation of myeloid stem cells to red blood cells. This was produced by kidney and fetal liver cells. The molecular weight is approximately 30,400 daltons and it is 166 amino acid polypeptide chain. The protein contains two disulfide bonds and four sites of glycosylation. The disulfide bonds and glycosylation are very much essential for the biological activity. It acts as mitogenic substance and enhances survival of cells in culture condition and prevents programmed cell death.

4.12 EYE DERIVED GROWTH FACTOR (EDGF)

This is having 14,000 daltons molecular weight. The exact action on the culture cells is not yet discovered, but it is known to have positive effect on the cells in cultural condition.

4.13 PLATELET-DERIVED ENDOTHELIAL CELL GROWTH FACTOR (PD-ECGF)

The platelet-derived endothelial cell growth factor (PD-ECGF) is a newly described growth factor. This is an acidic single-chain polypeptide with molecular weight of 45KD. It stimulates only endothelial cells and not fibroblasts. This acts as a specific endothelial cell mitogen.

4.14 INSULIN-LIKE GROWTH FACTOR-I (IGF-I)

Insulin-like Growth Factor-I is secreted from the liver into blood circulation. It is structurally related to insulin. This growth factor involved in a process regulated by pitutary growth hormone (GH) and so it mediates the growth promoting activity of growth hormone. This factor is earlier known as Somatomedin C, because of its mediation activity of growth hormone activities particularly on bone growth. Later it was found that IGF-I has autocrine and paracrine action along with its initially noticed endocrine action on both IGF-I is mitogenic in nature and has effect on fibroblasts, osteoblasts, smooth muscle cells, fetal brain cells, neuroglial cells and erythroid progenitor cells. This IGF-I exerts its effect exclusively through the IGF-I receptor (IGF-IR).

The IGF-I receptor, similar to insulin receptor (IR) on the cell membrane has intrinsic tynosine kinase activity. Because of its structural similarity IGF-I binds to the insulin receptor but at lesser affinity than does by insulin itself. It is also known to regulate angiogenesis and stimulates protein synthesis.

The IGF-I is 7.5 kDa polypeptide compound of 70 aminoacids. This shares 48% homology with proinsulin and highly conserved between species, with 100% identity among human, bovine and procine IGF-I. The IGF-I has 3 major effects on cell growth:

1. It signals mitosis in a different varieties of cells.
2. It is a very essential factor for establishing and maintaining cells in a transformed phenotype.
3. It gives the protection against programmed cell death or apoptosis, both *in vitro* and *in vivo*.

4.15 INSULIN-LIKE GROWTH FACTOR-II (IGF-II)

Insulin-like Growth Factor-II is also known as multiplication stimulating activity (MSA). The IGF-II is almost exclusively expressed in embryonic and neonatal tissues. After the birth the traceable amount of IGF-II level falls down significantly. Because of this reason it is considered as foetal growth factor necessary for overall growth and development of foetus. The IGF-II shows its bioactivities through its binding to IGF-II receptor (IGF-IIR).

This growth factor is 7.5 kDa polypeptide chain having 67 aminoacid residues.

SUMMARY

Presently it is well-known that *in vitro* cultured animal cells must require blood serum for their proper growth and proliferation. Serum contains several proteins which generally aid in the growth and proliferation of animal cells. Recently these serum proteins were separated in pure form and identified their chemistry. These include epidermal growth factor (EGF), endothelial cell growth factor (ECGF), fibroblast growth factor (FGF), platelet derived growth factor (PDGF), nerve cell growth factor (NGF), transforming growth factor (TGF), eye derived growth factor (EDGF), erythropoietin (EPO), interleukin-1, interleukin-2, granulocyte colony stimulating factor (G-CSF) and etc.

At present a particular growth factor is used to stimulate the particular type of animal cells instead of using whole serum. This is because specific mitogenic effect of individual growth factor was very well understood in culture condition. And also as serum may contain infective viruses and immunoglobulins, which are known to suppress the growth of many animal cell cultures. Hence, it helps to formulate a defined medium usage in the culture of specific animal cell type.

EXERCISE

1. Name different types of growth factors for animal culture. (BU April 2001)
2. Explain application of interleukins and erythropoietin.
3. Write a note on transforming growth factor.
4. What are the applications of growth factors in culture?
5. Explain the role of growth factors in animal cell proliferation. (BU Oct. 2001)
6. Write a short note on erythropoietin. (BU April 2001)
7. What is meant by mitogen?
8. Give the composition of serum and its application in animal cell culture.
9. Name types of cells induced by platelet derived growth factor.
10. Describe effects of interleukin 1 on target cells.
11. Write a short notes on fibroblast growth factor.

(BU April 2003)

5

Bioreactors for Large Scale Culture of Cells

5.1 INTRODUCTION

Bioreactor can be defined as a vessel or container used to culture the cells at controlled environmental condition for the production of some commercially valuable product. The bioreactor is usually of large volume vessels varies from one litre to 10,000 litres. These bioreactors has provisions for maintaining the aeration, stirring to achieve medium and cell mixing, contamination control and replacement of used medium or used medium with cells.

Animal cells produce macromolecules of medical interest that are not available by any other chemical methods or microbial source. The only available source of such substances previously was urine, animal tissue or even human tissue when human-specific molecules such as growth hormones were required. Later various cell lines have been identified for such purposes. In 1925, on the recommendation of the Oncological Council and the National Cancer Research Institute a cell bank was founded in the USA (American Type Culture Collection; (ATCC), Rockville, Maryland). In 1979 ATCC lists 400 cell lines and 400 species, including about 200 human fibroblast lines with special genetic features. Thus, now it is possible to establish animal culture system for the production of varieties of compound both in laboratory scale and large scale.

Before starting large scale cell culture system, the conditions such as cell type, source of cells (embryonic or adult), subculture, serum free growth, medium adoptability, gas phase requirement, culture system and capacity to grow in suspension cultures are to be properly determined and standardized. Otherwise the large scale production cannot be achieved with greater success.

Large-scale cell cultures are used for the production of various biomedical compounds such as interferons, interleukins, hormones, antibodies, enzymes and vaccines (Table 5.1). The bioreactors up to 10,000 litres are generally used for the production purposes. The large-scale culture of cells can be done by three methods such as (1) monolayer cultures, (2) suspension cultures and (3) immobilized cell cultures.

5.2 MONOLAYER CULTURES

Monolayer cultures system are used for anchorage dependent cells. Increase in cell concentration in culture is based on the principle of increasing the surface area available for the growth by using plates, ceramics, spirals and microcarrier particles. Some of the monolayer cultures are as follows.

Table 5.1: Various biomedical compounds produced in animal cell culture system

Biomedical compound	Type of cells	Used for
Currently in Therapeutic Use		
Human growth hormone (HGH)	Mouse mammary tumour cell line	Dwarfism
Tissue plasminogen activator (tPA)	Chenase hamster ovary (CHO) cell line	Thrombolysis
Erythropoietin*	Chenase hamster ovary (CHO) cell line	Anemia
Blood clotting factor VIII*	CHO cell line	Hemophilia
In Advanced Stages of Development		
Antithrombin VIII	CHO cell line	Anticoagulant
Blood factor VIII	Baby hamster kidney (BHK)-21 cell line	Hemophilia
Blood factor IX	CHO cell line	Hemophilia-B
Interleukin-2 (Il-2)	Lymphocytes	Cancer therapy
Human recombinant antibodies	CHO cell line and Mouse myeloma lines	Infections, various cancers
Hepatitis-B surface antigen (HBSAg)	CHO cell line and C-127 cell line	As vaccine against Hepatitis-B
Interferons (IFN)	Lymphocytes	Tumours
Plasminogen activator (including chimeric and modified)	CHO cell line and *Spodoptera frugiperda*	Thrombolysis
CD4 Protein	Mammalian cell line	AIDS vaccine
gp160 protein	Mammalian cell line	AIDS vaccine
gp24 protein	Mammalian cell line	AIDS vaccine
Interleukin-1 receptor antagonist	Mammalian cell line	Severe sepsis
Atrial natriuretic	Mammalian cell line	Kidney failure
Ciliary neurotropic factor	Mammalian cell line	Amyotrophic sclerosis
DNase	Mammalian cell line	Cystic fibrosis

* Approved for marketing in India.

Roux bottle

It is a flat glass bottle having more internal surface area. This was more often used in the laboratory for the production purpose (Fig. 5.1). Roux bottle is kept stationary after pouring the medium, thus it provides space approximately 175-200 cm^2 internal surface area for the cell attachment on the medium and it occupies around 750-1000 cm^3 area.

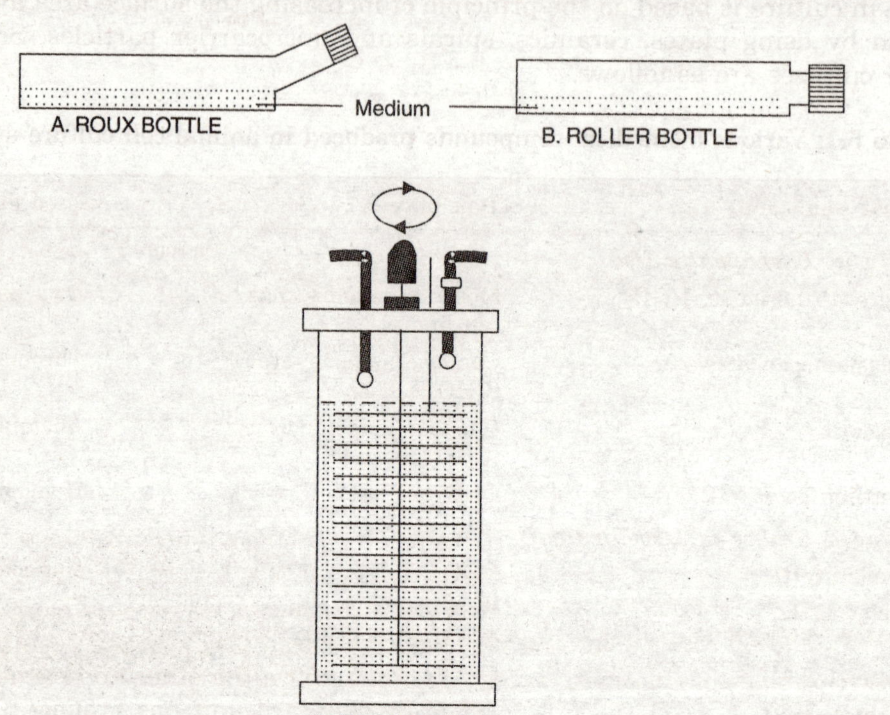

Fig. 5.1. Figures showing three types of culture vessels used for large scale animal cell cultures. (A) Roux bottle, (B) Roller bottle, (C) Stack plate vessels.

Roller bottle

The glass tube/bottle is used which is rotated in special incubators at 2 to 4 rpm. If the cells are nonadherent, they will be agitated by the rolling action, but will remain in the medium. If the cells are adhesive, they will gradually attach to the inner surface of the bottle and grow to form monolayer (Fig. 5.2). This provides an inner surface area of 600-1500 cm^3 area. A further development to this is made by semiautomatic exchange of medium. This helped in an inoculation with 4×10^5 cells/ ml in 100 ml of medium, a yield of 8×10^8 cells per flask can be obtained within six days.

The roller bottle culture has four advantages over static Roux bottle culture: 1) the increase in surface area, 2) the constant but gentle, agitation of the medium, 3) the increased ratio of the medium's surface area to its volume, which allows gas exchange to take place at an increased rate through the thin film of medium over cells not actually submerged in the deep part of the medium and 4) large numbers of cells or for collecting supernatant medium this method is most economical.

Later several modifications of roller bottle further manifold increase in the useful surface area. Such changes include introduction of glass tubes, introduction of polyester spirals or corrugated (ribbed) bottle surface or mechanics of rotations on roller racks or roller drum. These modified surfaces are known to offer the equivalent of up to 100 bottles.

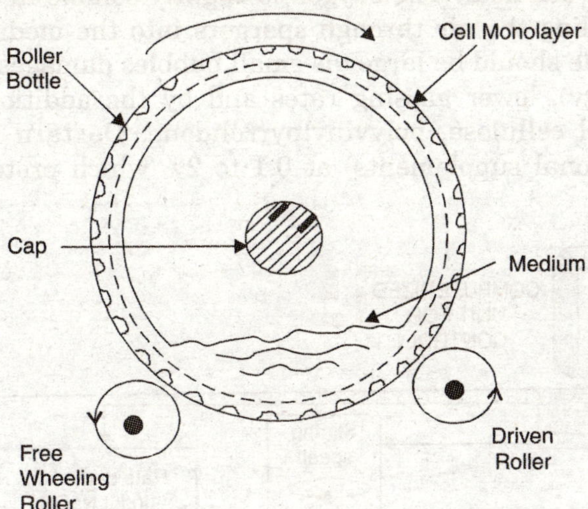

Fig. 5.2. Roller bottle culture. The cell monolayer (dotted line) is constantly bathed in liquid, but is submerged only for about one-fourth of the cycle, enabling free gas exchange.

Stacked plate vessel

In this vessel number of chambers are stacked one above the other and each chamber is interconnected by the channels (Fig. 5.1C). The standard unit has around 10 chambers stacked one above the other and have interconnecting channels. This makes it possible that the condition present in one chamber will be going to all the chambers. The total volume of the vessel is 12.5 liters and each chamber has around 600 cm^3 surface area. This type of culture system to gives good results with respect to production.

In the monolayer culture system by one or the other reason it is difficult to get good production when compared to the other described methods.

5.3 SUSPENSION CULTURES

In this method cells are grown in the liquid medium without anchorage. Here medium must be continuously stirred to keep the cells in suspension and to make culture homogeneous. The various types of stirrers range from simple magnetic stirrers, flat blade turbine impellers to that complex airlift stirrers or perfusion of medium (taking medium out by hydraulic energy and giving back to the culture vessel). The proper mixing can be done by making alteration in the design of stirrer paddle or by using multiple impellers.

The suspension cultures are monitored by using pH, oxygen, CO_2 and glucose electrodes that read from the culture *in situ* and by assaying the utilisation of nutrients, such as glucose and amino acids or the buildup of metabolism such as lactate or ammonia (Fig. 5.3). The stirring operation should be done without causing damage to the cells. Because animal cells are highly fragile as these cells are not like plant cells or microbial cells having rigid cell wall.

Another factor, which influences the growth and production, is the sufficient amount of oxygen supply without affecting the cells. The mean oxygen consumption by cultured cells is about 6 mg/10^6 cell/hour. The oxygen is slightly soluble in medium at the rate of 17 mg/cm^2/hour. Bubbling the air through spargers into the medium can do the proper aeration. The air bubble should be large (as small bubbles damages plasma membrane by its high surface energy), lower gassing rates and by the addition of methyl cellulose, sodium carboxymethyl cellulose, polyvinylpyrrolidone, Dextarn 50000, Ficol 400 and polyglycol (non-nutritional supplements) at 0.1 to 2% which protects from shear forces and bubbles.

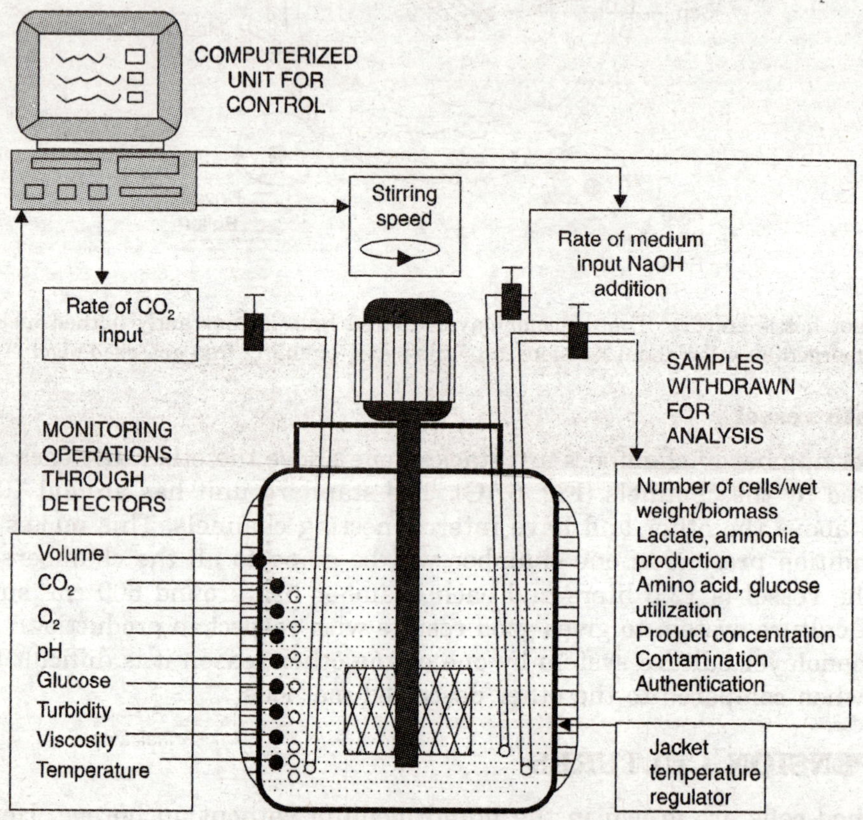

Fig. 5.3. Bioreactor process control. Schematic representation of a paddle-stirred bioreactor with direct-reading probes on the left, feeding to a control unit (top left) that stores the data and also regulates conditions within the bioreactor. A sampling port on the right withdraws the cell suspension from the bioreactor for analysis.

There are three types of suspension cultures a) stirred bioreactors, b) continuous flow reactors and c) airlift bioreactors.

Stirred Bioreactors

These types of bioreactors are closed vessels with fixed culture volume. The help of motor-driven stirrers along with variation in design to stir the medium. These bioreactors are glass (smaller) or stainless steel (larger) culture vessels of 1 to 1000 or even 8000 liter capacity (Fig. 5.4). But normally 20 liter capacity culture vessels are appropriate because it is easy to handle, autoclave and agitate the culture medium without any problem. Stirring of such cultures is done slowly with a magnet encased in a glass pendulum or with a large surface area paddle. The stirring speed should be between 30-100 rpm, sufficient to prevent cell sedimentation.

The stirring speed should not create shear forces that would damage the cells. These types of bioreactors are used for the production of interferons, interleukin, hormones, monoclonal antibodies and vaccines.

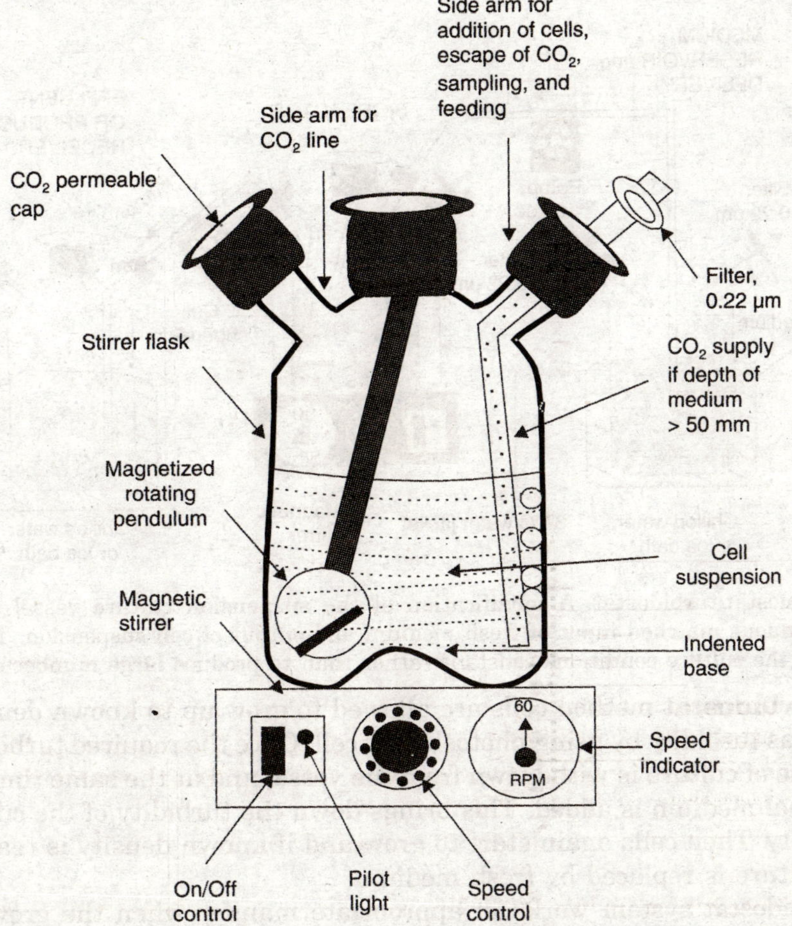

Fig. 5.4. Stirred animal cell culture. Diagrammatic representation of a large stirred flask suitable for volumes upto 8 liters.

Continuous flow bioreactors

In these types of bioreactors culture was maintained continuously by adding fresh medium after a specific period of growth. They may be of chemostat or turbidostat method of culture system. Initially both the methods of cultures started as batch culture but later became continuous culture (Fig. 5.5).

In the **chemostat** method the inoculated cells grow to the maximum density in the medium when some nutrient (e.g. vitamin) becomes growth limiting. At a constant rate fresh medium is added after 24-48 hours of growth and at an equal rate the culture content is withdrawn. When rate of growth equals the rate of cells withdrawn, the culture is in a steady state and both the cell density and medium composition remain constant. In this condition one of the growth component/nutrient is used at a lower concentration to make it growth limiting. But chemostat is the less efficient or controllable at the cell's maximum growth rate, hence the steady state growth rates in them are much lower than the maximum.

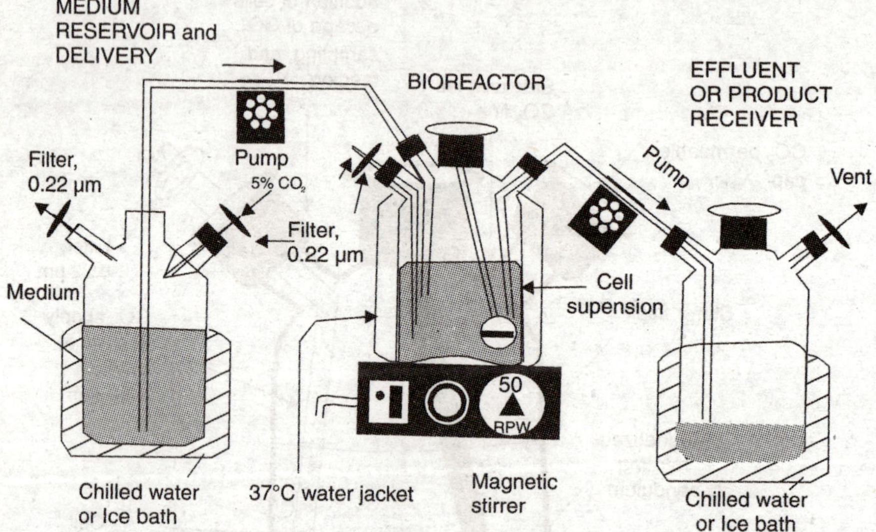

Fig. 5.5. Chemostat/turbidostat. A modification of the suspension culture vessel in Fig. 5.4, with continuous matched input of fresh medium and output of cell suspension. The objective is to keep the culture conditions constant rather than to produce large number of cells.

In the **turbidostat** method cells are allowed to grow up to known density which will be measured as turbidity by using photoelectric cell. Once the required turbidity is achieved a fixed volume of culture is withdrawn from the vessel and at the same time equal volume of fresh normal medium is added. This brings down the turbidity of the culture or lowers the cell density. Then cells again start to grow and if known density is reached, the fixed volume of culture is replaced by fresh medium.

This turbidostat system works in appropriate manner when the growth rate of the culture is close to the maximum for the cell line. The continuous flow bioreactors give continuous sources of cells and are most suitable for production formation (interferons, hormones and vaccines). In animal culture systems it is always necessary to use a two-stage system in which first stage supports the cells growth, while second stage helps in the product formation.

Airlift bioreactors

In this type of bioreactors both aeration and agitation is done by introduction of air bubbles at the bottom of the reactor. The bioreactor has an inner draft tube through which the air bubbles and the aerated medium rise because aerated medium is lighter than non-aerated medium. This results in mixing of the culture as well as aeration. The air bubbles moves to the top of the medium and passes out through an outlet. The cells and medium that lift out of the draft tube moved outside the tube and are recirculated (Fig. 5.6). In this method oxygen supply is more efficient but scaling up poses certain problems. Bioreactors of 2-90 liters capacity are commercially available, but generally 20,000 liters capacity bioreactors are being used for the monoclonal antibodies production (Fig. 5.6).

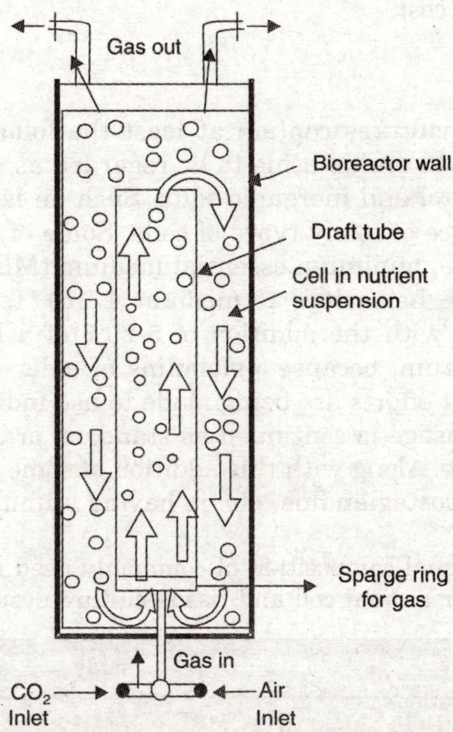

Fig. 5.6. Air-lift fermentor. In an air-lift fermentor, the bioreactor consists of two concentric cylinders, with the inner cylinder being shorter at both ends than the outer, thereby creating an outer and inner chamber. The bottom of the inner chamber carries a sintered steel ring through which 5% CO_2 in air is bubbled. The bubbles rise, carrying the cell suspension with them. CO_2 is vented from the top, and displacement ensures the return of the cell suspension with them.

5.4 IMMOBILIZED BIOREACTORS

In this method cells are generally made encapsulated in a polymeric matrix by adsorption, covalent linkage, cross-linking or entrapment. The gelatin, polylysine, alginate and agarose are used as matrix. This makes cells fixed in one position or immobilized. Immobilization effectively protects cells from mechanical damage in large bioreactors, allows medium to diffuse freely into the matrix and to the cells, the products freely comes out of the cells into the medium, the product formation (hormones, immunochemicals and enzymes) prolonged for longer periods when compared to the suspension cultures. The bioreactors of capacity ranges from 1-3 litres are commercially available.

Immobilization of cells in the cultures are known to have several advantages over other methods in the following ways.

1. Higher cell densities can be maintained (50-200 × 10^6 cells/ml).
2. Stability and longevity of cultures is more.
3. Protect the cells from shear forces due to medium flow.
4. This system is applicable for both suspension and monolayer cultures.
5. Less dependence of cells at higher densities on external supply of growth factors, which saves culture cost.

5.5 MEDIA

All media for animal cell cultures contain at least the following components: a hexose (glucose or galactose), the 13 amino acids to be regarded as essential, 7 to 8 vitamins of the B series, choline, inositol and inorganic salts. Such an isotonic nutrient solution can be used for the maintenance of many types of cells. Some of these category of media are basal medium Eagle (BME), minimum essential medium (MEM), Dulbecco's modification of MEM (DME), McCoy 5A, Leibovitz's 15 medium (L 15) (Table 5.2).

These media are used with the addition of 5-20% of a homologous or heterologous serum (preferably fetal serum, because γ-globulins of older animals serum inhibits the growth). Even though great efforts are being made to use individual growth factors of the serum, still whole serum usage is continued as standard practice, as it is the most variable and expensive additive. Along with this addition of some hormones such as glucagon, hydrocortison, estradiol, prostaglandins, etc. is having stimuli effect on cultured cells.

Table 5.2. Nutritional composition of commonly used media (with serum), for animal cell and tissue culture system

Nutrient	mg/lt								
	Eagle's MEM	Dulbecco's modification	Ham's F12	CMRL 1066	RPMI 1640	Medium 199	L15	Fischer's	Waymouth MB 752
Amino acids									
L-alanine	–	–	8.90	25.0	–	25.0	225	–	–
L-arginine (free base)	–	–	–	–	200	–	550	–	–
L-arginine-HCl	126	84.0	211	70.0	–	70.0	–	15.0	75.0
L-asparagine	–	–	–	–	50.0	–	–	–	–
L-asparagine H_2O	–	–15.0	–	–	–	260	11.4	–	
L-aspartic acid	–	–13.3	30.0	20.0	30.0	–	–	60.0	
L-cysteine (free base)	–	–	–	–	–	–	120	–	61.0

(Contd.)

Nutrient	mg/lt								
	Eagle's MEM	Dulbecco's modification	Ham's F12	CMRL 1066	RPMI 1640	Medium 199	L15	Fischer's	Waymouth MB 752
L-cysteine	24	48.0	–	20.0	50.0	–	–	–	15.0
L-cysteine 2Na	–	–	–	–	–	23.7	–	23.7	–
L-cysteine HCl.H_2O	–	–	35.1	260	–	0.0987	–	–	–
L-glutamic acid	–	–14.7	75.0	20.0	66.8	–	–	150	–
L-glutamine 292	584	146	100	300	100	300	200	350	–
Glycine	–	30.0	7.50	50.0	10.0	50.0	200	–	50.0
L-histidine (free base)	–	–	–	–	15.0	–	250	–	128
L-histidine HCl.H_2O	42.0	42.0	21.0	20.0	–	21.9	–	81.1	–
L-hydroxyproline	–	–	–	10.0	20.0	10.0	–	–	–
L-isoleucine	52.0	105	3.94	20.0	50.0	20.0	125	75.0	25.0
L-leucine	52.0	105	13.1	60.0	50.0	60.0	125	30.0	50.0
L-lysine. HCl	73.1	146	36.5	70.0	40.0	70.0	93	50.0	240
L-methionine	15.0	30.0	4.48	15.0	15.0	15.0	75.0	100	50.0
L-phenylalanine	33.0	66.0	4.96	25.0	15.0	25.0	125	67.0	50.0
L-proline	–	–	34.5	40.0	20.0	40.0	–	–	50.0
L-serine	–	42.0	10.5	25.0	30.0	25.0	200	15.0	–
L-threonine	48.0	95.0	11.9	30.0	20.0	30.0	300	40.0	75.0
L-tryptophan	10.0	16.0	2.04	10.0	5.0	10.0	20.0	10.0	40.0
L-tyrosine	36.0	72.0	5.40	40.0	20.0	–	–	–	40.0
L-tyrosine. 2Na	–	–	–	–	–	49.7	373	74.6	–
L-valine	47.0	94.0	11.7	25.0	20.0	25.0	100	70.0	65.0
Vitamins									
L-ascorbic acid	–	–	–	50.0	–	0.050	–	–	17.5
Biotin	–	–	0.0073	0.010	0.200	0.010	–	0.010	0.02
D-Ca-pantothenate	1.00	4.00	0.480	0.010	0.250	0.010	1.00	0.500	1.00
Calciferol	–	–	–	–	–	0.100	–	–	–
Choline chloride	1.00	4.00	14.0	0.500	3.00	0.500	1.00	1.50	250
Folic acid	1.00	4.00	1.30	0.010	1.00	0.010	1.00	10.0	0.40

(Contd.)

Nutrient	mg/lt								
	Eagle's MEM	Dulbecco's modification	Ham's F12	CMRL 1066	RPMI 1640	Medium 199	L15	Fischer's	Waymouth MB 752
i-inositol	2.00	7.20	18.0	0.050	35.0	0.050	2.00	1.50	1.00
Nicotinamide	1.00	4.00	0.04	0.025	1.00	0.025	1.00	0.50	1.00
Pyridoxal. HCl	1.00	4.00	0.04	0.025	–	0.025	–	0.50	–
Riboflavin	0.10	0.40	0.038	0.010	0.20	0.010	–	0.50	1.00
Thiamin. HCl	1.00	4.00	0.34	0.010	1.00	0.010	–	1.00	10.0
Vitamin B_{12}	–	–	1.36	–	0.005	–	–	–	0.20
Pyridoxine HCl	–	–	0.062	0.025	1.00	0.025	–	–	10.0
Cholesterol	–	–	–	0.200	–	–	–	–	–
Para-amino benzoic acid	–	–	–	0.050	1.00	0.050	–	–	–
Nicotinic acid	–	–	–	–	–	0.025	–	–	–
Menaphthone Sodium bisulphite $3H_2O$	–	–	–	–	–	0.019	–	–	–
Dl-Â tocopherol PO_4.2Na	–	–	–	–	–	0.01	–	–	–
Vitamin A acetate	–	–	–	–	–	0.115	–	–	–
Riboflavin	–	–	–	–	–	–	0.10	–	–
PO_4.2Na Thiamine mono PO_4, $2H_2O$	–	–	–	–	–	–	1.00	–	–
Inorganic salts									
$CaCl_2$ (anhyd.)	200	200	–	200	–	–	–	–	–
$CaCl_2.2H_2O$	–	–	44.0	–	–	186	186	91.0	120
$Fe(NO_3)_3.9H_2O$	–	0.10	–	–	–	0.10	–	–	–
KCl	400	400	224	400	400	400	400	400	150
KH_2PO_4	–	–	–	–	–	60.0	60.0	–	80.0
$MgCl_2.6H_2O$	–	–	122	–	–	–	–	–	240

(Contd.)

Nutrient	mg/lt								
	Eagle's MEM	Dulbecco's modification	Ham's F12	CMRL 1066	RPMI 1640	Medium 199	L15	Fischer's	Waymouth MB 752
$MgSO_4.7H_2O$	200	200	–	200	100	200	400	121	200
NaCl	6,800	6,400	7,599	6,799	6,000	8000	8000	8,000	6,000
$NaHCO_3$	2,200	3,700	1,176	2,200	2,200	350	–	1,125	2,240
$Na_2H_2PO_4.H_2O$	140	125	–	140	–	–	–	78.0	–
Na_2HPO_4 (anhyd)	–	–	–	–	–	47.5	190	60.0	–
$Na_2HPO_4.7H_2O$	–	–	268	–	1,512	–	–	–	566
$CuSO_4.5H_2O$	–	–	0.00249	–	–	–	–	–	–
$FeSO_4.7H_2O$	–	–	0.834	–	–	–	–	–	–
$ZnSO_4.7H_2O$	–	–	0.863	–	–	–	–	–	–
$CaNO_3.4H_2O$	–	–	–	–	100	–	–	–	–
Other components									
D-glucose	1,000	4,500	1,802	1,000	2,000	1,000	–	1,000	5,000
D-galactose	–	–	–	–	–	–	9.000	–	–
Lipoic acid	–	–	0.21	–	–	–	–	–	–
Phenol red	10.0	15.0	12.0	20.0	5.00	17.0	10.0	5.00	10.0
Sodium pyruvate	–	110	110	–	–	–	550	–	–
Hypoxanthine	–	–	4.10	–	–	0.30	–	–	–
Linoleic acid	–	0.084	–	–	–	–	–	–	25.0
Putrescine 2HCl	–	–	0.161	–	–	–	–	–	–
Thymidine	–	–	0.73	10.0	–	–	–	–	–
Cocarboxylase	–	–	–	1.00	–	–	–	–	–
Coenzyme A	–	–	–	2.50	–	–	–	–	–
Deoxyadenosine	–	–	–	10.0	–	–	–	–	–
Deoxycytidine HCl	–	–	–	10.0	–	–	–	–	–
Deoxyguanosine	–	–	–	10.0	–	–	–	–	–
Diphosphopyridine nucleotide									

The pH optimum (6.9-7.8) should be known for each cell line. This may be helpful in saving half the quantity of serum used in the cultural condition. Earlier for pH control use of 5% CO_2 in the air space with 0.22% sodium hydrogen carbonate ($NaHCO_3$) are followed. Recently, use has been made of non-toxic organic buffers e.g., HEPES (N-2-hydroxyethylpiperazine-N'-2-ethanesulfonic acid) in combination of up to 850 mg $NaHCO_3$/liter and 2% of CO_2 in the external air space. Because, the traditional concentration of CO_2 is above the optimum requirement by the cell culture.

The use of antibiotics in media is disputed but a mixture of 20 to 200 units of penicillin/ml and 10 to 100 μg of streptomycin/ml is added almost as routine and helps against the majority of bacterial contamination. A proper routine test for Mycoplasmas must be carried out. According to one estimate these organisms contaminate almost the 60% of cell lines. Mycoplasmas do not lead to the immediate death of the cells but changes the nutritional requirements and cell properties.

Thus, it is very important to test for mycoplasma at regular intervals. Often for every one to three months in the cell stock cultures. For this purpose direct culture test, PCR method and fluorescent staining techniques are used.

Similarly testing should be carried out for identification of endogenous and contaminant viruses. Especially with human cell lines tests are necessary for screening human immunodeficiency virus (HIV) and hepatitis viruses (HBV).

SUMMARY

The vessel or container used to culture the cells at controlled environmental conditions for the production of some commercially valuable product is known as bioreactor. The volume of the bioreactor usually ranges in between 1 litre and 1000 litres. These bioreactors provided with temperature regulator, pH regulator, agitator to mix the culture medium and devices to supply air, fresh medium and to remove the used medium. Many cell lines are available for the production of therapeutic proteins. The American type Culture Collection Centre lists around 400 cell lines. Bioreactors are being used for the production of various therapeutic proteins, which include interferons, interleukins, hormones, antibodies etc.

Animal cell lines can be cultured by three methods. They are monolayer cultures, suspension cultures and immobilized cell cultures. Monolayer culture systems are used for anchorage dependent cells and for production of small scale therapeutic proteins. These culture systems can be maintained in Roux bottle or roller bottle or streaked plate vessel. In suspension culture system the cells are grown in the liquid medium. The stirred bioreactors, continuous flow bioreactors and air-lift bioreactors are generally used to maintain suspension cultures. Suspension cultures are used to produce interferons, interleukins, hormones, monoclonal antibodies and vaccines. In immobilized cell culture systems gel encapsulated cells are used. The immobilized cell culture systems have several advantages over other methdos. Immobilized cell culture systems are basically employed for the production of enzymes, immunochemicals and hormones.

EXERCISE

1. What is meant by monolayer cultures?
2. Explain different types of monolayer culture system.
3. Write a note on stirred bioreactors.
4. Briefly explain Airlift bioreactor.

5. Mention advantages of immobilized bioreactors.
6. Write a short note on bioreactors. (BU Oct. 2002)
7. Differentiate between chemostat and turbidostat.
8. With neat diagram explain roller bottle culture system.
9. Describe suspension culture systems.
10. Define immobilization and outline the advantages of immobilized bioreactor.

Part II
Immunology

6

History of Immunology

6.1 INTRODUCTION

The immunology as branch of science started with the notice of two major functions of immune response *i.e.*, recognition and response. Immune recognition is remarkable for its specificity. The immune system is able to recognize small chemical differences that distinguish one foreign material or organism from another. In addition, the immune system is able to recognize and discriminate between foreign molecules and the body's own cells and proteins. As these organisms or material is recognized, the immune system allows the participation of a variety of cells and molecules to mount an appropriate and suitable response and is known as effector response. This occur to eliminate or neutralize the effect of foreign materials or organism. This complex of events are not understood completely in the beginning. This has undergone a slow process of knowledge development.

Beginning of immunity observation

The branch of immunology grew out of the careful thinking and observations made by the different people throughout the world. **Thucydides** made the first recorded work on immunity in 430 BC. He said that in Athens only those people who has recovered from the plague are able to take care of plague-affected people without any problem of infections. The first practical attempt to induce immunity was made by **Chinese and Turkish people** in 15th century. They had a practice of snuffing in to nostrils or keeping dried smallpox pustules in the skin. This technique is known as **variolation.** A British Lady in 1718 **Mary Wortley Montagu** gone with her husband who is a British ambassador to Constantinople show the native people doing variolation and done the similar thing to her children. This gave effective protection to the children. The birth and evolution of the germ theory of disease, the study of the mechanisms of immunity and the possible conquest of infectious diseases began almost simultaneously. Interest in immunology grew out of the everyday evidence that those individuals who survived; scarred and disfigured from the dreaded disease, small pox never did contract the infection again. As early as 1773, **Voltaire** documented an ancient Chinese custom where dried and powdered small pox scabs were inhaled, much like snuff, in an attempt at preventing the disease. Although it took several years before bacteria and viruses were fully characterized, theorists were already elucidating the mechanisms of host defense against infectious diseases and their possible prevention. Later in 1798 **Edward Jenner** an English physician practiced the similar type of variolation after observing the milkmaids who had exposed to cowpox were not able to infected by small pox afterwards. This practice spread rapidly throughout Europe. In 1885 **Louis Pasteur** first administered the vaccine to a human, a boy called Joseph Meister and thus vaccination has become practice.

Beginning of the cellular immunity

The cells involvement in immunity reactions takes its birth later. **The phagocytic theory** was the first to be developed in the 1880s, a bold and imaginative concept the fruit of **Elie Metchnikoff's** deep knowledge of biology. Metchnikoff hypothesized that the basis of inflammation was the cellular reaction and that vascular and nervous reactions were only of secondary importance. He further postulated that these migrating cells which were able to move in order to meet an enemy were the major guardians of health against bacterial infections. It was around this time, that **Louis Pasteur** reported the treatment of his first two patients with rabies using early vaccines. In 1888, Metchnikoff left Russia to continue his work on phagocytosis at Pasteur's Institute in Paris. In 1908, he shared the Nobel Prize with **Paul Ehrlich** for his early contributions to the understanding of inflammation.

After Metchnikoff died in 1916, other workers who had long cherished the theory that **soluble substances** in blood were also bactericidal, reported their observations. **Behring, Nuttall** and **Nissen** contributed substantially to the concept of soluble or **humoral factors** but found that these factors were not always bactericidal. Finally in 1903, **Almroth Wright** with his disciple **Stewart Douglas,** observed that a humoral component which they designated as **opsonin,** could render bacteria susceptible to phagocytosis. They thus forged a link between the two major theories of immunity. Since the time humoral and cellular components were shown to be interwoven in an intricate pattern, the complex chemicals associated with immunity have been the continuing subject of years of research. Large group of researchers started their work on immunology.

This paves the way for the further understanding of the immunology. Mere empirical methods of producing or increasing immunity were not sufficient, the whole basis of immunity needed questioning. Emil von Behring found that certain diseases were the expression of the action of toxins that could be neutralized by antitoxins. It could not be ignored that the blood of animals contained certain preformed bactericidal agents. **Emil von Behring** and **Kitasato** in 1890, inoculated animals with toxins of diphtheria and tetanus, to produce a neutralizing antitoxin serum. They introduced passive immunization into modem medicine and for this, von Behring was awarded the French legion of honour and the Nobel Prize in 1901. Behring maintained cordial ties with Pasteur—one of his great heroes, and he established a continuing friendship with Metchnikoff that began in 1888. This friendship was unique because the two men occupied opposite poles in explaining basic aspects of immunity—and yet enjoyed the stimulus of intellectual debate. This forms the basis of the successful achievements for the both group.

Meantime even though notably successful in passive immunization against diphtheria and tetanus, von Behring failed abysmally when he tried to extend these principles to passive immunization against tuberculosis. At this time, Robert Koch too had to taste failure in his attempts at providing a successful vaccine against tuberculosis. The prevention of tuberculosis was not to be found by the Germans. It was in France that **Albert Calmette** and **Camille Guerin** developed an effective vaccine for tuberculosis by the methods of attenuation so dear to Pasteur and disdained by Koch. This provided good theoretical idea for them.

They developed the first effective vaccine for tuberculosis. For establishment of a live but attenuated strain of *Mycobacterium tuberculosis* they had to continually transfer or sub-culture the organism till virulence was lost. This odyssey took them thirteen years and 230 transfers. Later, on Jan 5, 1921 it was completely avirulent even at high doses for all animal species. The **Bacille Calmette Guerin (BCG)** was thus born—it did not induce the formation of tubercles by intravenous, intraperitoneal or subcutaneous inoculation or even by ingestion, but formed an effective prophylactic against tuberculosis. This gave a world that suffered much from the tuberculosis, a trend of relief.

Complements and its identification

As more and more study on blood components started **Pfeiffer** in 1894-95, discovered the phenomenon of in-vivo cytolysis of *Vibrio cholerae* when the organism and the immune serum interacted intra peritoneally in the guinea pig. From these early experiments, the nature and function of complement mediated cytolysis was elucidated by Pfeiffer and later by **Buchner** and **Bordet**. Jules Bordet was awarded the Nobel Prize in 1920 for his pioneering work on complement, for his discovery of the whooping cough agent and for his enunciation of the diverse aspects of antigen-antibody reactions and the blood coagulation system. Around this time Gruber and Durham described the diagnostic value of the agglutination reaction, when agar cultures were mixed with their corresponding immune sera. The same year **Ferdinand Widal,** the Algerian born son of a French army surgeon described the agglutination reaction for typhoid fever—a test that is still widely used and bears his name. This formed the effective diagnosis of the infective disease.

Identification of blood groups and its usage

The technological and medical advancement made gave the way to describe the details of the blood system. In 1667, **Jean Baptiste Denis**, physician to Louis the XIV, performed what is considered to be the first transfusion of blood in man. The experiment was not well received by his fellow-physicians who were more accustomed to blood-letting than reversing the flow. Till the early 1900s the practice of transfusion of blood from man to man remained a risky, unpredictable business, when at the turn of the century, **Karl Landsteiner** discovered the blood groups. Landsteiner received the Nobel Prize in 1930 for elucidating the blood groups and for his work on the Rh factor.

As serotherapy developed, it soon became evident that antigen-antibody reactions *in vivo* could have some harmful effects and even produce death. Anaphylaxis, the most dramatic manifestation of hypersensitivity was first described by **P. Portier** and **Charles Richet** in 1902. In 1913 Charles Richet Received the Nobel Prize in recognition of his work on anaphylaxis. This started to observe the various compounds having such a negative effect in the entirely different immunological conditions by immunologists. Later, **P.G.H. Gell** and **R.R.A. Coombs** classified hypersensitivity in to four major types.

Two of the most vital immunological hypotheses made in the 1800s went undetected for several decades. They were rediscovered only in the 1940s and form the basis of the study of immunology today. Elie Metchnikoff, suggested that phagocytes were the prime detectors of foreign material; we trace the origins of cellular immunity to this hypothesis. Paul Ehrlich proposed the pre-existence of receptors (which he called toxophores) on the living cell, that reacted with toxins. Excess receptors were liberated into the circulation as antibodies—the molecular basis of humoral immunity, as we know it today, is remarkably close to this novel but original concept. Thus, the foundations of modern immunology, as we recognize it today, were laid by several European scientists over the last hundred or more years.

The twentieth century saw phenomenal progress in the understanding of immunological concepts. This formed the large number of new observations in a greater look. How antigens direct the various immunological processes, was the chief point of contention among early theorists who had to choose between the instructive and the selective theories. In the past 30 years, the biological basis of the immune response was, in part, clarified when the theory of clonal selection of antibody formation found acceptance. This theory proposed, in essence, by Paul Ehrlich almost a 100 years ago, fell out of favour with the introduction of the template theory. The first comprehensive attack against the template theory was launched in 1955 by Niels Kaj Jerne, who recalled certain criticisms made by Macfarlane Burnet against the template theory. It, however, remained for Burnet to draw together the new conceptualization and in 1957, he asserted that "each cell and its clones can produce just one kind of receptor" to explain the specificity of the immune response and the exponential rise in antibody production following contact with an antigen.

The secondary response is more powerful because antigenic memory leads to rapid clonal expansion during subsequent contact with the same antigen. We know today, that during this second exposure, the binding ability of antibody improves and we also know this to be the result of affinity maturation and somatic hypermutation. The clonal selection theory also explained tolerance as the deletion or suppression of an entire clone of cells which occurred before or soon after birth; or even much later in some instances. **Macfarlane Burnet** was awarded the Nobel Prize in 1960, which he shared with **Peter Medawar.** In 1984, **Niels Jeme** was also awarded the Nobel Prize for his theoretical contributions to immunology, the most fundamental of which was his role in developing the concept of clonality and for his description of the idiotype network in the regulation of immune responses.

In 1960 **S.A. Berson** and **Rosalyn Yallow** described the radioimmunoassay (RIA) for the determination of insulin and anti-insulin complex in the blood. This has given the way for identification of various serum proteins, cytokines, growth factors and leucocyte activating factors. **Engvall** and **Perlmann** and **van Weeman** and **Schuurs** in 1971 discovered independently the enzyme linked immunosorbent assay (ELISA) that has made enormous benefit for the study and identification of pathogenic agents, cell surface receptors, signaling molecules, tissue typing and immunity changes.

Study of antibody and its structure

Humoral immunity got a breakthrough when the antibody structure was discovered. The elucidation of antibody structure is credited to **Rodney R. Porter** and **Gerald M. Edelman** who shared the Nobel Prize for their discovery in 1972. In 1959, Edelman showed that the immunoglobulin molecule had 4 polypeptide chains: 2 of each kind and that each could be separated by chemical means. He called them light and heavy chains because of their size. At the same time Porter showed that the molecule could be cut into 3 different pieces (Fab × 2; Fe × I), by enzymes that cleave polypeptides. In 1970, Edelman showed that chemical differences responsible for the specificity of antigen binding were embodied in the amino acid sequences of their variable regions at the upper, outer arms of the antibody molecule. All subsequent work have confirmed his conclusions with some elaborations. In 1970 **T.T. Wu** and **E.A. Kabat** demonstrated the presence of hypervariable regions. **Dreyer** and **Bennet** in 1965, suggested that the germ line contains many variable (V) region genes and one constant (C) region gene, which combine to yield an immunoglobulin molecule. As a cell matures, it selects one V gene out of many and combines it with the one C region gene. This theory was attractive; yet it required that the cells have some means of rearranging genes in somatic cells. Evidence that immunoglobulin genes do undergo somatic recombination, but in more complicated ways than **Dreyer** and **Bennet** suggested, was found in 1970 by **Hozumi** and **Tonegawa** then at Basel, Switzerland.

They demonstrated that V and C genes were far apart in embryonic cells but much closer to each other in plasma cells. They also delineated the mechanism of shuffling of the many gene segments. Tonegawa was awarded the Nobel Prize in 1987, for elucidating the mechanism by which the immune system generates an almost limitless variety of antibodies. Somatic hypermutation, as another major source of the tremendous diversity of antibody specificity generated, was demonstrated by **Weigert** and **Cohn** in 1970.

The production of monoclonal or monospecific antibodies by somatic cell hybridization of antibody forming cells and continuously replicating cell lines was called the technique of hybridoma formation. This technique described by **Georges Kohler** and **Cesar Milstein** in 1975, has enabled immunologists to prepare virtually unlimited quantities of antibodies that are chemically, physically and immunologically completely homogenous. For this invaluable contribution to immunology they were awarded the Nobel Prize in 1984.

Transplantation immunology

As much of the organs function and structure was understood, it diseases are controlled. Transfer of cells, tissues or organs from one individual to another is known as transplantation. Immunobiology of transplantation and tolerance was an active field of research in the 1940s and 50s. In 1953, **Medawar, Brent** and **Billingham** performed a series of dazzling experiments to explain the mechanisms that enable the host to recognize and destroy foreign cells. Medawar showed conclusively that such a mechanism was immunological. He suggested that specific individuality markers were associated with every cell of an organism and that an animal could be immunized against foreign grafts by first injecting it with cells derived from immunocompetent tissues of the donor. Earlier Ray Owen had shown, in his classic experiment with dizygotic twin calves, that dissimilar blood group antigens of one calf were tolerated by the other twin calf. Drawing from this experiment, Medawar demonstrated that an inbred strain of mice-strain A could be made to tolerate skin grafts from another inbred strain—strain B, by injecting the strain A animal soon after birth with strain B spleen cells. With this experiment, he proved that/ resistance (or the lack of it) to foreign tissue is an immunological phenomenon. Peter Medawar was awarded the Nobel Prize (with Bumet) in 1960.

Existence of markers of biological individuality—what we now know as the histocompatibility antigens was first suggested by **Gorer** in 1937, who demonstrated that loci determining blood group antigens and those controlling tumour rejection in mice were distinct. With **Snell** in 1948 he showed that there were multiple alleles in the mouse H-2 locus and that this locus was genetically complex. The H-2 locus was shown to control the phenotype expression of certain cell surface markers in mice. In 1950 **Dausset** identified the HLA (Human Leukocyte Antigen) locus or it is sometimes known as MHC (Major Histocompatibility Complex). **Benacerraf** showed that genes of the HLA determining loci may control immune responses. Snell, Dausset and Benacerraf were awarded the Nobel Prize in 1980. Recent elucidation of the structures of some HLA molecules by Bjorkman and Wiley in 1987, using X-ray crystallography, have revealed that the molecules of the major histocompatibility complex (MHC) bind to antigenic peptides and present these peptides to the T cell receptor.

First glimpsed in experiments by **Allison** and **Kappler,** and **Reinherz** the T cell receptor was studied and characterized by **Tak W. Mak** and **Mark M. Davis** in 1984. These workers cloned and sequenced a gene expressed and rearranged in T cells but not in B cells. Their analysis showed that many of the T cell receptor sequences were homologous to those of immunoglobulin genes.

Immunology continues to fascinate, ignite and frustrate researchers the world over. A better understanding of blood components, pathobiology, tumour immunology, autoimmunity and responses to immunization could help develop strategies for protection against many devastating and major crippling diseases. Eventually, what has evolved as a precise and powerful tool created by nature to ensure the continued survival of the species, could perhaps be manipulated to yield a better quality of life. With the present technological advancements made in different fields of biology may change the understanding of immunology in a better and effective way.

SUMMARY

The earliest documented reference of immunity was observed by Thucydides. From that point onwards Edward Jenner and Louis Pasteur contributed to development of immunology by their society needed studies. von Behring, Metchnikoff, Richet, Landsteiner, Burnet, Porter, Edleman, Yalow, Snell, Dausset, Benacerraf, Milstein, Kohler, Jerne, Tonegawa, Thomas, Murray, Doherty, Zinkernagel made their significant contributions to the development and understanding of the immunology in the scientific methods. This made the present day wonders in all branches of science, especially medicine and human health.

EXERCISE

1. Write a note on history of immunology.
2. Define variolation.
3. Explain observations of Edward Jenner and Louis Pasteur.
4. Mention the importance of Milstein and Kohler contributions.
5. What are the contributions of Porter and Edelman?
6. Explain advantages of RIA and ELISA for immunology development.

7. What do you mean by MHC molecules?
8. Who first discovered the anaphylaxis?
9. Name the immunologists who got the Nobel Prize.
10. Describe the work of Karl Landsteiner in immunology.
11. Define transplantation.
12. The V and C regions of the immunoglobulin and their discoverer.

7

Immune System and Immunity

7.1 INTRODUCTION

Immunology is the study of specific immunity and how the immune system responds to specific infectious agents. In simple way immunology can be defined as study of immunity to infection.

The vertebrate organisms have remarkable ability to fight against infections caused by varieties of organisms and cells. This resistance shown by the host towards infectious disease is known as immunity (*immunis* meaning "exempt", literally means free from burden).

7.2 THE IMMUNE SYSTEM AND IMMUNITY ALONG WITH HISTORICAL PERSPECTIVES

The immune system is similar to other body systems (respiratory or reproductive systems) in having number of different cell types, tissues and organs. Many of these cells organized in to lymphoid organs or glands. The infection from microbes can occur at any site of the body, the immune system has moving force of cells in the blood stream. These cells are prepared to attack the invading microbe whenever it enters the body. Even though cells of immune system are separated from each other they maintain communication through cell contact and molecules secreted by them.

First reference of immunity was seen in 430 B.C. from **Thucydides**, when describing plague in Athens he wrote only those who have recovered from plague could nurse sick because they would not contract the disease second time. In fifteenth century the first recorded crude attempt to induce immunity done by Chinese and Turks. In 1718 Lady Mary **Wortley Montagu** by observing variolation (the process of inhaling the dried crusts from smallpox pustules or inserting into small cuts in the skin), she followed same method for her own children. English physician **Edward Jenner**, in 1798 made first scientific attempt of immunization. He extracted the contents of a pustule from the arm of cowpox-infected milkmaid and injected it to an eight-year-old boy. When he subsequently inoculated with smallpox, the boy showed no symptoms of the disease. This technique of inoculating with cowpox to protect against smallpox spread rapidly throughout Europe.

Louis Pasteur carried out further work on immunization. He showed that if culture of chicken cholera bacteria were allowed to age for two to three months, the bacteria cause mild cholera and allowed the chickens' recover. He then found that fresh cultures of bacteria failed to produce cholera that has been inoculated with old attenuated (weaken) cultures. Louis Pasteur called this attenuated culture as **vaccine** (vacca meaning "cow").

In 1881 **Louis Pasteur** extended the same method to anthrax bacillus (*Bacillus anthracis*). One group of sheep he vaccinated with heat attenuated bacteria and kept another group without vaccinated. When both group inoculated with the fresh bacteria, only vaccinated sheep lived whereas all unvaccinated sheeps died. This forms basis for beginning of immunology. Later in 1885 he vaccinated first human, Joseph Meister a young boy who had rabies. Later the boy recovered from the disease and became a custodian at the Pasteur Institute, Paris.

The research work of **Emil von Behring** and **Shibasaburo Kitasato** gave first insight into the mechanism of immunity earning Nobel Prize in 1901. They demonstrated that serum (liquid left over after coagulation) from animals previously immunized with diphtheria could transfer the immunity to unimmunized animals. **Elvin Kabat** in 1930s showed a fraction of serum known as gamma globulin was responsible for all these activities. This gamma globulin is now called **immunoglobulin** and active molecules in immunoglobulin is known as **antibodies**. As antibodies are present in the body fluid or humors this type of immunity is called humoral immunity.

Elie Metchinkoff in 1883 showed that certain white blood cells known as phagocytes were responsible for immunity by ingesting microbes and other foreign materials. Merrill Chase further confirmed this in 1940s by transferring immunity against the tuberculosis bacteria in guinea pigs.

Types of Immunity: There are two types immunity based on the immunity shown before exposure to foreign substances (antigen) or later. They include innate (non-specific) immunity and adaptive (specific) immunity.

7.3 INNATE (NON-SPECIFIC) IMMUNITY

It is the resistance that an individual possesses by virtue of his genetic makeup and inherently present in the host. It doesn't require prior contact with the microorganisms. This acts as first line of defense. The various nonspecific defense mechanisms include physical barriers (skin, mucous membrane), chemical barriers (gastric juices, salivary juices), biological barriers (microflora) and general barriers.

7.4 ADAPTIVE (SPECIFIC) IMMUNITY

It is the resistance obtained by recognizing and selectively eliminating specific foreign microorganisms and molecules. This acts as second line of defense. It requires prior contact with the microorganisms and has long time memory. This leads to a more rapid removal of the microbe on its second and third time of entry.

Acquired immunity refers to the type of specific immunity a host develops after exposure to a suitable antigen or after transfer of antibodies or lymphocytes from an immune donor. Acquired immunity can be obtained actively or passively by natural or artificial means.

Naturally acquired active immunity occurs when an individual's immune system contacts an antigenic stimulus such as an infection during normal activities. The immune system responds by producing antibodies and sensitized lymphocytes that inactivate or destroy the antigen. The immunity produced can be either lifelong (e.g. chicken pox, measles) or last for only a few years (e.g. tetanus).

Naturally acquired passive immunity involves the transfer of antibodies from one host to another. For example, some of pregnant women's antibodies pass across the placenta to her foetus. If the female is immune to diseases such as polio or diphtheria, the placental transfer also gives the foetus and newborn immunity to these diseases. Certain other antibodies can pass from the female to her offspring through colostrum. This kind of immunity generally lasts only a short time.

Artificially acquired active immunity results when an animal is given an antigen preparation to induce the formation of antibodies and activated lymphocytes. A vaccine is a preparation of killed microorganisms or inactivated bacterial toxins that are administered to an animal to induce immunity artificially.

Artificially acquired passive immunity results when antibodies that have been produced either in an animal or by specific methods *in vitro* are introduced into a host. Although this type of immunity is immediate, it is short lived and lasts only a few weeks to a few months. For example, Botulinum anti toxin is produced in a horse and is given to a human suffering from botulism food poisoning.

Based on the type of materials or cells involved immunity can be classified into two types, namely, humoral (antibody–mediated) immunity and cellular (cell-mediated) immunity. Both these type are branches of specific immunity.

Humoral (antibody–mediated) immunity humor means body fluid. This type of immunity is obtained by the action of soluble proteins called antibodies that occur in the body fluids and on the plasma membrane of B lymphocytes. Circulating antibodies bind to bacteria, toxin or any foreign proteinaceous material and make them neutralized or marking them for degradation by various mechanisms.

Cellular (cell-mediated) immunity this type of immunity is obtained from the action of specific kinds of T lymphocytes that directly attack cells infected with viruses or parasites, transplanted cells or organs and cancer cells. T lymphocytes can lyse these cells or release chemicals (cytokines) that enhance specific and nonspecific defenses (phagocytosis, inflammation).

7.5 ORGANS AND CELLS OF THE IMMUNE SYSTEM

The immune system has many different types of organs and tissues that are present throughout the body. The blood vessels and lymphatic vessels connect these organs and forms functional system as a whole.

Organs of the Immune System

The organs involved in immune system can be divided into two types based on their function as primary and secondary lymphoid organs (Fig. 7.1).

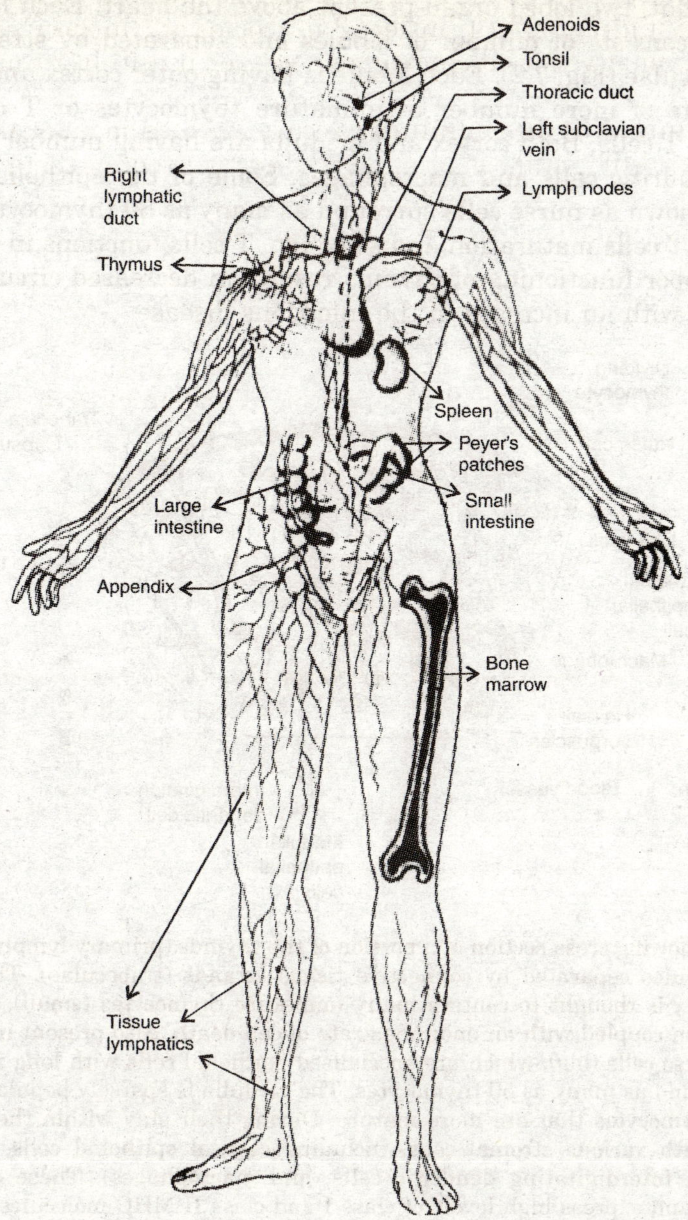

Fig. 7.1. The location of human lymphoid system in the body. The primary organs (bone-marrow and thymus); secondary organs and tissues involved in immune response. These structurally and functionally diverse lymphoid organs and tissues are interconnected by the blood vessels (in figure it is not shown) and lymphatic vessels through which lymphocytes circulate.

Primary lymphoid organ: The primary (or central) lymphoid organs are thymus and bone marrow in case of mammals. They provide appropriate microenvironment for the development and maturation of lymphocytes (a type of white blood cells/leukocytes).

The **thymus** is a flat, two lobed organ present above the heart. Each lobe is surrounded by a capsule and consists of number of lobules and separated by strands of connective tissue called trabeculae (Fig. 7.2). Each lobule is having outer cortex and central medulla. The cortex consists of more number of immature thymocytes or T cells and medulla contain only a few T cells. Both cortex and medulla are having number of epithelial cells, interdigitating dendritic cells and macrophages. Some of the epithelial cells present in the outer cortex known as nurse cells surround as many as 50 thymocytes. Thymus gland provides a site for T cells maturation and selection. T cells functions in cellular immunity of the body. Improper functioning of thymus results in decreased circulating T cells and cellular immunity with an increase in the infectious disease.

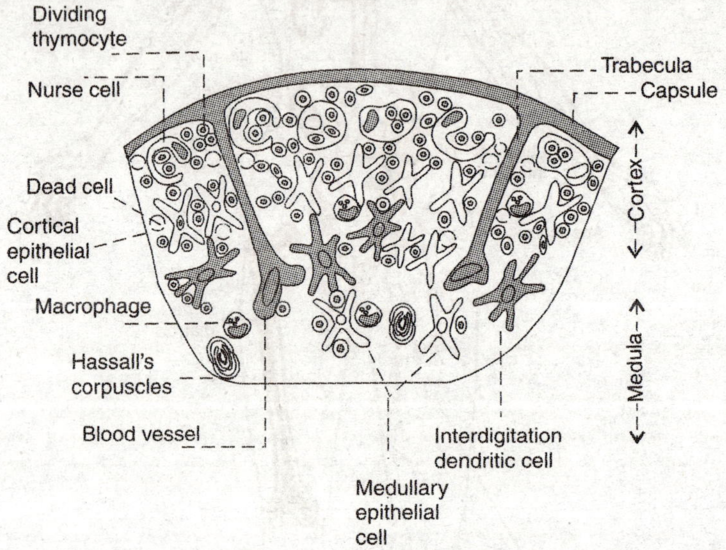

Fig. 7.2. Diagram showing cross-section of a portion of the thymus (primary lymphoid organ), indicating several lobules separated by connective tissue strands (trabeculae). The densely populated outer cortex is thought to contain many immature thymocytes (small), which undergo rapid proliferation coupled with an enormous rate of cell death. Also present in the outer cortex are thymic nurse cells (big), which are specialised epithelial cells with long membrane extensions that surround as many as 50 thymocytes. The medulla is sparsely populated and is thought to contain thymocytes that are more mature. During their stay within the thymus, thymocytes interact with various stromal cells, including cortical epithelial cells, medullary epithelial cells (tan), interdigitating dendritic cells, and macrophages). These cells produce thymic hormones and express high levels of class I and class II MHC molecules. Hassalls corpuscles, found in the medulla, contain concentric layers of degenerating epithelial cells.

In mammals the **bone marrow** acts as primary lymphoid organ, whereas in birds a specialized organ called **bursa of Fabricius** (a gut associated lymphoid tissue) helps in maturation, division and diversification of B cells. B cells produce various types of antibodies and functions in the humoral immunity of the body.

Secondary lymphoid organ: The secondary lymphoid organs are lymph nodes, spleen and mucosal-associated lymphoid tissue (MALT). The MALT is present in Peyer's patch (small intestine), tonsils, adenoids, appendix lining of upper airways, bronchi and genital tract. The **lymph nodes** are encapsulated bean shaped tissue consists of lymphocytes, macrophages and dendritic cells. Blood liquid part called **plasma** has the capacity to seeps out through the capillaries but some liquid portion remains in the surrounding tissue spaces, this fluid is called **lymph**. This lymph is collected by lymphatic capillaries and then by larger lymphatic vessels (Fig. 7.3). Lymphatic vessels finally attaches to the lymphnode by afferent lymphatic vessel. Lymphnodes anatomically having three regions (Fig. 7.4). The outermost layer cortex contains B cells, dendritic cells and macrophages. These cells aggregates together and forms primary follicle. After antigen exposure it converts into secondary follicle with a centre structure called germinal centre. Beneath the cortex isthe paracortex having T cells and dendritic cells. In the center medulla, having plasma cells, which are engaged in antibodies production. The lymphnodes help in

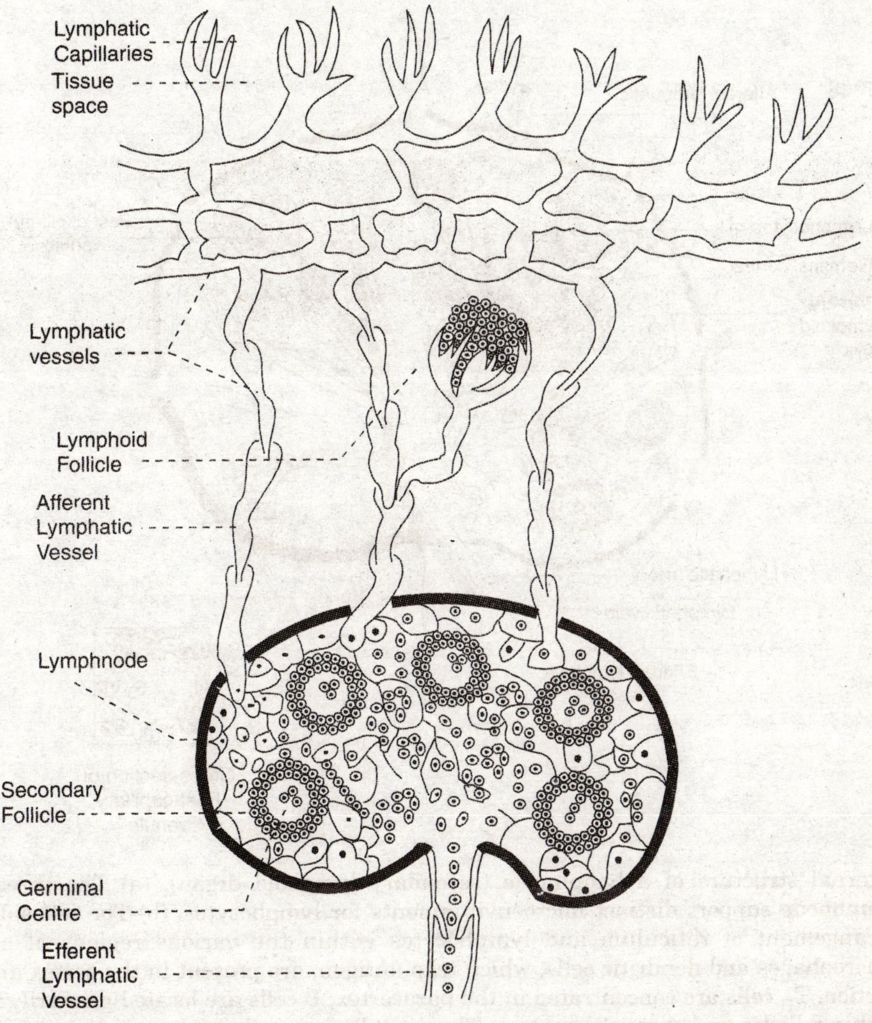

Fig. 7.3. Arrangement of lymphatic vessels. Small lymphatic capillaries opening into the tissue spaces pick up interstitial tissue fluid and carry it into progressively larger lymphatic vessels; which carry the fluid, now called lymph, into regional lymphnodes. As lymph leaves the nodes, it is carried through larger efferent lymphatic vessels, which eventually drains into the circulatory system at the thoracic duct or right lymph duct.

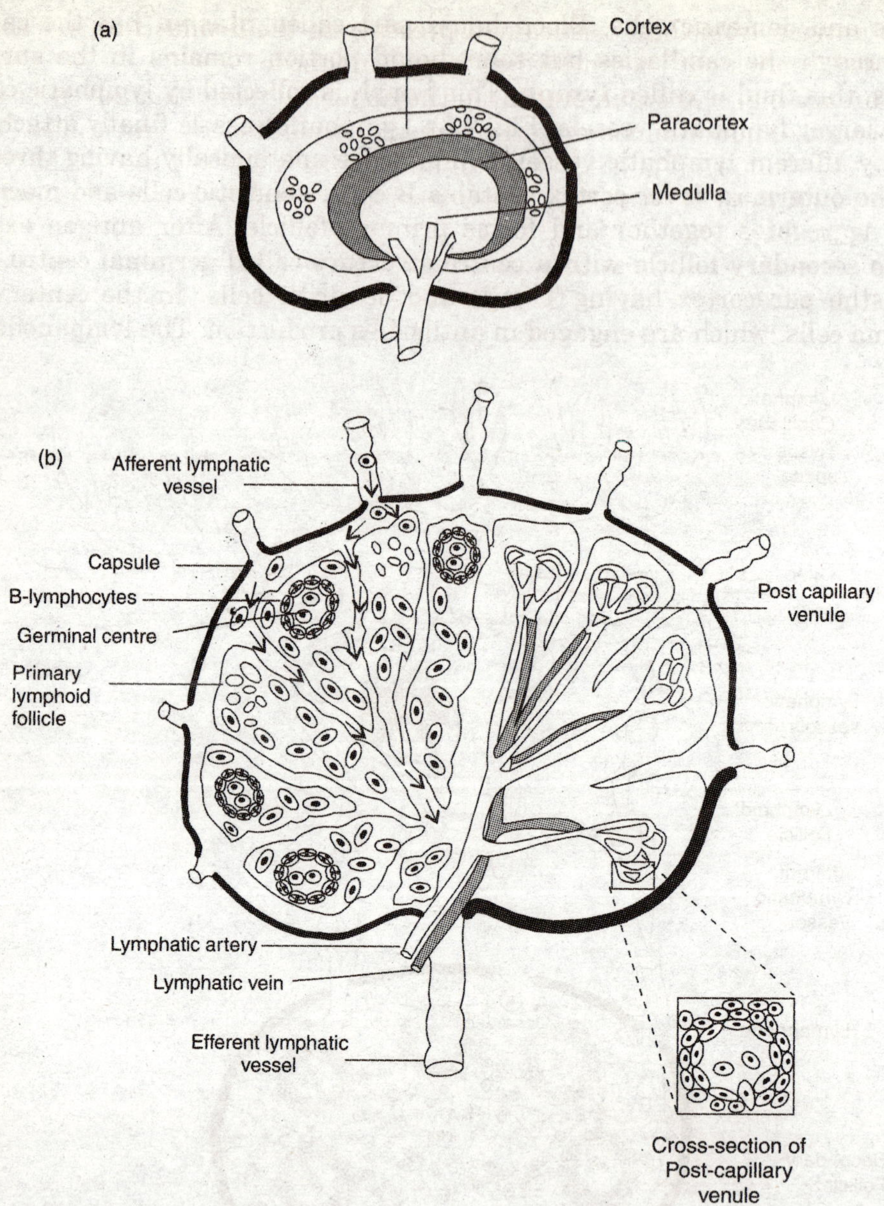

Fig. 7.4. Internal structure of a lymphnode (secondary lymphoid organ). (a) The three layers of a lymphnode support distinct microenvironments for lymphocytes. (b) The left side depicts the arrangement of reticulum and lymphocytes within the various regions of a lymphnode. Macrophages and dendritic cells, which trap antigen, are present in the cortex and paracortex portion. T_H cells are concentrated in the paracortex; B cells are located primarily in the cortex, within follicles and germinal centers. The medulla is populated largely by antibody-producing plasma cells. Lymphocytes circulating in the lymph are carried into the node by afferent lymphatic vessels; they either enter the reticular matrix of the node or pass through it and leave by the efferent lymphatic vessel. The right part of (b) shows the lymphatic artery and vein and the postcapillary venules. Lymphocytes in the circulation can pass into the lymphnode from the post-capillary venules by a process called extravastation (inset).

trapping the antigen from local tissues and presents to the T and B cells. The efferent lymphatic vessel carries out the antibodies from lymphnode.

The **spleen** is a big, oval organ present in the upper left abdominal cavity. It is having outer capsule and inner number of compartments separated by trabeculae (Fig. 7.5). Compartments are composed of outer red pulp containing many red blood cells (site of old and defective RBCs are removed) with macrophages and white pulp containing spleenic artery and T cells. The white pulp surrounds the spleenic artery and forms periarteriolar lymphoid sheath having large number of T cells. Both red and white pulp are separated by diffuse marginal zone rich in B cells. Spleen filter blood and traps blood-borne antigens.

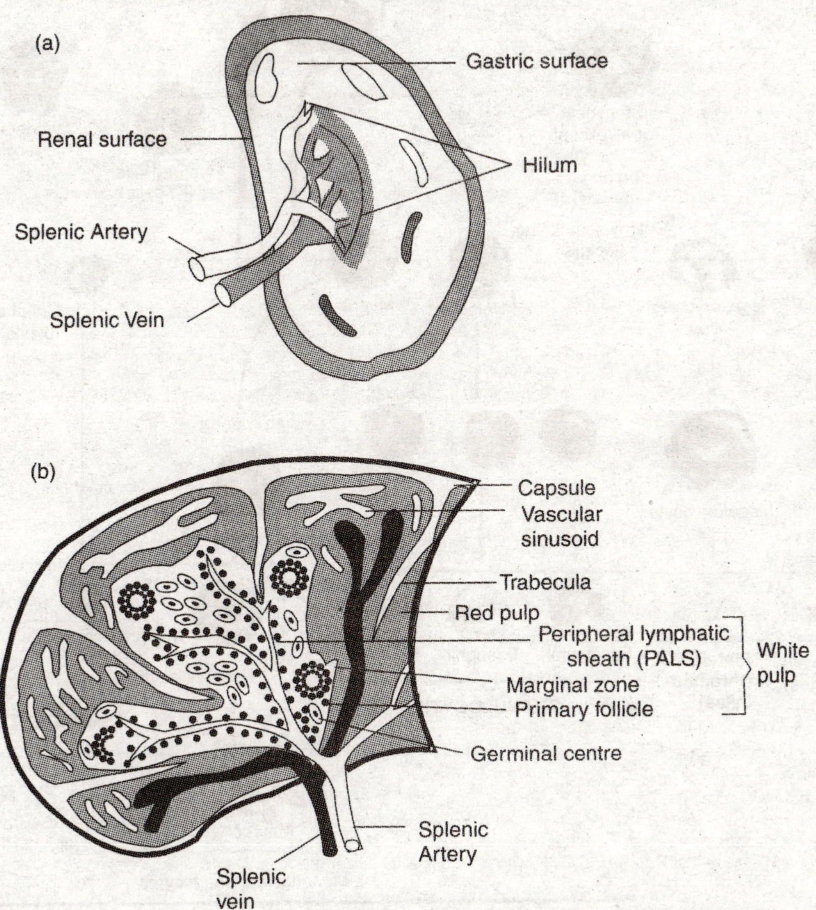

Fig. 7.5. Structure of the spleen (secondary lymphoid organ). (a) The spleen is approximately 5 inches long in adults, and is the largest secondary lymphoid organ. This organ is specialised for trapping blood-borne antigens. (b) Diagrammatic cross-section of the spleen. The splenic artery pierces the capsule and divides into progressively smaller arterioles, ending in vascular sinusoids that drain back into the splenic vein. The erythrocyte-filled red pulp surrounds the sinusoids. The white pulp forms a sleeve, the periarteriolar lymphoid sheath (PALS) around the arterioles; this sheath is made up of numerous T cells. Associated with the PALS is the marginal zone, an area rich in B cells that contains lymphoid follicles that can develop into secondary follicles containing germinal centers.

The digestive, respiratory and urinogential mucous membranes are the main entry spots for infectious pathogens. This part is generally protected by a group of organized lymphoid tissues and is named as **mucosal-associated lymphoid tissue (MALT)**. These tissues vary from loosely organized clusters of lymphoid cells in Lamina propria of intestinal villi to well developed tonsils, appendix and Peyer's patches (present in submucosa). MALT will be having large number of antibody producing plasma cells.

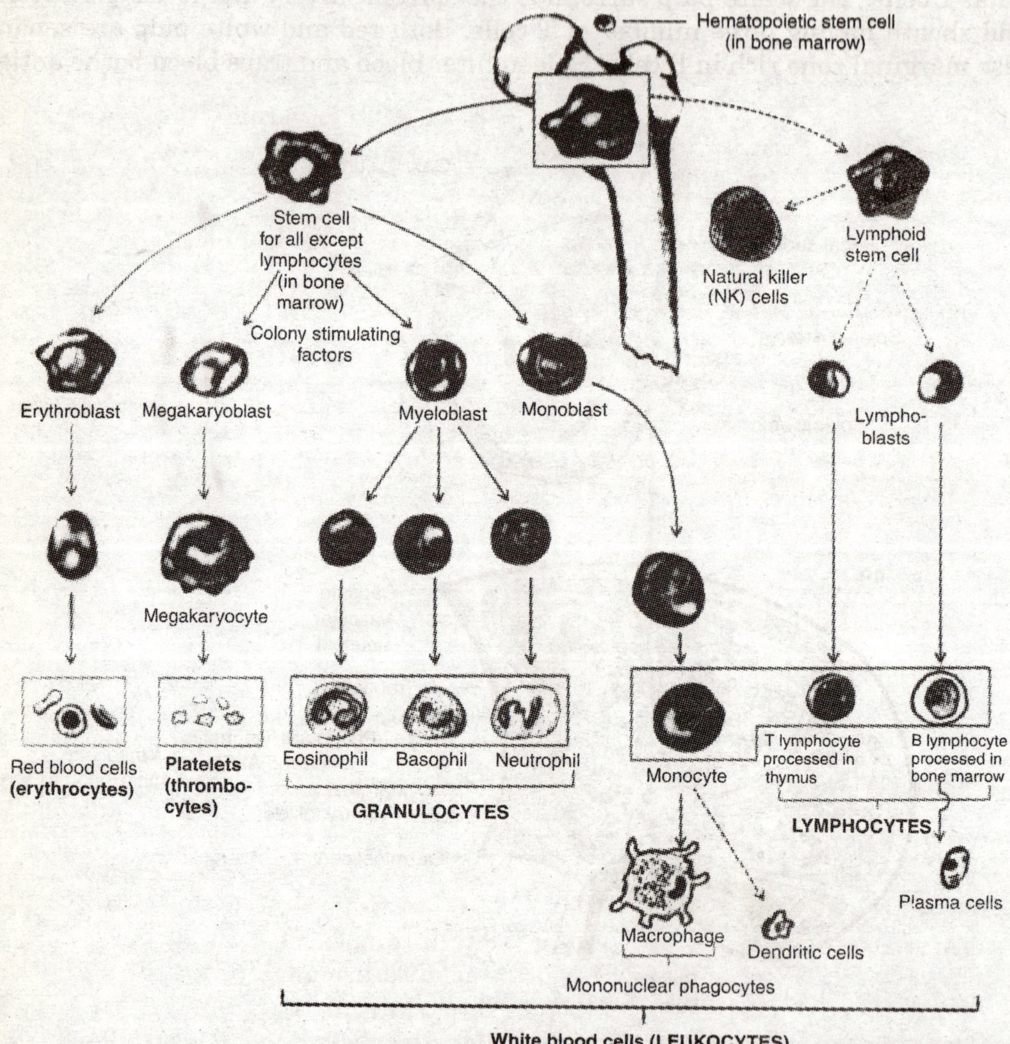

Fig. 7.6. The different types of blood cells in human. Pluripotent stem cells in the bone marrow divide to form two cells lineages: (1) the lymphoid stem cell give rise to B cells that become antibody-secreting plasma cells. T cells that become activated T cells, and natural killer cells. (2) The common myeloid progenitor cell gives rise to the granulocytes (neutrophils, eosinophils, basophils), monocytes that give rise to macrophages and dendritic cells, are as unknown precursor that gives rise to mast cells, megakaryocytes that produce platelets, and the erythroblast that forms erythrocytes (red blood cells). The dashed lines indicate steps that have not been very well studied.

7.6 CELLS OF THE IMMUNE SYSTEM

Leukocytes or WBCs are the major cells involved in immune response (Table 7.1). All blood and lymph cells derived from stem cells, which are pluripotent (capacity to develop into many cell types) in nature (Fig. 7.6). These cells are known as hematopoietic stem cells. Later these cells differentiate into two types, namely, myeloid stem cells and lymphoid stem cells. Myeloid stem cells give rise to RBCs, platelets, neutrophils, basophils and eosinophils and monocytes. Lymphoid stem cells give rise to B cells (B lymphocytes), T cells (T lymphocytes) and natural killer cells. These lymphocytes are the major cells of the immune system involved in all the activities of immunity (diversity, specificity, memory and self and non-self recognition). All the cells except RBCs and platelets referred as WBCs (leukocytes). The neutrophils, basophils and eosinophils are rich in granular materials in cytoplasm and hence they are known as granulocytes. The monocytes, B cells (B lymphocytes), T cells (T lymphocytes) and natural killer cells are not rich in granular materials and are known as agranulocytes.

Table 7.1: Blood cell count of a normal adult person

Type of Cell	Cells/mm^3	Percentage
Red blood cells	5×10^6	–
Platelets	2.5×10^5	–
White blood cells	7.3×10^3	–
Basophil		< 1%
Eosinophil		1–3%
Monocytes		1–6%
Lymphocytes		20–40%
Neutrophil		50–70%

7.7 AGRANULOCYTES

B cells (B lymphocytes): The B is derived from *bursa* of Fabricius in birds or *bone* marrow the site of maturation of the cells (Fig. 7.7). This cell types is always having surfaced, membrane bound antibodies (~ 1.5×10^5 antibodies per B cell). These cells after coming in contact with the antigen differentiate into antibodies producing plasma cells and memory B cell. Plasma cells, which lack membrane-bound antibody, synthesize and secrete one of the five classes of antibody. All clonal progeny from a particular B cell produce antibody with the same antigen binding specificity. The mature B cells contain number of membrane bound receptor molecules known as **cluster of differentiation (CD)**. Along with these number of membrane bound cluster of differentiation (**CD**), and B cells also have the capacity to presenting the antigen to immune system hence called antigen presenting cell (APC). These cells contain $CD_{40,}$ $CD_{45,}$ $CD_{35,}$ $CD_{21,}$ CD_{80} and CD_{86} molecules. B cells are known to bind the soluble antigens. The major function of B cells in immune system is the synthesis of the specific antibodies and the immunity obtained through the antibodies is known as humoral immunity (*humor* means *body fluid* contained by antibodies). It constitutes 5-15% of the circulating lymphoid cells.

T cells (T lymphocytes): The T is derived from *Thymus*, the site of maturation of the cells. These cells express various receptor molecules of the group CD_3, CD_4, CD_8, CD_{28}, CD_{45} (Cluster of Differentiation) (Fig. 7.7) and T cell receptor molecules (TCR) for antigen binding. T cells are known to bind to the antigens, which are processed and present on the self-cells. There are three sub types identified in the T cells, namely, T helper (T_H) cells, T cytotoxic (T_C) cells and T suppressor (T_S) cells.

T helper cells (T_H) interact with B cells to make antibodies and interact with phagocytes to destroy the pathogens. The T_H cells has CD_4^+ surface molecules and recognizing antigen bound to class II major histocompatability complex (MHC) molecules. T cytotoxic (T_C) cells having $CD8^+$ surface markers are responsible for elimination of host cells harboring viruses and other pathogens.

T suppressor (T_S) cells are known to suppress the functions of B and T cells, but still dispute is going on between immunologists regarding its exact function. The immunity obtained through the various T cells is known as cellular immunity.

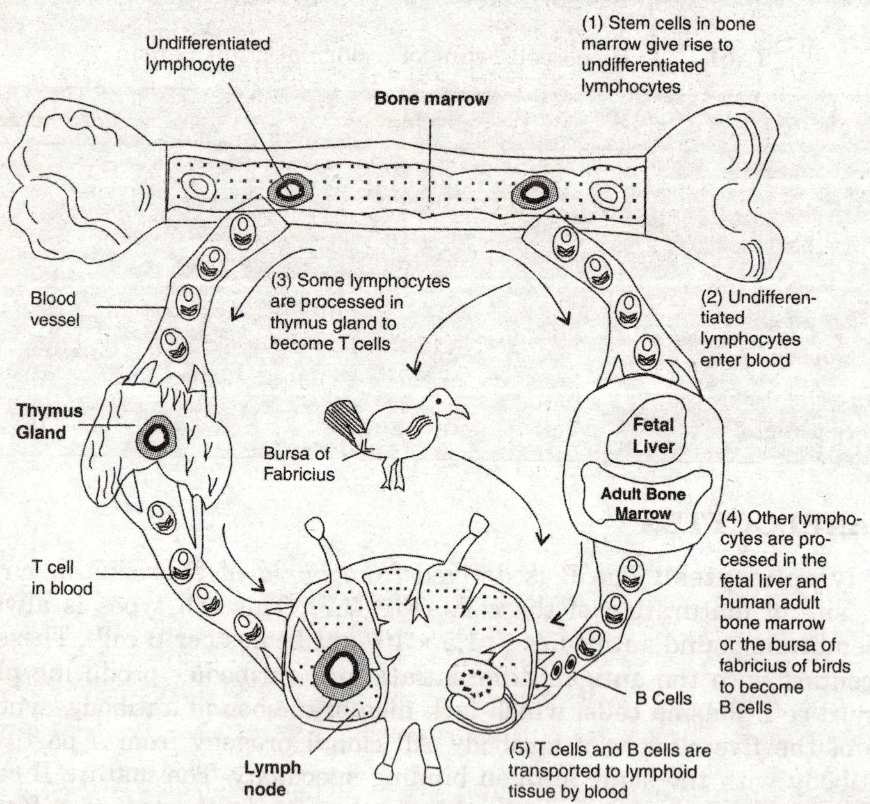

Fig. 7.7. Diagrammatic representation of lymphocyte development in lymphoid organs: Bone marrow produces undifferentiated lymphocytes, which, after processing develop into T and B cells.

Natural killer cells (NK cells): The lymphocytes that are not having any type of surface receptors and fail to synthesize antibodies are called null cells. Many of the null cells that are large, granular lymphocytes are called natural killer cells. It accounts for up to 15% of blood lymphocytes. These cells constitute 5%-10% of lymphocytes in human peripheral blood. The NK cells destroy tumor cells and virus infected cells. It also produces gamma interferon (proteins involved in protection against viral infections) and bears CD_2, CD_{16}, CD_{56}, killer activatory receptor (KAR) and killer inhibitory receptor (KIR) molecules on the surface.

Monocytes: Monocytes are mononuclear cells and function as phagocytic cells in blood. During phagocytic activity they engulf particles including infectious agents and finally destroy them. These cells contain characteristic horse shoe-shaped nucleus and are larger than most circulating lymphocytes. Monocytes come out of the blood vessels and differentiate into **macrophages** in the specific tissues. Such as alveolar macrophages in the lung, histiocytes in connective tissues, kupffer cells in the liver, mesangial cells in the kidney, microglial cells in the brain and osteoclasts in bone (Fig. 7.8).

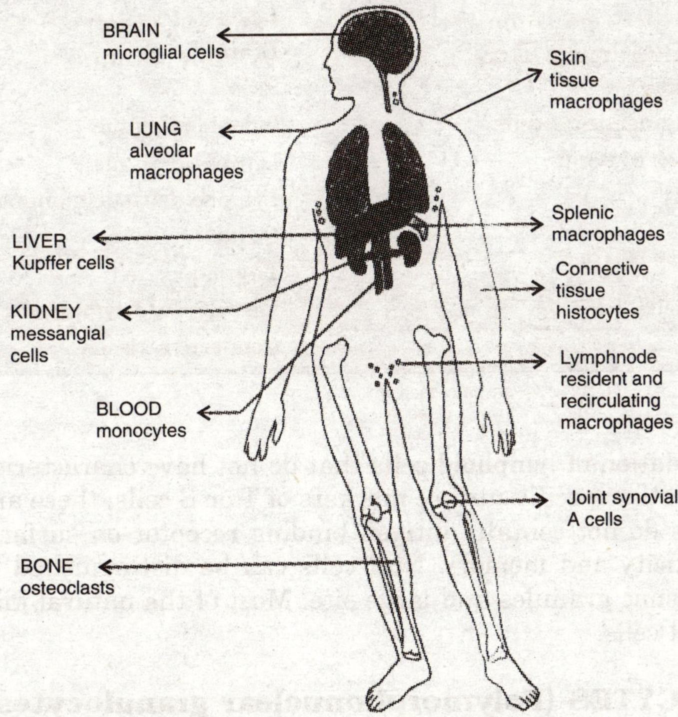

Fig. 7.8. The various Monocyte-Macrophage System in human body. This system consists of tissue (within the liver, spleen, and lymphnodes) having "fixed" or immobile phagocytes that have separate names depending on their location in the body.

Antigen presenting cells (APCs): These are heterogeneous group of leucocytes, which stimulate the immune response (Table 7.2). Some of them induce the activity of T_H cells and few communicate with other leucocytes. Antigen presenting cells are primarily found in the skin, lymphnodes, spleen and thymus. These cells after recognition of antigen they bind to the respective antigens and present to the immune system for further action. The Langerhans' cells, interdigitating cells, dendritic cells, B cells and macrophages can act as APCs, which are known to have class II MHC molecules. Follicular dendritic cells can differentiate between self cells and foreign cells. These APCs don't bear class II MHC molecules on the surface, but binds to antigen via CD21, CD35 (complement receptors that attaches complement associated immune complex) and Fc receptors. Recently another type of APC has been identified and they are called germinal centre dendritic cell (GCDC). They are involved in phagocytosis and activated by α-interferon, heat shock proteins and tumor necrosis factors that are released in response to microbial infection.

Table 7.2: Different types of antigen presenting cells and their location

Cell type	Location
Monocytes	Blood
Macrophages	Tissue, spleen, lymphnode
Kuffer cells	Liver
Microglia	Brain
Langerhan's cells	Skin
Interdigitating dendritic cells	Lymphoid tissue
Follicular dendritic cells	Lymphoid tissue
B cells	Lymphoid tissue/immune reaction site
Astrocytes	Brain
Follicular cells	Thyroid
Endothelial cells	Vascular and lymphoid tissue
Fibroblasts	Connective tissue

Null cell

These are the population of lymphoid cells that do not have characteristics of either T or B cells. As they lack the specific surface markers of T or B cells, these are called null cells. Because these cells do not contain antigen binding receptor on surface, they lack exact immunologic specificity and memory. Null cells can be distinguished from them by the presence of cytoplasmic granules and large site. Most of the natural killer cells belong to the category of null cells.

7.8 GRANULOCYTES (Polymorphonuclear granulocytes)

Granulocytes have irregular shaped nuclei and hence they are called Polymorphonuclear granulocytes. The nucleus are of different shapes and consists of two to five lobes. The cytoplasmic matrix has granules that contain reactive substances that kill microorganisms and enhance inflammation. These cells are produced in the bone marrow and are short-lived (2-3 days). Three types of granulocytes have been identified, namely, neutrophils, eosinophils and basophils (Fig. 7.9).

Basophils: These are having an irregular-shaped nucleus with two lobes and granules stain bluish-black with basic dyes. Basophils consist of histamine, prostaglandins, serotonin leukotrienes in their granules and releases after stimulation by antigens. These compounds (pharmacologically active) act on blood vessels to change the diameter and are termed as vasoactive agents. Basophils have receptor for immunoglobulin E molecules and they are coated with IgE antibodies. Once coated, antigens can trigger the basophils to secrete vasoactive agents and play a major role in allergenic response such as asthma, hay fever and eczema. The basophils constitute the less than 1% of total circulating white blood cells.

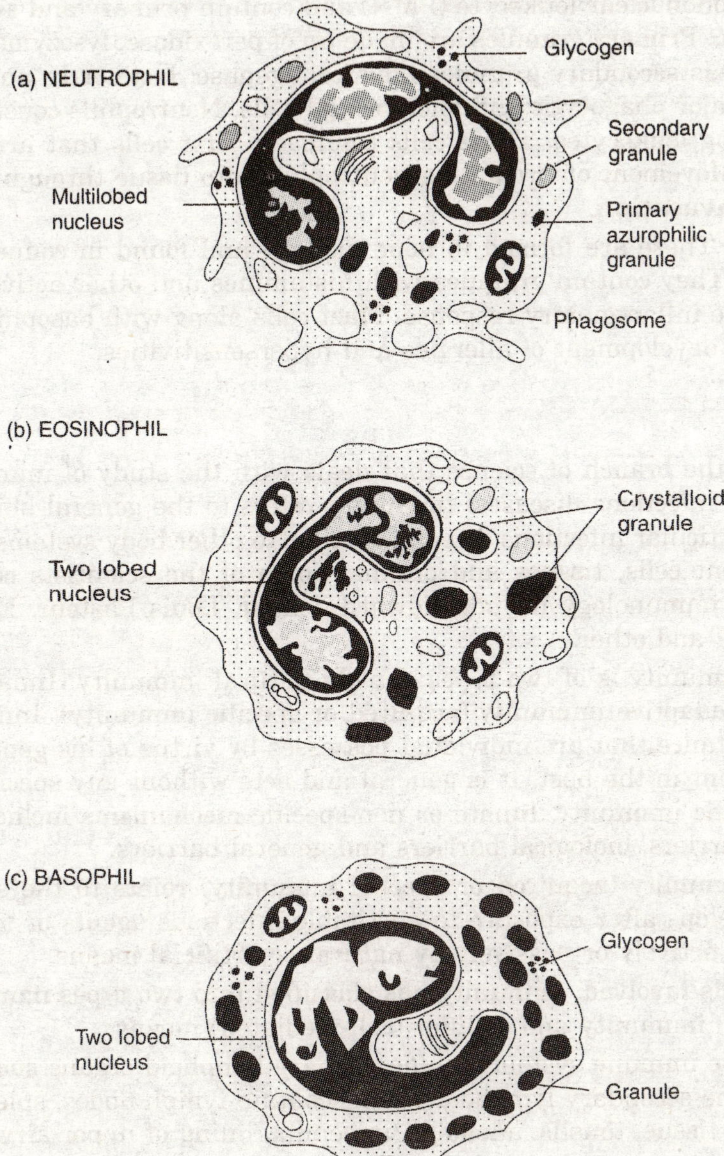

Fig. 7.9. Drawing showing typical morphology of granulocytes. Notes differences in the shape of the nucleus and in the number and shape of cytoplasmic granules.

Eosinophils: These cells have a two lobed nucleus connected by a slender thread of chromatin and granules stain red with acidic dyes. They can migrate from blood stream into tissue spaces and acts as phagocytic cells. Eosinophils play a major role in defense against parasitic organisms such as protozoan and helminthes by secreting cationic proteins and reactive oxygen metabolites which damage plasma membrane of the parasitic organisms. They constitute 1% to 3% of the total leukocytes.

Neutrophils: These cells have three to five lobed (multi lobed) nucleuses, hence also called poly morphonuclear leukocyte (PMN) and contain primary and secondary granules in the cytoplasm. Primary granules are made up of peroxidase, lysozyme and various lytic enzymes, whereas secondary granules have collagenase, lactoferrin and lysozyme. Thus they act as a major phagocytic and microbicidal cells. Neutrophils constitute 50% to 70% of the circulating leukocytes. These cells commonly first cells that arrive at the site of inflammation. Movement of circulating neutrophils into tissue through blood capillary is known as **extravasation**.

Mast cells: These are formed in bone marrow and found in connective tissue after differentiation. They contain granules with histamines and other active substances that contribute to the inflammatory response. Mast cells along with basophils play an important role in the development of allergies and hypersensitivities.

SUMMARY

Immunology is the branch of science that deals with the study of immunity of the host system towards infectious diseases. Immunity refers to the general ability of the organism to resist particular infectious disease. Similar to other body systems, immune system also has different cells, tissues and organs. Some of the scientists contributed to the development of immunology include Edward Jenner, Louis Pasteur, Emil von Behring, Elie Metchinkoff and other.

Basically immunity is of two types namely natural immunity (Innate or non-specific immunity) and adaptive immunity (acquired or specific immunity). Innate immunity refers to the resistance that are individual possesses by virtue of his genetic make-up and inherently present in the host. It is general and acts without any specificity hence come under non-specific immunity. Innate or non-specific mechanisms include physical barriers, chemical barriers, biological barriers and general barriers.

Adaptive immunity (acquired or specific immunity) refers to the specific immunity that a host develops after exposure to a suitable infectious agents or foreign antigen. It can be obtained actively or passively by natural or artificial means.

Based on cells involved, immunity has classified into two types namely humoral (antibody-mediated) immunity and cellular (cell-mediated) means.

Organs of the immune system include primary lymphoid organs such as thymus and bone marrow. The secondary lymphoid organs contain lymph nodes, spleen, mucous associated lymphoid tissue, tonsils, adenoids, appendix, lining of upper airways, bronchi and genital tract.

Cells involved in immune system include granulocytes and agranulocytes. The agranulocytes are B lymphocytes, T lymphocytes, natural killer cells, monocytes, antigen presenting cells and null cells. The granulocytes are basophils, eosinophils, neutrophils and mast cells.

EXERCISE

1. Define immunity.
2. Distinguish between antigen and immunogen.
3. Explain different types of immunity.
4. Describe different types of organs of immune system. (BU April 2001, Oct. 2002)
5. Elucidate various types of cells of immune system.
6. What is meant by granulocytes and agranulocytes? (BU Oct. 2002)
7. Define immunity. Explain active and passive immunity. (BU April, 2002)
8. What is meant by innate immunity? Explain. (BU April, 2002)
9. Write a note on Elie Metchinkoff. (BU Oct. 2002)
10. Give a note on contributions of Louis Pasteur. (BU Oct. 2001)
11. Write an account of Natural Killer Cells. (BU April, 2001)
12. Briefly explain Thymus gland. (BU April, 2001)
13. Describe the role of lymphoid organs in immunity. (BU April, 2003)
14. Explain different barriers which constitute innate immunity. (BU April, 2003)
15. Explain role played by dendritic cells in immunity.
16. List a historical account of development of immunology.
17. Describe anatomical structure of spleen.
18. Write a note on granulocytes of immune system.
19. Illustrate the formation of different blood cells.
20. Discuss internal structure of lymphnode.

8

Antigen, Antibody and Their Structure

8.1 INTRODUCTION

An **antigen** (**anti**body **gen**erator) is defined as any substance, which induces immune system to produce an antibody. An antigen reacts specifically with antibody in an observable manner. Most antigens are large, complex molecules with a molecular weight generally greater than about 10,000 Daltons. Chemically antigens may be proteins, nucleoproteins, polysaccharides or glycolipids. The ability of a molecule to function as an antigen depends on its size and structural complexity.

Each antigen can have several antigenic determinant sites or **epitopes**. Epitopes are the regions or sites in the antigen that bind to the antigen binding site of a specific antibody or with a T cell receptor. Antibodies are formed most readily in response to determinant that project from the foreign molecule or to terminal residues of a specific polymer chain. The number of antigenic determinant sites (epitopes) on the surface of an antigen is its valence and it determines the number of antibodies binding with the antigen. If one epitope is present then the antigen is said to be monovalent, if two it is referred as bivalent and so on.

Many small organic molecules are not antigenic by themselves but can become antigenic, if they bond to a larger carrier molecules (such as protein). These molecules are referred as **haptens** (*haptein* means to grasp). Hapten cannot stimulate antibody formation, but can react with antibodies once formed for this reason they are also known as **partial antigen**. Penicillin is best example of hapten. Penicillin by itself is not antigenic, however when it combines with certain serum proteins of sensitive individuals it acts as antigen.

8.2 IMMUNOGEN AND ANTIGEN

Thus we can separate immunogen and antigen as different molecules. **Immunogen** is a molecule that provoke immune system to produce antibodies or other cellular response. Whereas **antigen** is a molecule which has the capacity to react with the antibody but cannot stimulate the antibody formation. Haptens are antigens, but not immunogens. Generally both immunogen and antigen are often used synonymously. **Immunogenicity** is the ability of the substance to induce a immune response to produce antibodies or cellular response and **antigenicity** is the ability of the substance to react with antibodies or cell surface receptors (cells of cellular mediated immunity). All molecules that have the property of immunogenicity also have the property of antigenicity. But all antigens are not immunogens.

Some compounds are known to act as immunostimulants i.e. enhance the immunogenicity of the antigens and such compounds are called **adjuvants**. Some of the example for adjuvants include aluminium hydroxide or phosphate, Freund's incomplete adjuvant (protein antigens in water oil emulsion) and Freund's complete adjuvant (Freund's incomplete adjuvant + Muramyl peptide). Adjuvants along with humoral responses they induces high degree of cellular immunity and prolongs the antigenic stimulus. Certain antigens provoke a drastic immune response by stimulating T cells non-specifically and such antigens are known as **superantigens**. The bacterial proteins are example for superantigens (staphylococcal enterotoxins).

Any substance to be an effective antigen is based on the following properties:

a. Foreigners
b. Molecular weight
c. Chemical composition and heterogenicity
d. Susceptibility to the tissue enzyme
e. Genotype of the immune system
f. Dosage and route of exposure
g. Capacity to bind to carrier molecules

8.3 ANTIBODIES

Antibodies or **immunoglobulin (Ig)**, on the other hand, are produced rapidly in response to an antigen and can recognize and bind to the antigen that caused its production. Antibodies are chemically glycoproteins and present in the blood serum, tissue fluids and mucosal surfaces of vertebrate animals. Serum glycoproteins are separated by electrophoresis and gammaglobulin fraction consists of heterogenous mixture of immunoglobulins. These immunoglobulins differ from each other in molecular structure, charge, amino acid composition and carbohydrate content.

Antibody structure

Immunoglobulins are large glycoproteins with molecular weights ranging from 150,000 to 900,000. Rodney R. Porter and Gerald M. Edelman discovered the immunoglobulin structure, for their work they got Nobel Prize in 1972. All immunoglobulins are having basic "Y" shaped structure and consists of two similar light chains (220 amino acids each, 25kDa) and two similar heavy chains (about 440-450 amino acids each, 50-70kDa) which are covalently interconnected by disulfide bond near the hinge region (Fig. 8.1). The stem region has polysaccharide side chain. This forms the monomeric structure of an antibody. Each arm of the Y shaped is composed of one light chain and one heavy chain, joined by the one disulfide bond. But stem consists of only two heavy chains joined by the two disulfide bonds. Both heavy (H) chain and light (L) chain contain two different regions. Constant regions

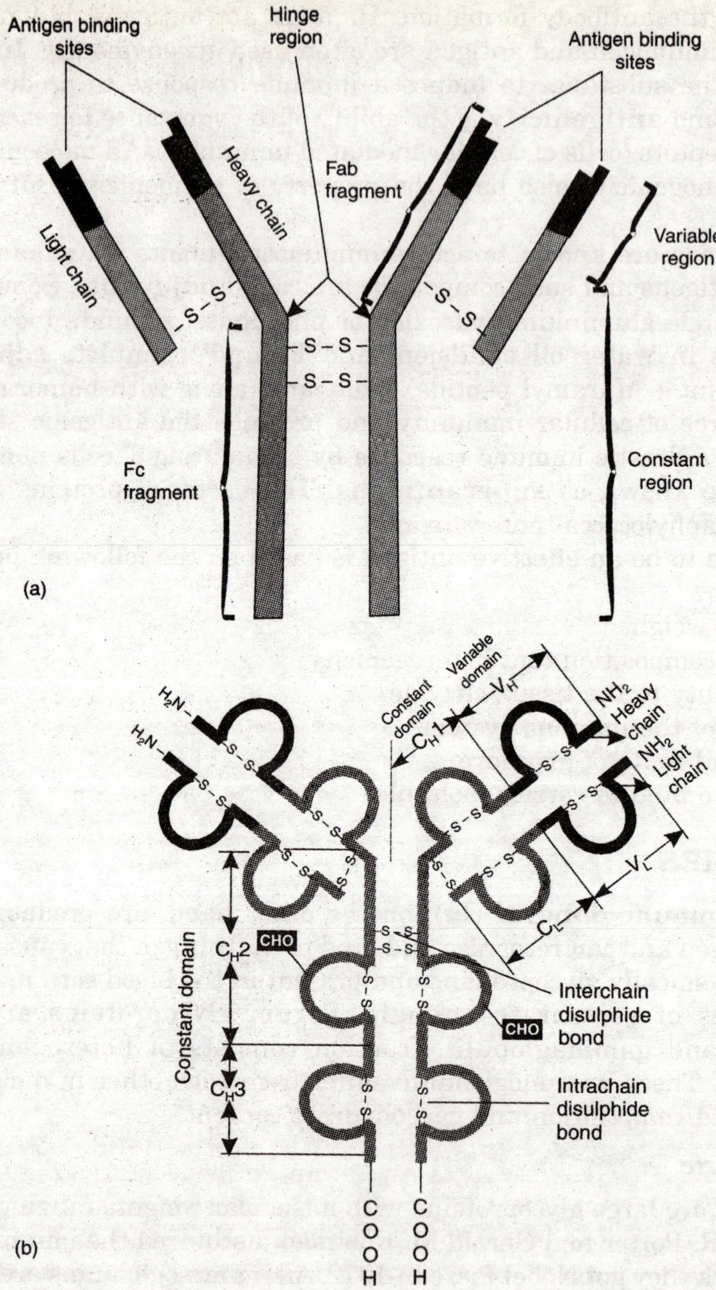

Fig. 8.1 a & b. Diagrammatic representation of an immunoglobulin (antibody) structure. (a) An immunoglobulin molecule showing two identical light chains and two identical heavy chains joined together by disulphide bonds. (b) The immunoglobulin unit structure shows intrachain disulphide bonds create loops that form domains. All light chains contain a single variable domain (V_L) and a single constant domain (C_L). Heavy chains contain a variable domain (V_H) and either three or four constant domains (C_HI, C_H2, C_H3 and C_H4). The variable regions (V_H, V_L), when folded together in three-dimensions, form the antigen-binding sites.

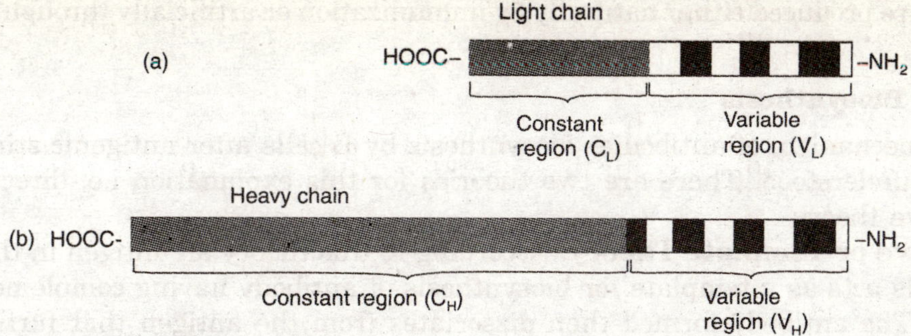

Fig. 8.1c. Constant and variable domains of an immunoglobulin. Location of constant (C) and variable (V) domains within (a) light chains and (b) heavy chains. The dark bands within V_L and V_H region represent hypervariable regions or complementarity-determining regions within the variable domains.

(**C_L** and **C_H**) have amino acid sequences that do not vary significantly between the antibodies of the same class. The variable regions (**V_L** and **V_H**) from different antibodies do have different sequences. It is the variable regions (**V_L** and **V_H**) that when folded together form the antigen binding sites.

The papain enzyme cleave each monomeric unit (Y-shaped structure) near hinge region into two fragments which bind to the antigen (Fab) and one fragment which does not bind to antigen but undergoes crystallization (Fc) (Fig. 8.1c). The Fc fragments are composed of only of constant regions (**C_H2-3**), whereas the Fab fragments have both constant (**C_H1**) and variable regions (**V_L** and **V_H**). Even within variable region some part has more variation and these regions are referred as hyper-variable regions (Fig. 8.1c) or complementarity-determining regions. Within each chain a loop is formed called domain by intrachain disulfide bonds consists of 60 amino acids.

The light chains are of two types i.e. kappa (κ) and lambda (λ) types. Its constant region determines the type of a light chain. Each antibody molecule is made up of either two kappa or two lambda light chains. The two separate genes encode the kappa (genes in human chromosome 2) and lambda (human chromosome 22) light chains. The gene of human chromosome 14 codes the heavy chain.

The Fab fragment is involved in binding of antigen, whereas the Fc fragment helps in binding to various cells of the immune system (mast cells, B cells, T cells), some phagocytic cells, or the first component of the complement system. Thus, immunoglobulins are bifunctional in nature. The simple binding of an antibody with an antigen usually does not cause total destruction of the antigen or of the microorganism, cell or agent to which it is attached. Rather the antibody serves to mark and identify the target for wide array of immunological attack and to activate nonspecific immune responses that can destroy the target by phagocytosis is termed opsonization. Immune destruction also is promoted by antibody-induced activation of the complement system.

Source of Antibodies

An antibody is an immunoglobulin whose formation is induced by the introduction of an antigen in an animal body. It reacts with the corresponding antigen specifically in some observable way, that each antibody has binding sites for an identified antigen. The need for pure homogeneous antibodies has increased dramatically in recent years. Currently antibodies are produced either naturally by immunization or artificially through hybridoma formation.

Antibodies Biosynthesis

The exact mechanism of antibodies biosynthesis by B cells after antigenic stimulation is completely understood. There are two theories for this explanation i.e. directive theory and selective theory.

Directive or Template Theory: According to this theory an antigen in the antibody forming cells acts as a template for biosynthesis of antibody having complementary configuration. The antibody formed then dissociates from the antigen that further acts as template for next antibody synthesis. It is considered that antigen brings about genetic changes in the cell so that it and its progenies continue to produce specific antibodies.

Selective Theory: In 1880, Paul Ehrlich first proposed this theory and hence is also known as Ehrlich side chain theory. According to this, antibody forming cells have antibody molecules as side chains on their surface and the antigen selects its corresponding side chain, attaches to it and finally knocks it down. This acts as a trigger and then the cell starts synthesizing similar side chains repeatedly, which are dislodged to free circulation and unite with the antigen.

8.4 CLASSES OF IMMUNOGLOBULINS (ANTIBODIES) AND THEIR SUBCLASSES

Immunoglobulins are classified into five types based on the antigenic character of heavy chains. These are Immunoglobulin G (IgG), Immunoglobulin M (IgM), Immunoglobulin D (IgD), Immunoglobulin A (IgA) and Immunoglobulin E (IgE). The heavy chains of an antibody are designated as γ heavy chain in IgG, μ in IgM, α in IgA, δ in IgD and ε in IgE. An individual person has all the five types of immunoglobulins (Table 8.1).

Immunoglobulin G (IgG): IgG is a monomer unit (Fig. 8.2). This is the major immunoglobulin (most abundant) in human serum and makes up 70 to 75% of the total immunoglobulin pool. IgG is present in blood plasma and tissue fluids. The IgG class acts against bacteria and viruses by opsonizing the invaders and neutralizing toxins. It is also one of the two immunoglobulin classes that activated complement by the classical pathway. IgG is the only immunoglobulin molecule able to cross the placenta (hence termed as **maternal antibody**) and provide naturally acquired immunity for the newborn and persists for several months.

Totally there are four IgG subclasses (IgG1, IgG2, IgG3, and IgG4) that vary chemically in their (γ heavy) chain composition and the number and arrangement of interchain disulfide bonds. About 65% of the total serum IgG is IgG1, and 23% is IgG2. Differences in biological function have been noted in these subclasses. For example, IgG2 antibodies are opsonic and develop in response to antitoxins. Anti-Rh antibodies are of the IgG1 or IgG3 subclass. IgG1 and IgG3 also bind best to monocytes and macrophages and activate complement most effectively. The IgG4 antibodies function as skin sensitizing immunoglobulins.

Immunoglobulin A (IgA): IgA is the second most abundant immunoglobulin and account for about 15% of the immunoglobulin pool. Some IgA is present in the serum as a monomer of two heavy and two light chains. Most IgA, however, occurs in the serum as a polymerized dimer joined together by a polypeptide J chain (Fig. 8.3). IgA has special features that are associated with secretory mucosal surfaces. IgA, when transported from the mucosal-associated lymphoid tissue to mucosal surfaces, acquires a polypeptide termed secretory protein and called **secretory antibody**.

The secretory IgA (sIgA), as the modified molecule is now called the primary immunoglobulin of the secretory immune system. This system is found in the gastrointestinal tract, upper and lower respiratory tracts, and genitourinary system. Secretory IgA is also

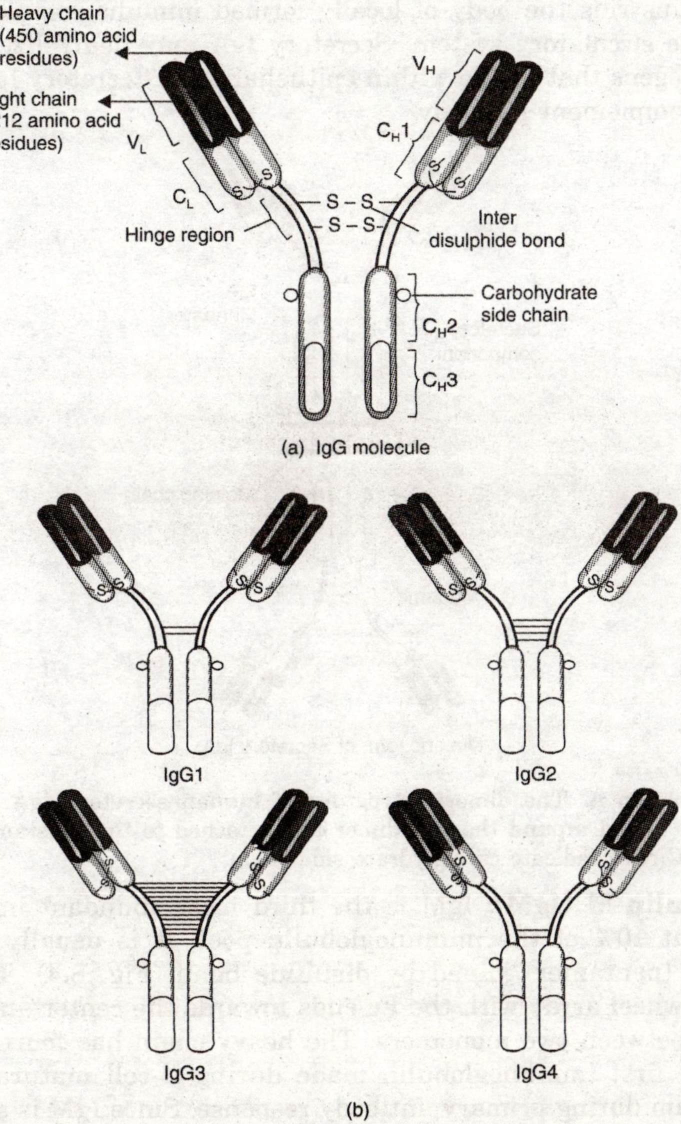

Fig. 8.2. Immunoglobulin G. (a) The basic structure of human IgG. (b) The structure of the four human IgG subclasses IgG1, IgG2, IgG3 and IgG3. Note the arrangement and numbers of disulphide bonds (shown as thin black lines). Carbohydrate side chain is represented by circles.

found in saliva, tears, and breast milk (*colostrum*). In these fluid and related body areas, sIgA by the formation of an immune barrier plays a major role in protecting surface tissues against infectious microorganisms. For example, in breast milk sIgA (maternal antibodies) helps to protect nursing newborns. In the intestine, sIgA attaches to viruses, bacteria, and protozoan parasites such as *Entamoeba histolytica*. This prevents pathogen adherence to mucosal surfaces and invasion of host tissues, a phenomenon known as **immune exclusion**. In addition, sIgA binds to antigens within the mucosal lamina propria, and the antigen-sIgA complexes are excreted through the adjacent epithelium into the gut lumen. This rids the body of locally formed immune complexes and decreases their access to the circulatory system. Secretory IgA may neutralize viruses and other intracellular pathogens that reside within epithelial cells. Secretory IgA also plays a role in the alternate complement pathway.

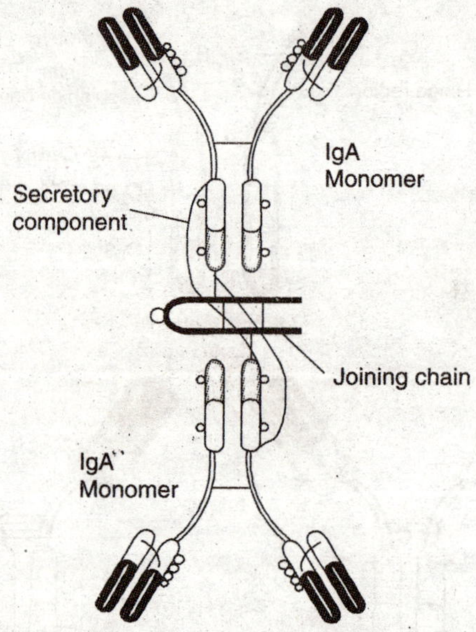

Fig. 8.3. Immunoglobulin A. The dimeric structure of human secretory IgA. Notice the secretory component wound around the IgA dimer and attached to the constant domain of each IgA monomer. Circles indicate carbohydrate side chain.

Immunoglobulin M (IgM): IgM is the third most abundant immunoglobulin and accounts for about 10% of the immunoglobulin pool. It is usually a polymer of five monomeric units (pentamer) joined by disulfide bond (Fig. 8.4). The monomers are arranged in a pinwheel array with the Fc ends towards the center and consist of special J (joining) chain between two monomers. The heavy chain has fourth constant domain (**C_{H4}**). IgM is the first immunoglobulin made during B-cell maturation and the first secreted into serum during primary antibody response. Since IgM is so large, it does not leave the bloodstream or cross the placenta. IgM agglutinates bacteria, activates complement by the classical pathway, and enhances the ingestion of pathogens by phagocytic cells. These classes also contain special antibodies such as red blood cell agglutinins and heterophile antibodies.

Although most IgM appears to be pentameric, around 5% or less of human serum IgM exists in a hexameric form. This molecule contains six monomeric units but seems to lack a J chain. Hexameric IgM activates complement up to twenty fold more effectively than does the normal pentameric form. It has been suggested that bacterial cell wall antigens such as gram-negative lipopolysaccharides may directly stimulate B cells to form hexameric IgM without a J chain. If this is the case, the immunoglobulins formed during primary immune responses are less homogenous than previously thought.

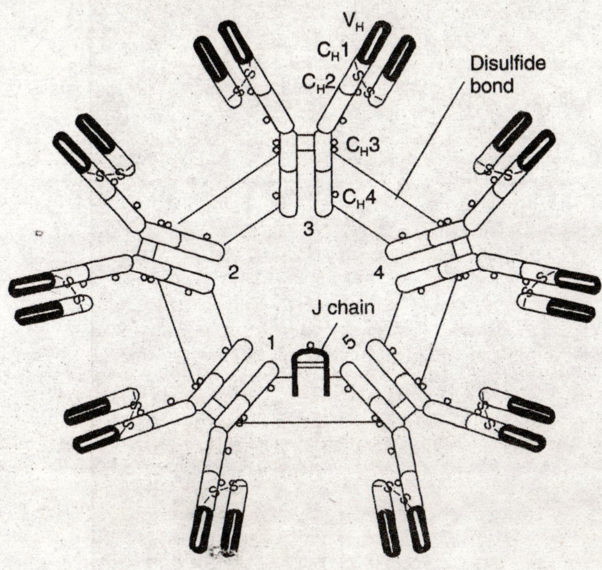

Fig. 8.4. Immunoglobulin M. The pentameric structure of human IgM. The disulphide bonds linking peptide chains are shown by line; carbohydrate side chains are indicated by dots. Note that 10 antigen-binding sites are present.

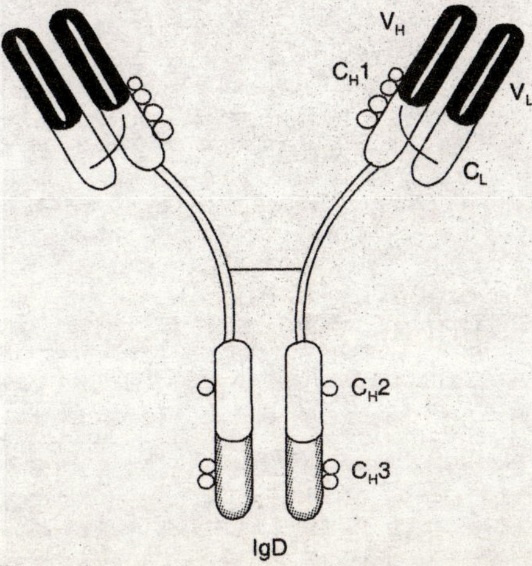

Fig. 8.5. Immunoglobulin D. The structure of human IgD. The disulphide bonds linking protein chains are shown by line; carbohydrate side chains are represented by circles.

Table 8.1: Physicochemical properties of human immunoglobulin classes

Property	Immunoglobulin Classes				
	IgG[a]	IgM	IgA[b]	IgD	IgE
Heavy chain	γ_1	μ	α_1	δ	ε
Mean serum concentration (mg/ml)	9	1.5	3.0	0.03	0.00005
Percent of total antibody	80-85	5-10	5-15	<1	0.002-0.05
Valency	2	5 (10)	(2-4)	2	2
Mass of heavy chain (kDa)	51	65	56	70	72
Mass of entire molecule (kDa)	146	970	160[c]	184	188
Placental transfer	+	–	–	–	–
Half-life in serum (day)[d]	23	5	6	3	2
Complemental activation					
Classical pathway	++	+++	–	–	–
Alternative pathway	–	–	+	–	–
Induces mast cell degradation	–	–	–	–	+
Major characteristics	Most abundant Ig in body fluids; neutralises toxins, opsonises bacteria, activates complement, maternal antibody	First to appear after antigen stimulation; very effective agglutinator; expressed as membrane-bound antibody on B cells	Secretory antibody; protects external surfaces	Present on B-cell surface; B-cell recognition of antigen	Anaphylactic-mediating antibody: resistance to helminths
% carbohydrate	3	7-10	7	12	11

[a]Properties of IgG subclass 1.

[b]Properties of IgA subclass I.

[c]sIgA = 360 – 400 kDa

[d]Time required for half of the antibodies to disappear.

Immunoglobulin D (IgD): IgD is an immunoglobulin found in very low concentration in the blood serum. It is highly sensitive to proteolytic enzymes and to heat. IgD has a monomer structure similar to IgG (Fig. 8.5). IgD antibodies do not fix complement and cannot cross the placenta, but they are abundant on the surface of B cells and bind antigens, thus signaling the B cell to start antibody production.

Immunoglobulin E (IgE): IgE makes up only 0.00005% of the total immunoglobulin pool. The classic skin-sensitizing and anaphylactic antibodies belong to this class. Most IgE is bound to the cytoplasmic membranes of basophils in blood and mast cells in connective tissue, where it mediates the release of histamine that causes immediate type allergic reactions. IgE molecules have four constant domains ($C_\varepsilon1$, $C_\varepsilon2$, $C_\varepsilon3$ and $C_\varepsilon4$) in heavy chain and two light chains (Fig. 8.6).

The Fc portion of the $C_\varepsilon4$ chain can bind to special Fc receptors on mast cells and basophils. When two IgE molecules on the surface of these cells are cross-linked by binding to the same antigen, the cells show degranulation. This degranulation releases histamine and other pharmacological mediators of anaphylaxis. It also stimulates eosinophils and gut hypermotility (increased rate of movement of the intestinal contents) that aid in the elimination of helminthic parasites. Thus, though IgE is present in small amounts, this class of antibodies has very potent biological capabilities.

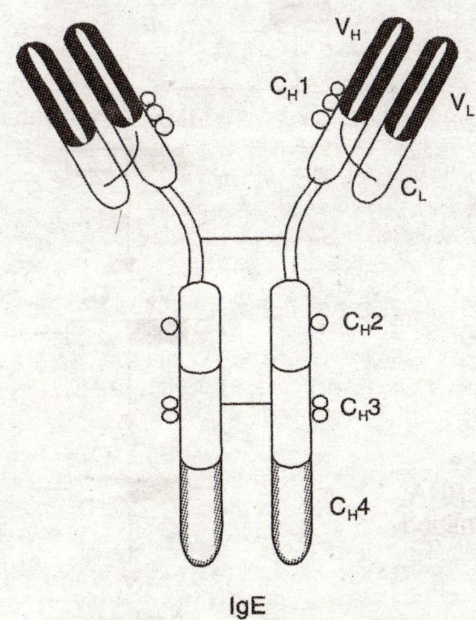

Fig. 8.6. Immunoglobulin E. The structure of human IgE. Circles indicate carbohydrate side chain.

8.5 THE BASIS OF ANTIBODY DIVERSITY

The most remarkable and unique property of antibodies is their diversity. According to current estimation each human or mouse can synthesize more than 10 million different kinds of antibodies. This remarkable diversity is generated due to rearrangement of antibody gene segments, somatic mutations and generation of different codons during antibody gene splicing.

Immunoglobulin genes are split or interrupted genes having many exons. Embryonic B cells contain a small number of exons, close together on the same chromosome, that determine the constant (C) region of the light chains (Fig. 8.7). Separated from them, but on the same chromosome, is a larger cluster of exons that determines the variable (V) region of the light chains. During B-cell differentiation, one exon for the constant region is spliced exon for the variable region. This splicing produces a complete light-chain antibody gene. A similar splicing mechanism also occurs to join the constant and variable exons of the heavy chains.

As the light chain genes actually consist of three parts and the heavy chain genes consist of four parts, formation of a final antibody molecule is slightly more complicated than previously mentioned. The germ line DNA for the light chain gene contains multiple coding sequences called V and J (joining) regions. During the differentiation of a B cell, a deletion (which is variable in length) occurs that fuse one V exon with one J exon. This DNA joining process is termed combinatorial joining since it can create many combinations of the V and J regions. When the light chain gene is transcribed, transcription

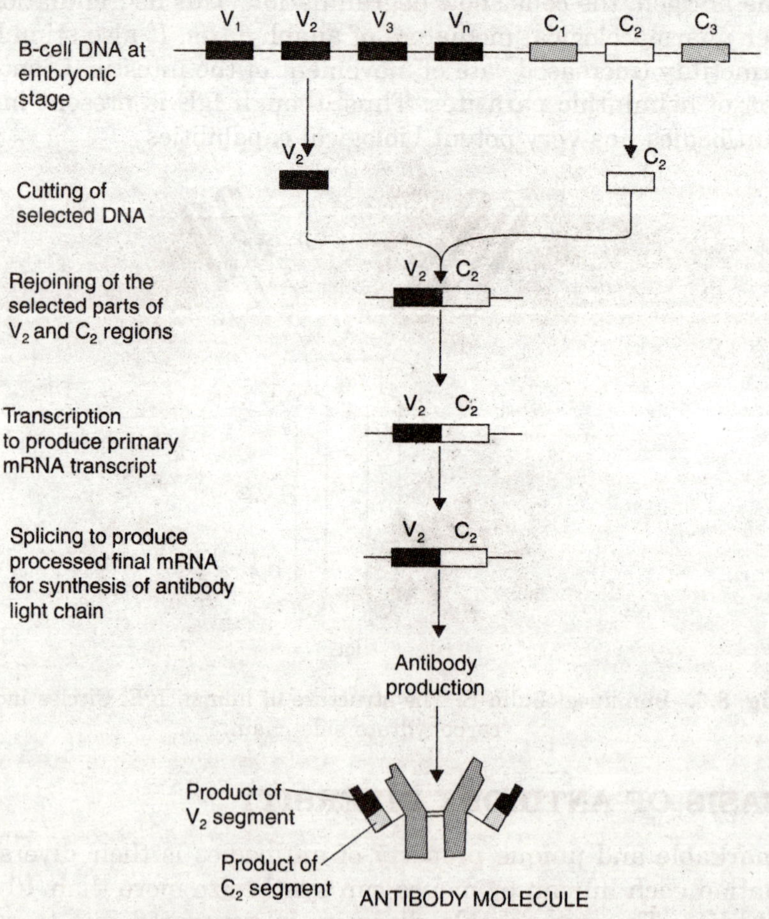

Fig. 8.7. Shuffling of genes to produce antibody diversity: Antibody diversity is partly the result of the shuffling of gene sequences that code for both heavy and light chains. This model shows the shuffling cutting, and splicing process used to produce an assembled light chain of the antibody molecule.

continues through the DNA region that encodes for the constant portion of the gene. RNA splicing subsequently joins the V-J and C regions creating mRNA.

The formation of a heavy chain gene through combinatorial joining occurs by means of DNA splicing of the heavy chain counterparts of V and J along with a third set of D (diversity) sequences (Fig. 8.8). Initially, all heavy chains have the μ type of constant region. This corresponds to antibody class IgM. Another DNA splice joins the VDJ region with a different constant region that can subsequently change the class of antibody produced by the B cell.

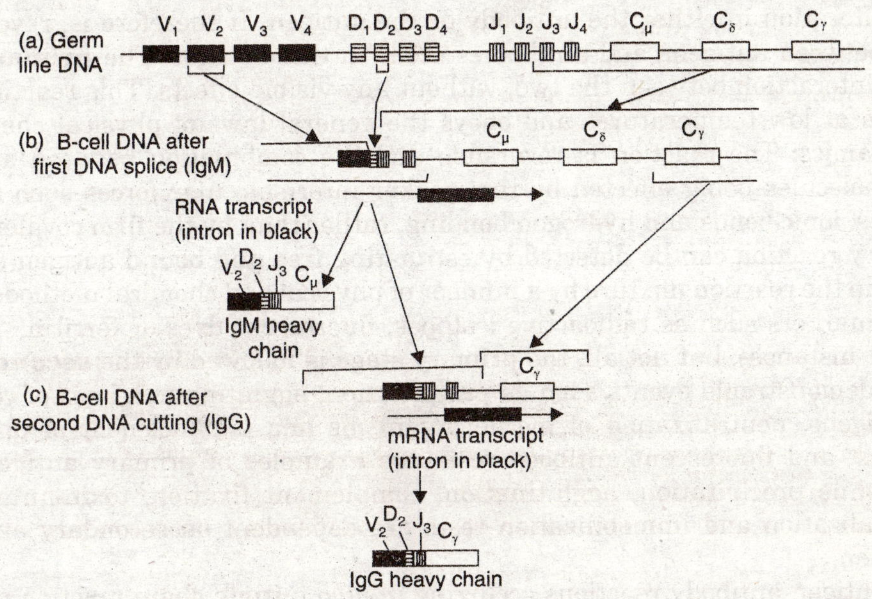

Fig. 8.8. The formation of a gene for the heavy chain of an antibody molecule.

The kappa (κ) light chains in mouse are formed from combinations of about 250-350 V_κ and 4 J_κ regions giving a maximum of approximately 1,400 different κ chains. The O chains have their own V_O and J_O regions but smaller in number than their κ counterparts (6 different O chains). The heavy chains have approximately 250-1,000 V_H, 10-30 D, and 4 J_H regions giving maximum 120,000 different combinations. Because any light chain can combine with any heavy, there will be a maximum of 2×10^8 possible κ chain antibody types.

The value 2×10^8 different antibodies is actually an underestimate because antibody diversity is further augmented by two processes:

1. The V regions of germ line DNA are susceptible to high rate of somatic mutation during B cell development in the bone marrow. These mutations allow B cell clones to produce different polypeptide sequences.
2. The junction for either VJ or VDJ splicing in combinatorial joining can occur between different nucleotides and thus one VJ splicing event can join the V sequence CCTCCC with the J sequence TGGTGG in two ways:

CCTCCC + TGGTGG = CCGTGG,

which codes for the amino acids proline and tryptophan; and

CCTCCC + TGGTGG = CCTCGG,

which codes for proline and arginine. Thus the same VJ joining could produce polypeptides differing in a single amino acid.

8.6 ANTIGEN-ANTIBODY INTERACTIONS

The antigen-antibody interaction is a bimolecular association similar to an enzyme-substrate interaction, with an important distinction: it does not lead to an irreversible chemical alteration in either the antibody or the antigen. It therefore is, reversible. The reactions between antigens and antibodies occur in three stages. The **primary stage** is the initial interaction between the two, without any visible effects. This reaction is rapid, occurs even at low temperatures and obeys the general laws of physical chemistry and thermodynamics. The reaction is reversible and the combination between antigen and antibody molecules being effected by the weaker intermolecular forces such as Van der Wall's forces, ionic bonds and hydrogen bonding, rather than by the firm covalent bonding. The primary reaction can be detected by estimating free and bound antigen or antibody separately in the reaction mixture by a number of physical and chemical methods, including the use of markers such as radioactive isotopes, fluorescent dyes or ferritin.

In most instances, but not all, the primary stage is followed by the **secondary stage** leading to demonstrable events such as precipitation, agglutination, lysis of cells, killing of live antigens, neutralization of motile organisms and enhancement of phagocytosis. Coomb's test and fluorescent antibody tests are examples of primary antigen-antibody reaction, while precipitation, agglutination, complement fixation, toxin-antitoxin reaction, neutralization and immobilization tests are dependent on secondary antigen-antibody reaction.

Some antigen-antibody reactions occurring *in vivo* initiate chain reaction that lead to neutralization or destruction of injurious antigens, or to tissue damage. These are the **tertiary reactions** and include humoral immunity against infectious disease as well as clinical allergy, xenotransplantation rejection and other immunological diseases.

The exquisite specificity of antigen-antibody interactions has led to the development of a variety of immunologic assays. These assays can be used to detect the presence of either antibody or antigen and have played vital roles in diagnosing diseases, monitoring the level of the humoral immune response and identifying molecules of biological or medical interest. These assays differ in their speed and sensitivity, some are qualitative and others are quantitative. In the body, reactions between antigens and antibodies form the basis of antibody-mediated immunity to infectious diseases, or to tissue injury in some types of hypersensitivity and autoimmune diseases. In the laboratory, they help in the diagnosis of infections, in epidemiological surveys, in the identification of infectious agents and of non-infectious antigens such as enzymes. In general, these reactions can be used for the detection and quantification of either antigens or antibodies. Antigen-antibody reactions *in vitro* are known as serological reactions.

GENERAL CHARACTERISTICS OF ANTIGEN-ANTIBODY REACTIONS

1. The reaction is specific, and antigen combine only with its homologous antibody

and vice versa. The specificity, however, is not absolute and 'cross reactions' may occur due to antigenic similarity or relatedness.

2. An entire molecule reacts and not fragments. When an antigenic determinant present in a large molecule or on a 'carrier' particle reacts with its antibody, whole molecules or particles are agglutinated.
3. There is no denaturation of the antigen or the antibody during the reaction.
4. During combination only the surface antigens participate.
5. Combination is firm, but reversible. The affinity and avidity of the reaction influence the firmness of the union. Affinity refers to the intensity of attraction between antigen and antibody molecules. Avidity is the strength of the bond after the formation of antigen-antibody complexes.
6. Both antigens and antibodies participate in the formation of agglutinates or precipitates.
7. Antigens and antibodies can combine in varying proportions, unlike chemicals with fixed valencies. Both antigens and antibodies are multivalent. Antibodies are generally bivalent, though IgM molecules may have 5 or 10 combining sites. Antigens may have valencies upto hundreds.

STRENGTH OF ANTIGEN-ANTIBODY INTERACTIONS

The noncovalent interactions that form the basis of antigen-antibody (Ag-Ab) binding include hydrogen bonds, ionic bonds-hydrophobic interactions and Van der Waals interactions.

Because the strength of each of these interactions is weak (compared with that of a covalent bond) a large number of such interactions are required to form a strong antigen-antibody interaction. Furthermore, each of these noncovalent interactions operated over a very small distance, generally about 1×10^{-7} mm (1 A) or a little less; consequently, a strong antigen-antibody interaction depends on a very close fit between the antigen and antibody. Such fits require a high degree of complementarity between antigen and antibody, and this requirement is the basis of the exquisite specificity that characterizes antigen-antibody interactions.

Antibody Affinity

The strength of the total noncovalent interactions between a single antigen-binding site on an antibody and a single epitope is the 'affinity' of the antibody for that epitope. Low affinity antibodies bind antigen weakly and tend to dissociate readily, whereas high-affinity antibodies bind antigen more tightly and remain bound longer. The association between a binding site on an antibody (Ab) with monovalent antigen (Ag) can be described by the equation

$$Ag + Ab \underset{k_{-1}}{\overset{k_1}{\rightleftharpoons}} Ag - Ab$$

where k_1 is the forward (association) rate constant

k_{-1} is the reverse (dissociation) rate constant

The ratio of k_1/k_{-1}, is the association constant Ka (i.e., k_1/k_{-1}= Ka), a measure of affinity. Because Ka is the equilibrium constant for the above reaction, it can be calculated from the ratio of the molar concentration of bound antigen-antibody complex to the molar concentrations of unbound antigen and antibody at equilibrium as follows:

$$Ka = \frac{\{Ag - Ab\}}{\{Ab\}\ \{Ag\}}$$

The value of Ka varies for different antigen-antibody complexes and depends upon both k_1, which is expressed in units of liters/mol/sec (L/mol/s), and k_1, which is expressed in units of 1/second. For small haptens, the forward rate constant can be extremely high; in some cases, k_1 can be as high as 4×10^8 L/mol/s. For larger protein antigens, however, k_1 is smaller, with values in the range of 10^5 L/mol/s.

Antibodies Avidity

The affinity at one binding site does not always reflect the true strength of the antibody-antigen interaction. When complex antigens containing multiple, repeating antigenic determinants are mixed with antibodies containing multiple binding sites, the interaction of an antibody molecule with an antigen molecule at one site will increase the probability of reaction between these two molecules at a second site. The strength of such multiple interactions between a multivalent antibody and antigen is called the 'avidity'.

The avidity of an antibody is a better measure of its binding capacity within biological systems (e.g., the reaction of an antibody with antigenic determinants on a virus or bacterial cell) than the affinity of its individual binding sites. High avidity can compensate for low affinity. For example, secreted pentameric IgM often has a lower affinity than IgG, but the high avidity of IgM, resulting from its higher valence, enables it to bind antigen effectively.

PRECIPITATION REACTIONS

When a soluble antigen combines with its specific antibody in the presence of electrolytes (NaCl) at a suitable temperature and pH, the antigen-antibody complex forms an insoluble precipitate. This reaction is called precipitation. The precipitates thus formed by antibodies are known as precipitins. When the precipitate remains suspended as floccules, the reaction is called flocculation. Precipitation reaction may occur in liquid medium or in semi-solid medium like agar gels, agarose or polyacrylamide.

Quantitative Precipitation in Liquid Medium

Antigen and antibody combine in varying proportions. When same amount of antiserum is mixed with increasing amounts of corresponding pure soluble antigen in test tubes, precipitation occurs most rapidly and abundantly in one of the middle tubes, in which the Antigen and the antibody are present in optimal or equivalent proportion. In the preceding tube, in which the antibody is in excess, and in the later tubes, in which the antigen is in excess, the precipitation will be weak or even absent. For a given antigen-antibody system, the optimal or equivalent ratio will be constant, irrespective of the quantity of the reactants. If the amounts of precipitate in the different tubes are plotted on a graph, the resulting curve will have three phases (Fig. 8.9):

(i) An ascending part (Prozone or Zone of antibody excess) (A), in which the uncombined antibody is present.

(ii) A peak (Zone of equivalence) (B), in which both antigen and antibody are completely precipitated and no uncombined antigen and antibody are present.

(iii) A descending part (post zone or zone of antigen excess) (C), in which all antibody has combined with antigen and some uncombined antigen is present.

This is called the zone phenomenon. The prozone is of importance in clinical serology, as sometimes sera rich in antibody may give a false negative precipitation of agglutination result, unless several dilutions are tested.

MECHANISM OF PRECIPITATION

Marrack (1934) proposed the lattice hypothesis to explain the mechanism of precipitation. This is supported by considerable experimental evidence and is now widely accepted.

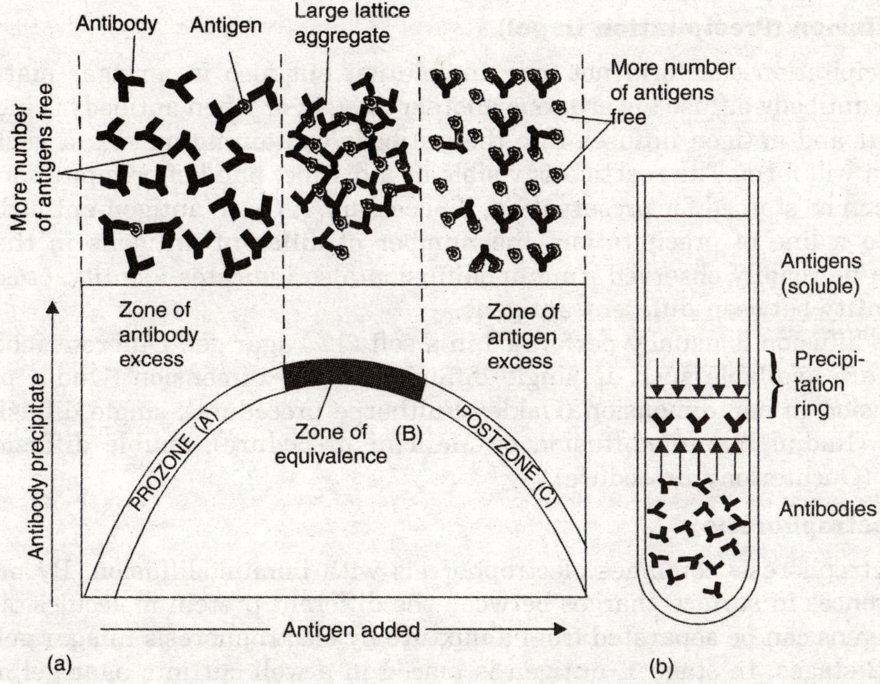

Fig. 8.9. Immunoprecipitation reaction. (a) Graph indicating a precipitation curve is based on the ratio of antigen to antibody. The zone of equivalence represents the optimal ratio for precipitation. (b) A precipitation ring test. Antibodies and antigens diffuse toward each other in a test tube. A precipitation ring is formed at the zone of equivalence.

According to this concept, multivalent antigens combine with bivalent antibodies in varying proportions, depending on the antigen-antibody ratio in the reaction mixture. Precipitation results when a large lattice is formed consisting of alternating antigen and antibody molecules. This is possible only in the zone of equivalence. In the zones of antigen or antibody excess, the lattice does not enlarge, as the valencies of the antibody and the antigen, respectively, are fully satisfied.

APPLICATIONS OF PRECIPITATION REACTIONS

Precipitation reactions are very sensitive and can detect as little as 1μg protein antigen. The precipitation test may be carried out either as a qualitative or quantitative test. The qualitative test is much more widely used to detect the presence of antigen and is of considerable value in:

1. Identification of bacteria, *e.g.*, detection of group specific poly-saccharide of *Streptococcus* (Lancefields's groupings).
2. Identification of bacterial components in infected tissues, *e.g.*, *B. anthracis* (Ascoli's thermoprecipitin test).
3. Detection of unknown antibody *e.g.*, VDRL and KT in syphilis.
4. Forensic application in identification of human blood and seminal stains.
5. Standardization of toxins and antitoxins.
6. Testing for food adulteration.

Immunodiffusion (Precipitation in gel)

Immunoprecipitation can form not only in solution but also in an agar matrix. When antigen and antibody diffuse towards one another in agar, or when antibody is incorporated into the agar and antigen diffuses into the antibody containing matrix, a visible line of precipitation will form. The reaction is visible as a distinct band of precipitation, which is stable and can be stained for preservation, if necessary. As each antigen-antibody reaction gives rise to a line of precipitation, the number of different antigens in the reacting mixture can be readily observed. Immunodiffusion also indicates identity, cross reaction and nonidentity between different antigens.

Immunodiffusion is usually performed in a soft (1%) agar gel. Different modifications of the test are available such as single diffusion in one dimension (Oudin procedure), double diffusion in one dimension (Oakley-Fulthorpe procedure), single diffusion in two dimensions (Radial immunodiffusion or Mancini procedure), double diffusion in two dimensions (Ouchterlony procedure).

Immunoelectrophoresis

Immunoelectrophoresis combines electrophoresis with immunodiffusion. By making use of the differences in surface charges between the different protein molecules of antigens, several antigens can be separated from a mixture by electrophoresis in agar gel. The test consists of 2 stages. In stage 1, antigen is placed in a well cut into agar gel on slide or plate and electric current is passed through the agar (1-2 hrs). The antigen components migrate in the electric field according to their charge and size and separate from each other. After separation of antigens electric current is stopped. In stage 2, a trough is cut in the gel parallel to the direction of migration of antigen and filled with antibody, and allowed to diffuse for an appropriate period of time. As the antigen and antibody diffuses towards each other, a series of precipitate arcs are formed where the antibodies encounter the antigen in optimal proportions. This technique is useful for determination of the presence or absence of serum proteins and detection of unusual patterns such as human myeloma protein.

In rocket electrophoresis, the antigen is electrophoresed through an agar gel containing Ab. This is a quantitative method. Antigen-antibody complexes precipitate to form cone-like structures (appearance of a rocket), the length of which corresponds to the concentration of antigen.

Counter Immunoelectrophoresis (CIE): It relies on movement of antigen and antibody in opposite direction. The antigen carrying a negative charge migrates towards the positive pole and the positive charged antibody towards the negative pole and vice versa. Antigen (specimen) and the homologous antibody are placed in separate wells in agar gel and through which electric current is passed. Line of precipitation appears at a point where the two meet in 30 to 60 min. This test is faster and 10 times more sensitive than double diffusion technique. CIE is useful in detection of hepatitis B Antigen and antibodies, DNA Abs in SIE etc.

AGGLUTINATION REACTIONS

When an antibody combines with a particulate antigen in presence of electrolytes at optimal pH and temperature, it results in visible clumping of particles. This is called agglutination. Antibodies that produce such reactions are called agglutinins.

Agglutination is more sensitive than precipitation for the detection of antibodies. Agglutination reactions are similar in principle to precipitation reactions; they depend on the cross linking of polyvalent antigens. Agglutination reaction occurs when antigen and antibody are mixed in optimal proportions. Just as an excess of antibody inhibits precipitation reactions, such excess can also inhibit agglutination reactions; this inhibition is called the prozone effect.

Because prozone effects can be encountered in many types of immuno-assays, understanding the basis of this phenomenon is of general importance. Several mechanisms can cause the prozone effect. First, at high antibody concentrations, the number of antibody binding sites may greatly exceed the number of epitopes. As a result, antibodies can bind antigen only univalently instead of multivalently. Antibodies (Abs) that bind univalently cannot cross link one antigen to another. Prozone effects are readily diagnosed by performing the assay at a variety of antibody (or antigen) concentrations.

When one is using polyclonal Abs, the prozone effect can also occur for another reason. The antiserum may contain high concentrations of Abs that bind to the antigen but do not induce agglutination; these antibodies, called 'incomplete' or 'monovalent' antibodies are often of the IgG class. At high concentrations of IgG, incomplete Abs may occupy all of the antigenic sites, thus blocking access by IgM, which is a good agglutinin. This effect is not seen with agglutinating monoclonal Abs.

Slide agglutination Test

This is a widely used method to detect unknown antigen (bacteria, red blood cells etc.). A fine suspension of antigen is made in saline on a slide. Small drop of antiserum is added and mixed properly and allowed to react for a minute. If specific antigens are present then it forms clumps within a few minutes. This method is routinely used in blood grouping, cross matching and bacterial identification.

Tube agglutination Test

Serum is serially diluted in saline in a series of tubes. An equal volume of antigen suspension (*e.g.* 0.5 ml of cell suspension) is added to all the tubes and incubated for known period. The highest dilution of serum at which agglutination occurs is recorded as antibody titer. This method is used for determination of antibody titer. It is routinely used for serological diagnosis of typhoid, paratyphoid, brucellosis and typhus fever.

Radioimmunoassay (RIA)

In this method generally antibodies are labeled with radioactive substances such as Iodine (I^{125} or I^{131}). This technique was first developed by two endocrinologists Berson, S.A. and Rosalyn Yallow (1960) for the determination of insulin and anti-insulin complexes in diabetics. They introduced RIA in clinical medicine in 1961 and since then it has been successfully utilized for quantitative assay of all hormones, drugs and Hbs antigen. For this pioneering work Yallow was awarded Nobel Prize in 1977. This is one of the most sensitive techniques used for detecting antigen and antibody in the sample.

Immunofluorescence

Coons and his coworkers (1942) showed that fluorescent dyes can be conjugated to antibodies and such labeled antibodies can be used to locate and identify antigens in tissues. Generally used fluorescence compounds are fluorescein isothiocyanate that shows blue green colour and lissamine rhodamine that impart orange red fluorescence.

Enyzme Linked Immunosorbent Assay (ELISA)

It is developed by Engvall and Perlmann and van Weeman and Schuurs independently in 1971. ELISA is a highly sensitive test used to detect pathogens, toxins, hormones or any antigenic agents at picogram levels in the sample. First antigen is attached to microtitre plate through antibody. Then enzyme conjugated (linked) antibodies specific for antigen was added. This specifically binds with the respective antigen. The most commonly used enzymes are horse radish peroxidase, alkaline phosphatase and β-galactosidase. Later substrate (colourless) solution was added. If antigen present in the sample enzyme linked antibodies binds tightly and will not be washed out in the washing step. After incubation period enzyme acts on colorless substrate and produces coloured products that can be visualized and measured.

SUMMARY

An antigen or immunogen is a foreign substance which can stimulate lymphocytes production. Each antigen has several antigenic determinant sites or epitopes that stimulate the production of antibodies by B lymphocytes. The epitopes can specifically react with the antibodies produced. Haptens are small organic molecules that are not antigenic by themselves but can become antigenic if bounded to a larger çarrier molecule. Adjuvants are the compounds such as alumonia hydroxide, which enhance the immunogenicity of the antigen.

Antibodies or immunoglobulins are a group of glycoproteins produced in response to an antigenic stimulation. They can recognize and specifically reacts with the antigens. All immunoglobulins have a basic structure composed of four chains of polypeptides (two light chain and two heavy chain) connected to each other by disulfide bonds. In humans five classes of immunoglobulins are present namely IgG, IgA, IgM, IgD and IgE. The template theory and selective theory explains the exact mechanism of antibody biosynthesis by B cell after antigenic stimulation.

Antibody diversity results from the rearrangement of the individual exons on the antibody coding chromosomes, somatic mutations, the generation of different codons during splicing and the independent of light and heavy chain genes.

Antigen-antibody reactions such as precipitation, agglutination etc. are very valuable for diagnosis of several infectious diseases. Precipitation can be defined as when the soluble antigen combines with its specific antibody in the presence of electrolyte (NaCl) at a suitable temperature and pH, antigen and antibody complex forms an insoluble precipitate. Immuno-diffusion, immunoelectrophoresis and coutner immunoelectrophoresis techniques are based on precipitation mechanism. Agglutination can be defined as when an antibody combines with a particulate antigen in the presence of electrolyte at optimal pH and temperature, it results in the visible clumping of particles. Based on agglutination mechanism the slide agglutination, tube agglutination, radioimmunoassay and enzyme linked immunosorbent assay techniques are widely used to detect unknown antigen, which are based on agglutination.

EXERCISE

1. Define antibody and write a note on structure of antibody molecule.
2. Mention different types of antibodies.
3. Describe antigen-antibody interactions.
4. What do you mean by affinity and avidity?
5. Define precipitation and explain precipitation reaction.
6. What is meant by agglutination? Mention its types.
7. What is meant by radio-immunoassay (RIA)? (BU Oct. 2002)
8. Write a note on immunoelectrophoresis. (BU Oct. 2001)
9. Explain properties of different immunoglobulins. Add a note on their significance. (BU Oct. 2001)
10. Give an account of any two antigen-antibody interactions. (BU April 2001)
11. Write a note on epitopes. (BU Oct. 2001) (BU April 2003)
12. Mention effect of adjuvants. (BU April 2001) (BU April 2003)
13. Briefly write a note on zone phenomenon. (BU Oct. 2002)
14. What are the factors that influence a compound to be an ideal antigen? (BU Oct. 2002)
15. Describe epitope in brief. (BU Oct. 2001)
16. Write a short note on ELISA. (BU April 2001) (BU April 2003)
17. What is meant by precipitation? Mention its type.
18. Explain in detail different precipitation pattern obtained in immunodiffusion. (BU April 2003)
19. Explain zone of equivalence with suitable diagram.
20. List out medical application of precipitation.
21. Give an account of antigenic determinants on immunoglobulins. (BU April 2003)
22. Discuss general characteristics of antigen-antibody reactions.

9

Humoral (Antibody-Mediated) Immunity

9.1 INTRODUCTION

Humoral immunity depends first on the ability of B cells (B lymphocytes) to recognize specific antigens and second on their ability to initiate responses that protect the body against foreign agents (antigens). In most cases the antigens are present on surfaces of infecting microorganisms. The most common response is the production of antibodies that will inactivate an antigen and lead to destruction of infecting microorganisms.

All B cells contain specific antibody/receptor on its membrane and can bind immediately to a specific antigen. This antigen binding activates or sensitizes the parent B cell and results in division of the cell to form more number of cells. Most of the resulted cells act as **plasma cells** and a few cells act as **memory cells** (Fig. 9.1). Plasma cells are large B lymphocytes that synthesize and release antibodies specific for a particular antigen. The activated plasma cells can produce as many as 2000 antibodies per second. The plasma cells have short life span of five to seven days. The memory cells can remain active for years or decades to instantly stimulate antibody production. These memory cells help in response to second and subsequent exposures to an antigen much more rapidly than to the first exposure.

9.2 ACTIVATION OF B CELLS

The activity of B cells can be influenced by certain T lymphocytes. If a B cell encounters large antigens with many epitopes, B cells generate plasma cells and antibodies without the help of T cells (Fig. 9.2). But most of the infective organisms have only few epitopes. In such cases T helper cells assist B cells by processing of antigens, so that B cells can respond to them. Macrophages also present antigens to B cells. The suppressor T cells block the activity of T helper cells and thus stop the antibodies production by the B cells.

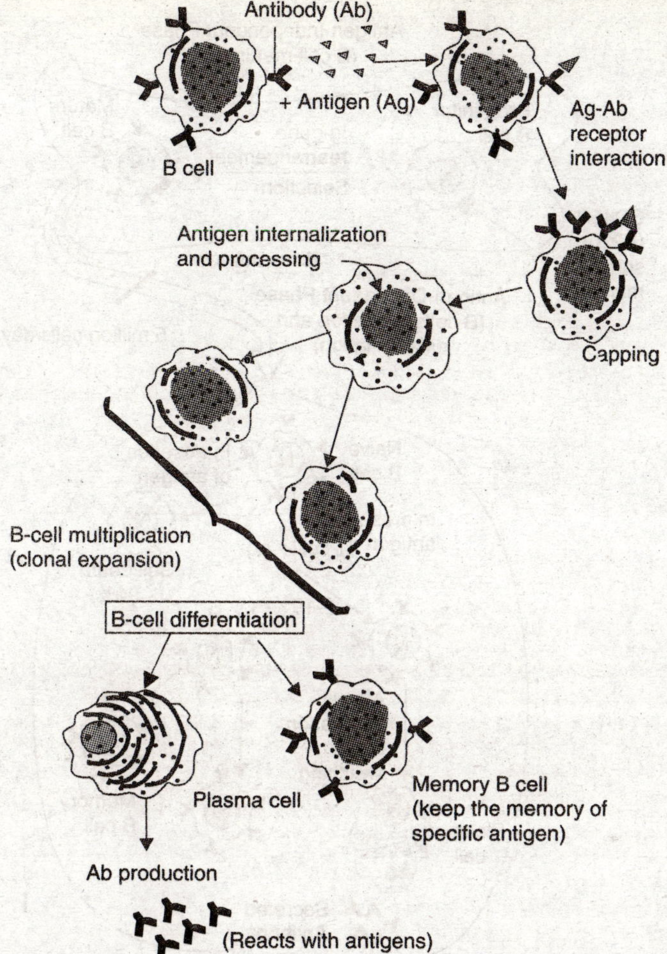

Fig. 9.1. The immune system can respond specifically to a myriad of possible antigens entering into body, whether they are individual molecules or are attached to pathogens and abnormal cells such as cancer cells. B cells or B lymphocytes constantly roam the body, particularly the blood and lymphoid tissues. Each B cell synthesises only one of the millions of possible antibodies and displays this antibody on its surface. When the antigen meets a B cell having a surface antibody of the proper specificity (top left), it complexes with the antibody and capping occurs. (Capping is the regional aggregation of antibodies on the surface of the cell following Ag-Ab interaction). The antigen is then internalised; the B cell swells and begins to divide rapidly, producing a B-cell clone. The activated B-cell clone differentiates into plasma cells and memory cells. Plasma cells form the specific antibody that immediately attacks the antigen that provoked its formation. Memory B cells persist in the body and boost the immune system's readiness to eliminate the same antigen if it presents itself in the future. Most antibody responses also require signals from T helper cells (not shown).

9.3 MOLECULAR EVENTS IN ACTIVATION OF B CELLS

The B cells directly produce antibodies when exposed to large number of antigens. In this method bacterial lipopolysaccharides (TI-1; **Thymus-independent antigen**-1) or

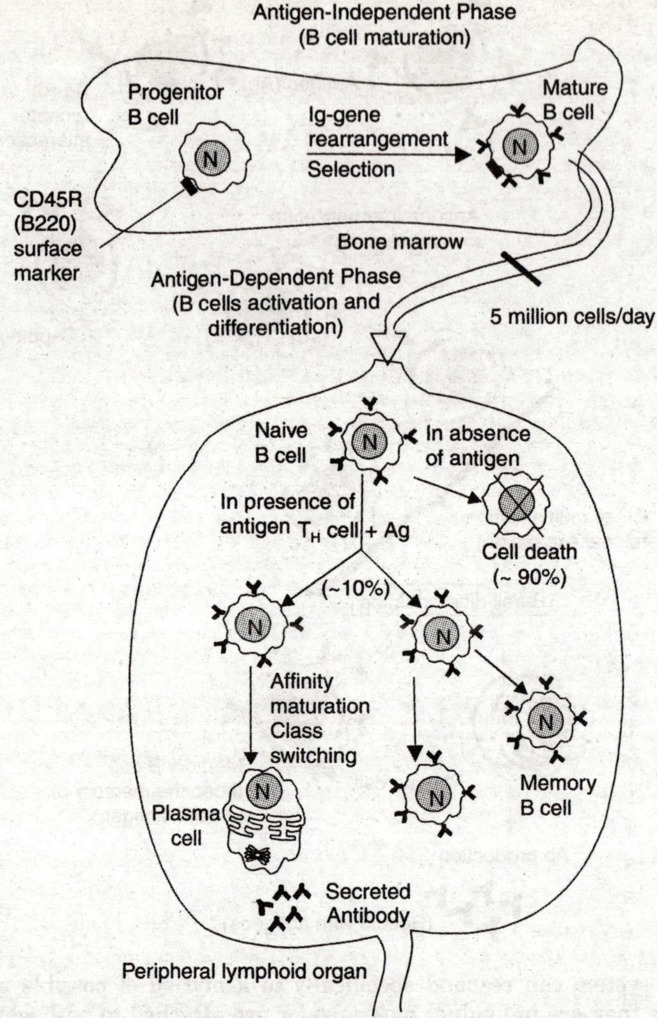

Fig. 9.2. An overview of B-cell development in immune system. During the antigen-independent maturation phase, immunocompetent B cells expressing membrane IgM and IgD are generated in the bone marrow. Only about 10% of the potential B cells reach maturity and exit the bone marrow. In the absence of antigen-induced activation, naive B cells in the periphery die within a few days. In the presence of soluble protein antigen and activated T_H cells, B cells are activated and proliferate within secondary lymphoid organs. Those bearing high affinity membrane-bound immunoglobulin differentiate into plasma cells and memory B cells, which may express different isotypes because of class switching. The numbers cited refer to B-cell development in the mouse, but the overall principles apply to humans as well.

bacterial flagellin/cell wall polysaccharides (TI-2) acts as an activator for B cells and both the signals are produced (Fig. 9.3). Here T cells will not help the B cells hence these antigens are referred as thymus-independent antigens. In case of **thymus dependent** activation first signal produced by cross linking of surface immunoglobulins on surface of B cells by thymus dependent (TD) antigen and second one by the interaction with T helper cells through CD40 (on the B cell) and CD40L (on T_H cells) with assistance by cytokines (Fig. 9.4).

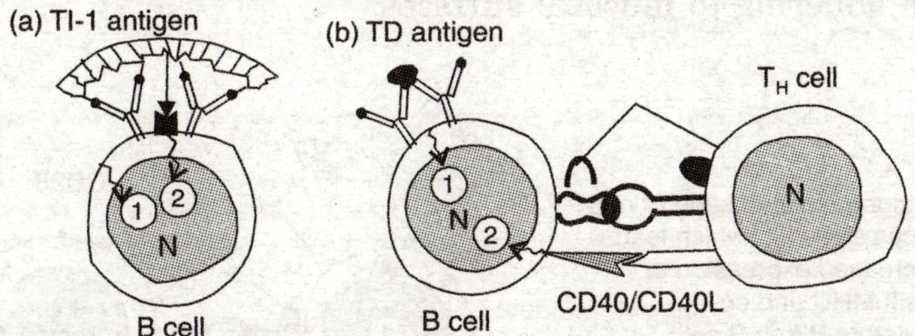

Fig. 9.3. An effective B-cell activation requires two distinct signals induced by membrane events. Binding of a type 1 thymus-independent (TI-1) antigen to a B cell provides both the signals. A thymus-dependent (TD) antigen provides signal 1 by cross linking immunoglobulin, but a separate interaction between CD40 on the B cell and CD40L on an activated T_H cell is required to generate signal 2.

9.4 PRIMARY AND SECONDARY RESPONSES

The initial exposure of antigen to B cells results in production of antibodies by plasma cells and within few days antibodies begin to appear in the blood plasma. This is known as **primary response** and antibodies concentration increases over a period of 1 to 10 weeks (Fig. 9.5). The first antibodies produced are IgM molecules, which can attack antigen directly. Then concentration of IgG production increases by B cells. Ultimately concentration of both IgM and IgG can become so less that cannot be detected in the plasma. The type of the antibody molecule produced by the plasma cells depends upon the type of the cytokines binds to the cells. Depending on the type of cytokines plasma cells differentiate to produce different class of antibodies (Fig. 9.6). The concentration of antibodies reaches highest around 10-17 days. But memory cells remain in lymphoid tissues. They generally do not participate in the primary response, whereas they retain their ability to recognize a particular antigen. These cells survive without dividing for many months to many years.

When the same antigen recognized by the memory cells in later stage results in the antibody production at a rapid rate. This is known as **secondary response or anamnestic response.** The presence of memory cells makes the secondary response much faster than the primary response. During this some memory cells divide rapidly to produce plasma cells and others remain as memory cells. Plasma cells synthesize and release rapidly large quantities of antibodies. Similar to primary response, IgM are the first antibodies produced and later replaced by larger quantities of IgG molecules than in the primary response. Thus, the secondary response is characterized by a rapid increase in antibodies most of which are IgG (Fig. 9.5). The concentration of antibodies reaches highest in around 2 to 7 days.

Humoral immunity based on antigen-antibody reactions is most useful in giving the protection against bacterial infections. It is also known to neutralize toxins and viruses that have not yet invaded host cells. The defensive mechanism of humoral immunity depends on recognizing antigens associated with pathogens. For the bacteria to colonize on surfaces or for viruses to infect host cells, these pathogenic agents must first attach to surfaces. The IgA antibodies in tears, nasal secretions, saliva and other body fluids react with antigens on the pathogenic microorganisms. They coat bacteria and viruses and prevent them from adhering to mucosal surfaces.

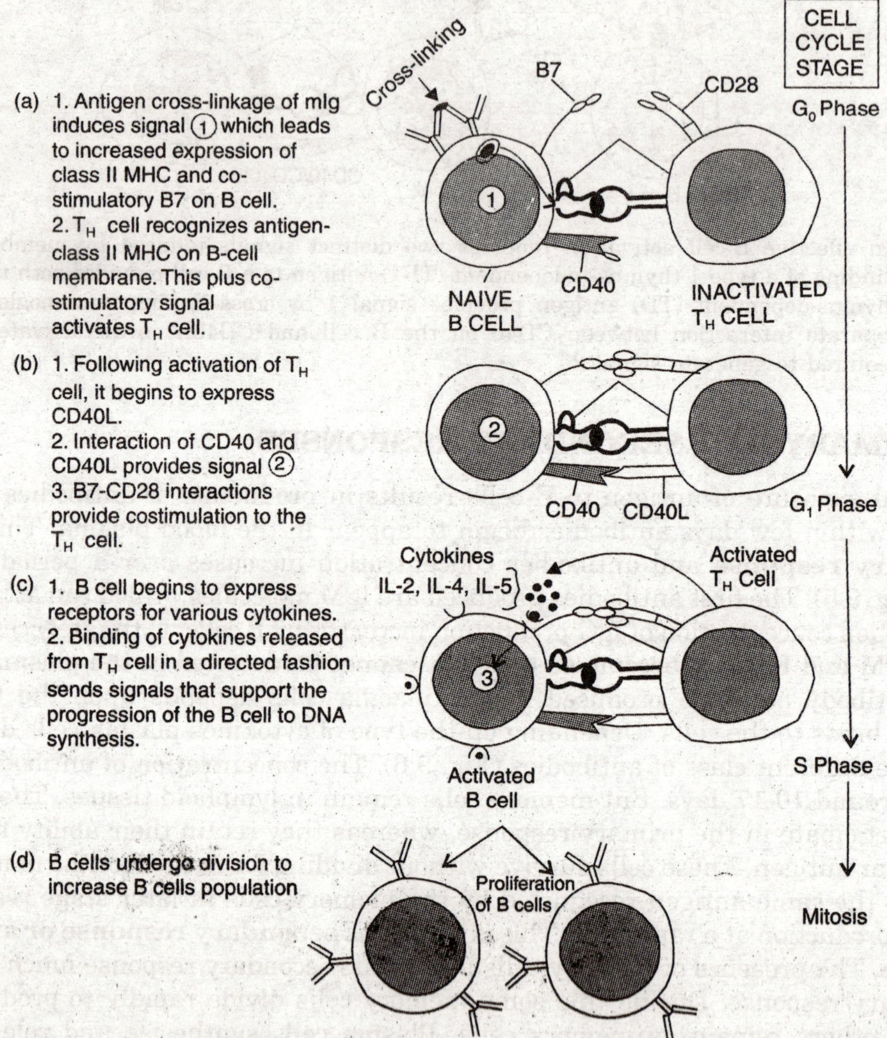

Fig. 9.4. The sequence of events in B-cell activation by a thymus-dependent antigen. The cell-cycle phase of the B cell is indicated on the right side.

Microorganisms that escape IgA invade tissues and encounter IgE in lymphnodes and mucosal tissues. Gut-associated lymphoid tissues (GALT) release large quantities of IgE, which bind to mast cells to release histamine and other substances that initiate and accelerate the inflammatory process. Inflammatory process means accumulation of large quantities of blood, complement factors, macrophages/phagocytes and antibodies to the site of infection, which results in swelling, redness, pain and heat.

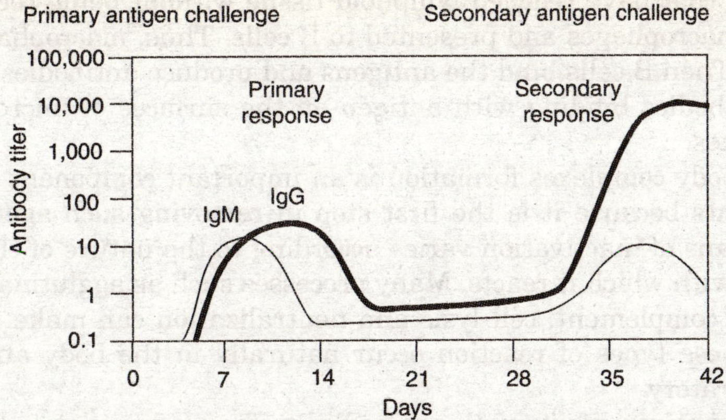

Fig. 9.5. Primary and secondary antibody responses after exposure to the antigen. Comparison of primary and secondary antibody responses. Note that the secondary response to the antigen is both faster and greater than the primary response against antigen. T-cell-mediated responses exhibit similar immunologic memory against the antigen.

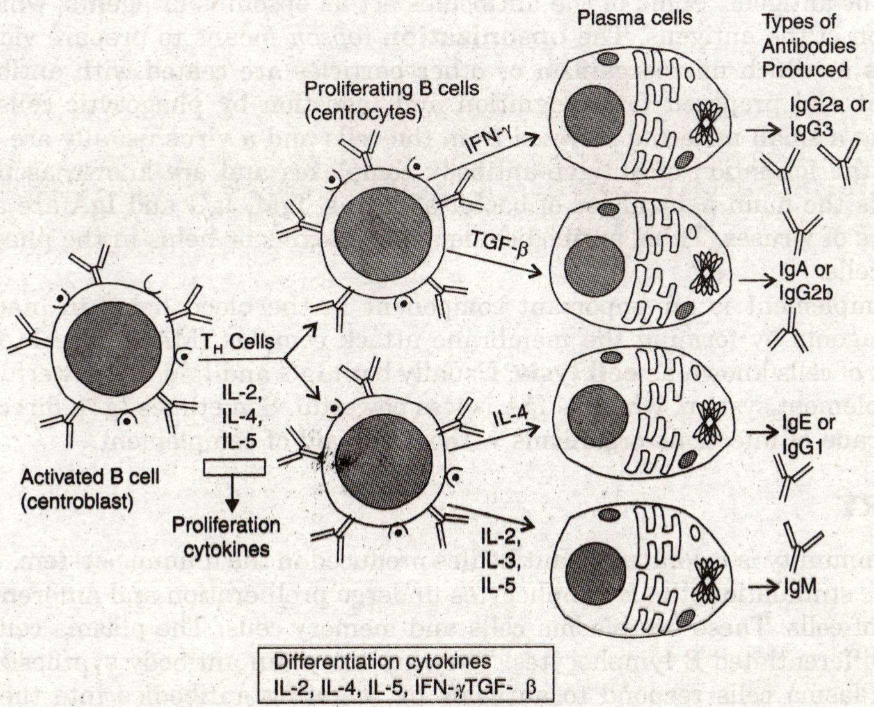

Fig. 9.6. The interactions of numerous cytokines with B cells generate signals required for proliferation and class switching during the differentiation of B cells into plasma cells. Binding of the proliferation cytokines, which are released by activated T_H cells, provides the progression signal needed for proliferation of activated B cells. The effects of the indicated cytokines have been demonstrated; however, similar or identical effects may be mediated by other cytokines. Class switching in the response to thymus-dependent antigens also requires the CD40-CD40L interaction, which is not shown here.

The microbes that have reached lymphoid tissue without being recognized by B cells are acted on by macrophages and presented to B cells. Thus, macrophage acts as antigen presenting cells. Then B cells bind the antigens and produce antibodies with the help of T helper cells. Antibodies binding with antigen on the surfaces of microbes form antigen-antibody complexes.

Antigen-antibody complexes formation is an important component of the inactivation of infectious agents because it is the first step in removing such agents from the body. However, the means of inactivation varies according to the nature of the antigen and the kind of antibody with which it reacts. Many processes such as agglutination, opsonization, and activation of complement, cell lysis and neutralization can make the inactivation of pathogens. All these types of reaction occur naturally in the body and can be made to occur in the laboratory.

If particulate antigen reacts with antibodies to form a complex is known as **agglutination**. Here bacterial cells acts as particles, which reacts with antibody and results in sticking together of microbes. IgM produces strong and IgG produces weak agglutination reaction with certain bacterial cells. This agglutination reaction produces visible reaction to the unaided eye and can be used for the laboratory test to detect the presence of antibodies or antigens. Some of the antibodies act as opsonins or agents, which helps in opsonization of the antigens. The **opsonization** (*opson* means to prepare victims for) is the process in which microorganism or other particles are coated with antibody and/or complement and prepared for recognition and ingestion by phagocytic cells. Bacterial toxins being a small molecule secreted from the cells and a virus usually are inactivated simply by the formation of antigen-antibody complexes and are known as **neutralization**. IgG is the main neutralizer of bacterial toxins. IgM, IgG and IgA are all effective neutralizers of viruses. Thus, antibodies neutralize toxins or helps in the phagocytosis of microbial cells.

The complement is an important component in the blood helps in inactivation of infectious agents by forming the membrane attack complex (MAC). The MAC helps in breakdown of cells known as **cell lysis**. Usually both IgG and IgM are powerful activators of the complement system, whereas IgA is less powerful. Sometimes IgM, directly lyse the cell membrane of infectious organisms without the aid of complement.

SUMMARY

Humoral immunity is mediated by antibodies produced in the immune system. In response to antigenic stimulation, the B lymphocytes undergo proliferation and differentiation into two types of cells. These are plasma cells and memory cells. The plasma cells are large and fully differentiated B lymphocytes. They are the main antibody synthesizing cells of the body. Plasma cells respond to antigens by secreting antibodies into the blood and lymph. The life span of plasma cells is five to seven days. The memory cells capable of producing antibodies on subsequent exposure to the same antigen. They circulate freely in blood and lymph and may live for years.

Activation of some B lymphocytes by antigens requires T lymphocytes help. These antigens that trigger B lymphocytes into immunoglobulin production with T lymphocytes co-operation. Such antigens are called T cell dependent antigens. Example for these includes bacteria, red blood cells etc. It is mediated by the cytokines produced by the

T helper cells. But some B lymphocytes do not require T lymphocytes help for their activation. These types of antigens that trigger directly B lymphocytes are known as T cell independent antigens. Example for this category include bacterial lipopolysaccharides, bacterial flagellin and cell wall polysaccharides.

Immune responses are of two types and they are primary immune response and secondary immune response. The antibody response to initial antigenic stimulus is called primary immune response and antibody response to the subsequent stimuli with the same antigen is known as secondary immune response. The secondary immune response is much faster than primary immune response. The first antibodies produced are IgM and later replaced by large quantities of IgG molecules.

EXERCISE

1. What is meant by primary response and secondary response?
2. Write a note on B lymphocytes.
3. What do you mean by humoral immunity?
4. How are B cells activated by antigens for antibody production?
5. Define oposonization.
6. Give an account on neutralization.
7. Give an account of 'B' immunocompetent cells. (BU Oct. 2001)
8. How do cytokines influence the class switching in B lymphocytes?
9. Explain mechanism of plasma cells production.
10. Describe molecular mechanism of B cells activation.
11. Differentiate between thymus dependent and thymus independent antigens.
12. Mention different cytokines helpful in class switching of plasma cells to produce different types of antibodies.
13. What is meant by capping?
14. Describe how bacterial and viral infections can be eliminated.
15. Compare primary and secondary antibody responses. (BU April, 2003).

10

Cellular (Cell-Mediated) Immunity

10.1 INTRODUCTION

In this type of immunity, directly immune cells are involved in protection of body, hence, it is referred as cellular immunity. In the immunity process antibodies alone are not enough to defense in certain circumstances, especially virus and other pathogens that have invaded host cells. This type of immunity is also involved in elimination of tumor (cancer) cells or transplantation rejection. Thus, cell-mediated immunity generally is helpful in killing intra-cellular pathogens, destruction of tumor cells, rejection of graft tissue, and hypersensitivity reaction to certain antigens (poison, drugs).

Cell-mediated immune response involves the differentiation and action of different T cells for killer activity and the production of chemical mediators called lymphokines. The major cells involved in lymphokines production are T helper ($\mathbf{T_H}$) cells. Cells involved in killing activity include cytotoxic T ($\mathbf{T_C}$) cells, activated macrophages, natural killer cells, antibody dependent killer cells.

The cell-mediated immune response begins usually with processing of an antigen by antigen presenting cells (macrophages, dendritic cells and B cells). These cells scavenge foreign materials (antigens or pathogens) from the blood, tissues and phagocytize such materials. These cells digest them in the cell interior and small pieces of the antigens bind to a special class of proteins called major histocompatibility complex (MHC) class II molecules. The MHC-antigen complex is then transported to the surface of the antigen presenting cells and displayed to circulating T helper cells. Before understanding T cells action and T cells-macrophages interaction some knowledge of major histocompatibility complex molecules.

10.2 MAJOR HISTOCOMPATIBILITY COMPLEX (MHC)

Major histocompatibility (*histo* means tissue; *compatibility* means ability to get along) complex is special group of proteins present on the cell surface. These proteins derived from collection of genes on chromosome 6 in humans (chromosome 17 in mice) and these surface antigens are referred as human leukocyte antigen (HLA) complex (H-2 complex in mice). Almost all human tissue cells contain MHC molecules on their plasma membranes. The MHC molecules can be divided into three classes: class I molecules are found on almost all types of nucleated body cells; class II molecules are present only on those leukocytes involved in T helper cell mediated immune response (macrophages, antigen presenting cells and B cells); class III molecules include various secreted proteins that have immune functions (complement component C2, C4a, factor B and tumor necrosis factor). The class I proteins are obtained from A, B, C locus of MHC, class II by DP, DQ, DR locus and class III by C4, BF, C2 locus on chromosome 6.

Class I MHC molecules contain a large glycoprotein **α** chain (45 kDa) associated noncovalently with smaller proteinaceous **β_2-microglobulin** (12 kDa) molecule (Fig. 10.1). The **α** chain consists of three external domain (α1, α2 and α3) and β_2-microglobulin contains only one external domain that is associated with α3 domains. This α chain (external domains) is anchored in the plasma membrane by its hydrophobic transmembrane segment and hydrophilic cytoplasmic tail. The α1 and α2 domains exposed outside and form the antigen binding pocket.

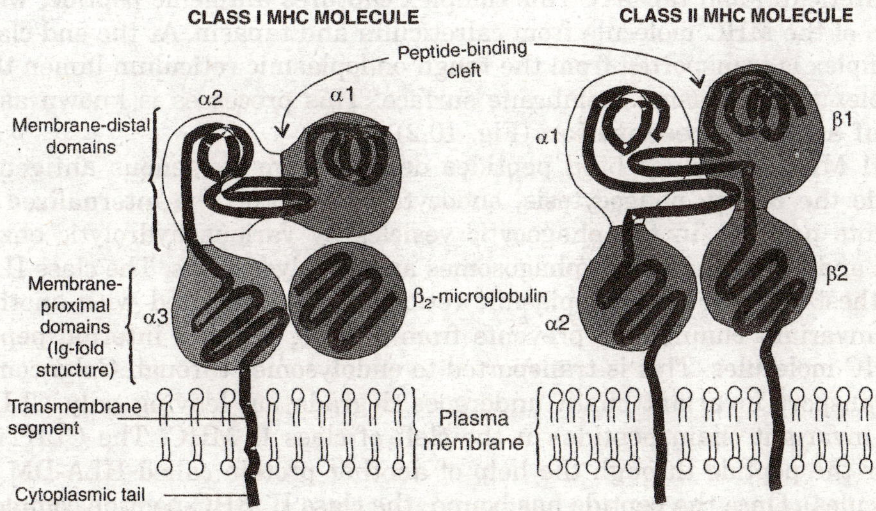

Fig. 10.1. Diagrammatic representation of a class I and class II MHC molecule showing structure of the external domains, transmembrane segment, and cytoplasmic tail on cell surface. The peptide-binding cleft is formed by the membrane-distal domains in both class I and class II molecules. The membrane-proximal domains possess the basic immunoglobulin-fold structure; thus, class I and class II MHC molecules are considered as members of the immunoglobulin superfamily.

Class II MHC molecules consist of two transmembrane polypeptide chains, a 33 kDa **α** chain and a 28 kDa **β** chain which associate by noncovalent interaction. Each chain contains two external domains in both **α** chain (α1, α2) and **β** chain (β1, β2). Both chains have transmembrane segment and cytoplasmic segment. The α1 and β1 domains exposed outside and form the antigen binding part.

10.3 ANTIGEN PROCESSING AND ANTIGEN PRESENTATION

Recognition of foreign antigen by a T cell requires that peptides derived from the antigen should display with MHC molecules on the membrane of cell. The formation of antigenic peptide from antigen by sequence of degradation events (by proteasome-multifunctional protease complex) in the cell is known as **antigen processing**. The degraded peptides then associate with MHC molecules within the cell interior and the peptide-MHC complexes are transported to the membrane surface, and they are displayed on the outer membrane surface known as **antigen presentation**.

Class I and class II MHC molecules bind peptides that have been processed in different intracellular compartments. Class I MHC molecules bind peptides derived from endogenous antigens that have been processed within the cytoplasm of the cell (normal cellular proteins, tumor proteins or viral and bacterial proteins produced within infected cells). The protein antigen is degraded in to peptides by proteasome and transported into rough endoplasmic reticulum lumen by a transport protein called TAP (transporters associated with antigen processing). The calnexin associated class I MHC α chain binds to β_2-microglobulin (calnexin, calreticulin and tapasin are molecular chaperones, which helps in maintaining protein stability). The β_2-microglobulin binding dissociates calnexin and binds to calreticulin and tapasin. This complex captures antigenic peptide, which allows dissociation of the MHC molecule from calreticulin and tapasin. At the end class I MHC-peptide complex is transported from the rough endoplasmic reticulum lumen through the Golgi complex to the plasma membrane surface. This processes is known as **cytosolic pathway of antigen presentation** (Fig. 10.2).

Class II MHC molecules bind peptides derived from exogenous antigens that are taken inside the cell by phagocytosis, endocytosis or both. The internalized antigen is degraded into peptides in the phagocytic vesicles by various hydrolytic enzymes with various pH and forms endosomes/phagosomes and endolysosomes. The class II MHC molecules synthesized in rough endoplasmic reticulum, is associated with another protein called the invariant chain. This prevents from binding of other internal peptide to the class II MHC molecules. This is transported to endolysomes through Golgi complex. During this transport invariant chain undergoes degradation leaving only CLIP (class II associated invariant chain peptide) in the cleft of class II MHC. The CLIP is later exchanged for the peptide through the help of another protein called HLA-DM (similar to MHC molecules). Once the peptide has bound, the class II MHC-peptide complex is transported to the plasma membrane surface. This processes is known as **endocytic pathway of antigen presentation** (Fig. 10.2).

The non-proteinaceous materials such as lipids, glycolipids are also known to have immunogenic capacity. A special type of receptors called CD1 receptors generally present these materials on the plasma membrane. The exact mechanism is not yet understood.

The T cells are known to bind to the antigen peptide, which are processed and present on the self-cells (antigen presenting cells). There are three sub-types identified in the T cells, namely, T helper (T_H) cells, T cytotoxic (T_C) cells and T suppressor (T_S) cells. T helper (T_H) cells and T suppressor (T_S) cells are involved in regulation of specific immune cells and hence called regulatory T cells. T cytotoxic (T_C) cells generally cause cytolysis and cells death of target cells (antigen containing cells), hence these cells are called effector T cells.

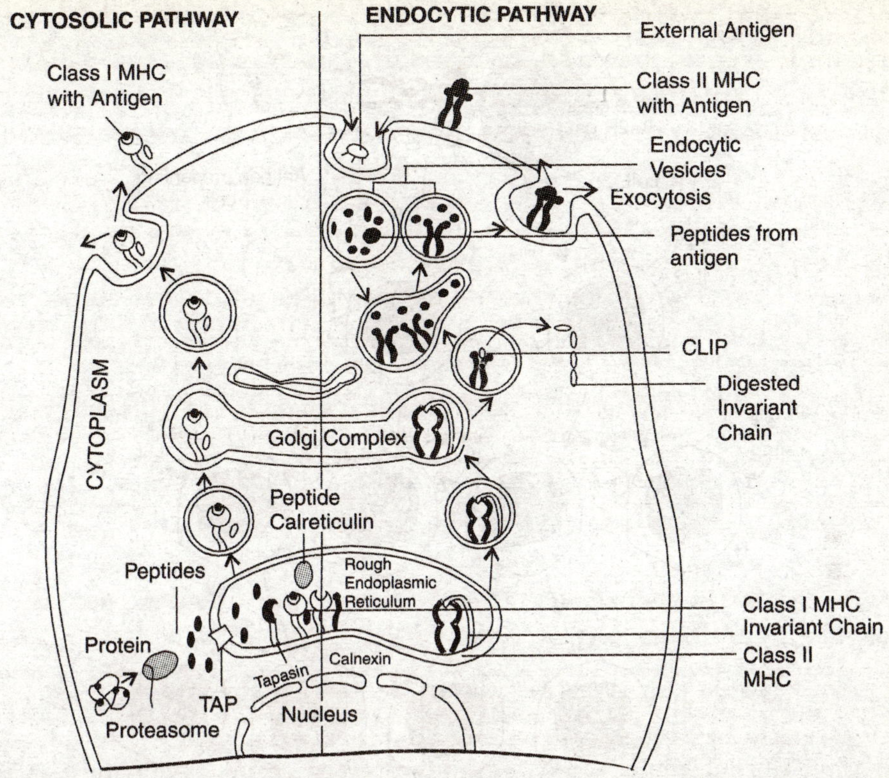

Fig. 10.2. Scheme showing separate antigen-presenting pathways for cytosolic (endogenous) and endocytic (exogenous) antigens. The mode of antigen entry into cells and the site of antigen processing appear to determine whether antigenic peptides associate with class I MHC molecules in the rough endoplasmic reticulum or with class II molecules in endocytic compartments.

10.4 ACTIVATION OF T CELLS

The naïve T cells are activated by the antigen-MHC complex, which generates memory T cells, regulatory T cells and effector T cells (T_H, T_C, and T_S cells) (Fig. 10.3). The T_H cells are also known as $CD4^+$ cells and two subtypes are noticed i.e. T_H 1 and T_H 2 cells. These subtypes cells produce and secrete a specific mixture of cytokines or lymphokines (Table 10.1). T_H 1 cells produce IL-2, IFN-γ and tumor necrosis factor that are involved in cellular immunity, especially macrophage activation (stimulate to destroy the pathogens by phagocytosis). The T_H 2 cells produce IL-4, IL-5, IL-10 and IL-13 that are involved in humoral immunity. T_H 2 cells interact with B cells (helpers for B cells) to make antibodies against antigen (Fig. 10.4).

10.5 MOLECULAR MECHANISM OF T_H1 CELLS ACTIVATION

The T helper (T_H1) cells require two signals (costimulation) to be stimulated by an antigen. The first signal is through interaction of TCR (T cell receptor)-CD3 complex (on T cells)

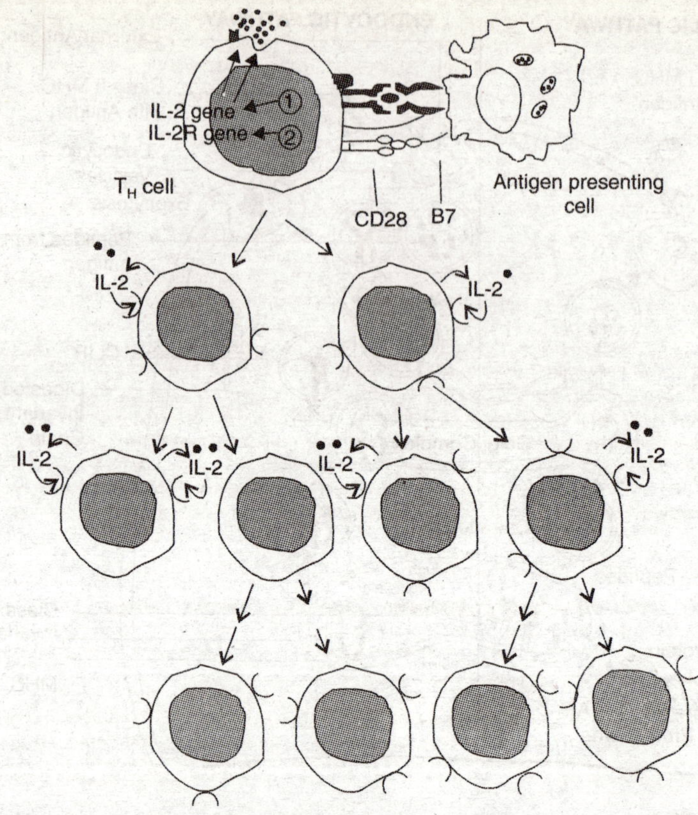

Fig. 10.3. Model showing activation of a T_H cell by both signal 1 and co-stimulatory signal 2 up-regulates expression of IL-2 and the high-affinity IL-2 receptor, leading to proliferation and differentiation.

Table 10.1: Production of lymphokines by T_H subsets and their effector functions in immune system

T_H Subsets	Lymphokines	Effector functions
T_H1	IL-3 GM-CSF IL-2 IFN-γ TNF-β	T_C, T_{DTH} cells production Opsonization Macrophage activation Inflammatory response Tissue injury
T_H2	IL-3 GM-CSF IL-4 IL-5 IL-10 IL-13	Chemotaxis of leucocytes B Cells activation of proliferation Eosinophils activation Allergic response Non-complement activation of TgG

and antigen bound to class II MHC on antigen presenting cells. The CD4 is also involved in the activation. The second signal for activation involves the binding of CD28 on T_H1 cell with a protein B7 (CD80) on macrophage (an APC). Both the activation signal goes to cytoplasm of T_H1 cells (Fig. 10.5) by following a signal transduction pathway in a cascade manner to show effector mechanism.

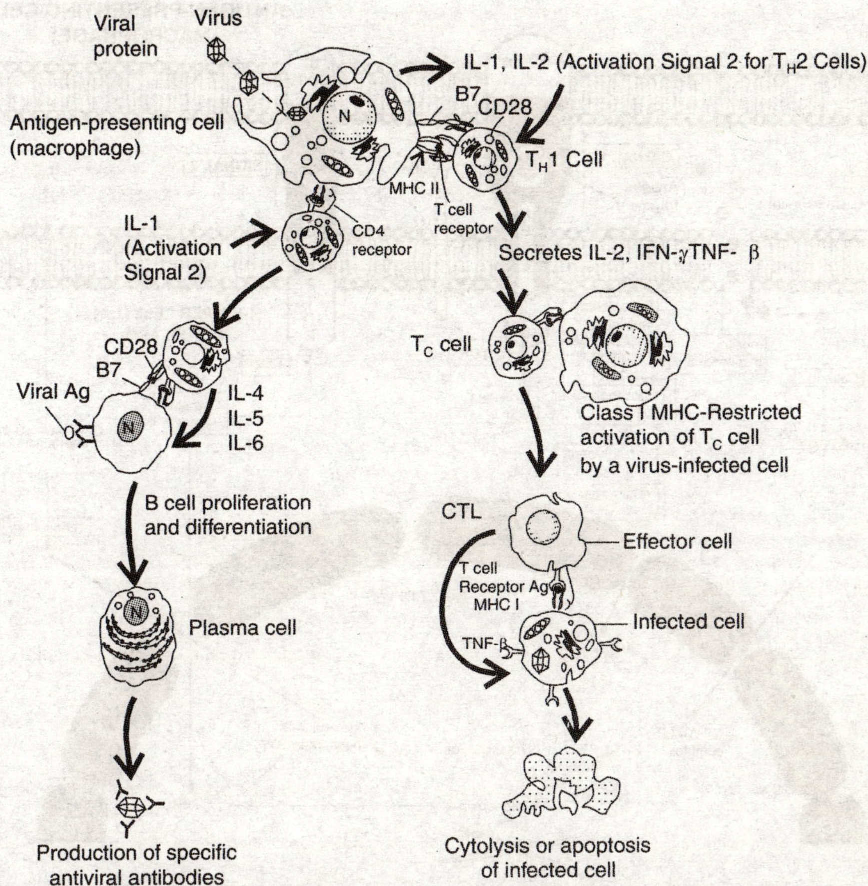

Fig. 10.4. The T_H-cell responses in cell-mediated immunity to antigen. A virus is phagocytosed by a macrophage and a small antigen fragment (peptide) presented to T-helper cells (T_HI and T_H2) in association with class II MHC molecules. Once activated (activation signals 1 and 2) the T_H2 cell secretes the cytokines IL-4, IL-5, IL-10 and IL-13 that cause B-cell proliferation and secretion of specific antiviral antibodies. Activation signals 1 and 2 also cause the T_H1 cell to secrete IL-2, IFN-γ, and TNF-β, IL-2 regulates the proliferation of cytotoxic T cells. Once the effector T_C cell proliferates, it attacks and causes lysis or programmed cell death (apotosis) of the virus-infected cell by either the Fas or perforin pathways.

First signal activates tyrosine kinase and it adds phosphate to phospholipase Cγ1 which cleaves phosphatidylinositol biphosphate to two products, diaceylglycerol and inositol triphosphate. The diaceylglycerol further activates the protein kinase C then it moves to nucleus where it catalyses the formation of a protein complex called AP-1 from separate components. The other product inositol triphosphate causes a calcium channel to open, thus calcium ions rush into the cytoplasm leading to further enzymatic activity and the activation of calmodulin, calcineurin and nuclear factor of activated T_H1 cells (NF-AT). Then NF-AT migrates into the nucleoplasm where it binds to the newly formed AP-1, forming complex NF-AT/AP-1 (transcription factor). This complex binds to specific gene in the DNA causing IL-2 mRNA to be transcribed. This messenger RNA moves out of the nucleus for protein (IL-2) synthesis.

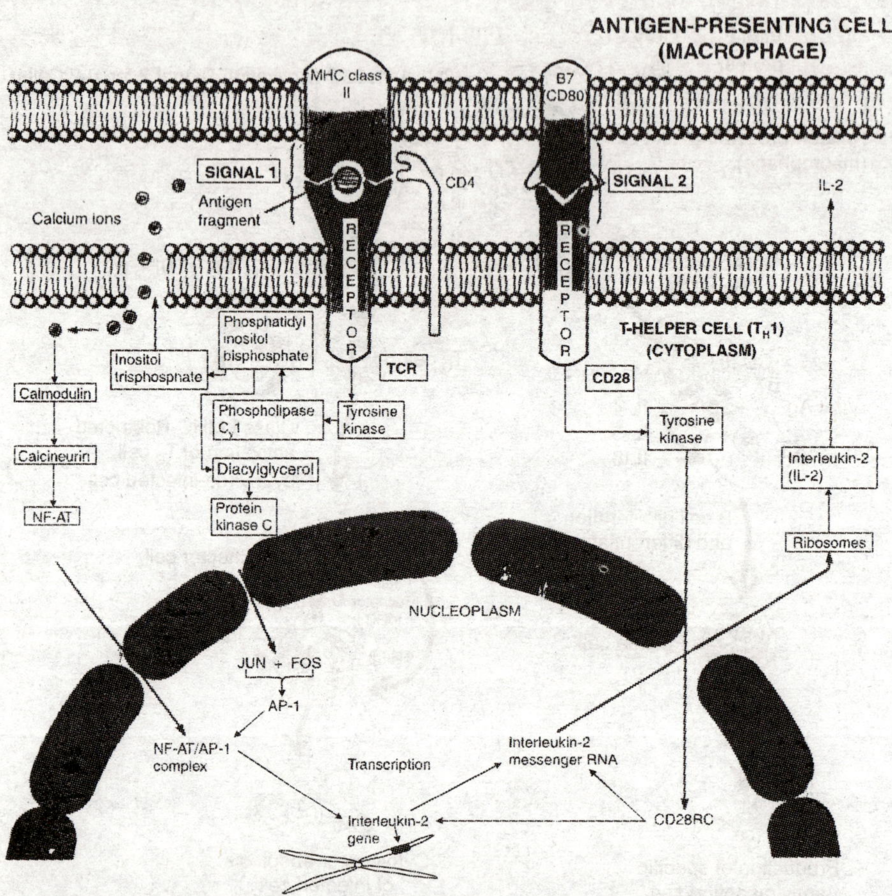

Fig. 10.5. Generation of two signals (costimulation) that are essential for T-helper (T_H1) cell activation. The signal (1) is the presentation of the antigen fragment by a macrophage or other antigen-presenting cell along with the MHC class II molecule to the T-helper cell receptor and CD4 protein. The signal (2) initiates when the macrophage presents the B7 (CD80) protein to the T-helper cell with its CD28 protein receptor. Both signals send information into the cytoplasm of the T-helper cell. The signal (1) causes interleukin-2 mRNA to be produced. The signal (2) boosts the production of interleukin-2 to effective concentrations. It is now known that the gene for interleukin-2 is under very tight regulation. It cannot be transcribed unless the NF-AT/AP-I complex are activated when the T-helper cell is stimulated through its antigen-specific receptor.

Second signal mediated by CD28 receptor and B7 molecule activates another tyrosine kinase result in the formation of transcription factor CD28RC, and also stabilizes the mRNA transcribed from the IL-2 gene. This increases the concentration of IL-2 molecules. The T_H1 cells activated by these two signals secretes large amount of IL-2, which further activates cytotoxic T cells. T_H1 cells also secrete IFN-γ, which is capable of activating macrophages cells and increasing microbicidal activity (Fig. 10.6).

The T helper (T_H2) cells also require two signals (costimulation) to be stimulated by an antigen. The first signal is through interaction of TCR (T cell receptor)-CD3 complex (on T cells) and antigen bound to class II MHC on antigen presenting cells along with CD4. This activated T_H2 cells express CD40L and reacts with CD40 on B cell and provides the second signal to the B cells for expression of cytokines receptors. Binding of B cell receptor (B7) and T cell receptor (CD28) also stimulates the activation. After binding to cytokines (IL-2, IL4 and IL-5) released from T_H2 cells, B cells divides to plasma cells and memory cells. Thus, B cells get activated by the T_H2 cell for the increased antibody production.

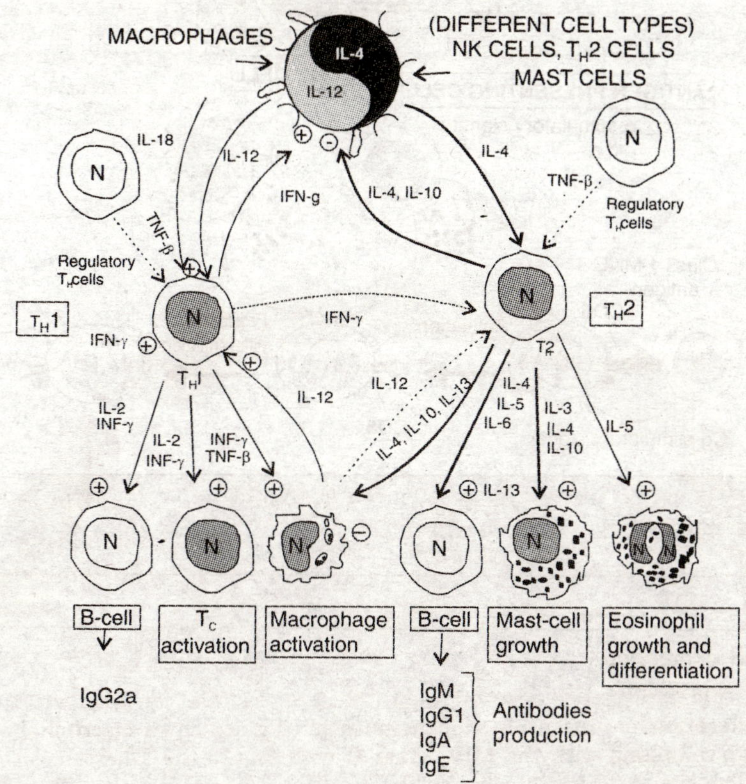

Fig. 10.6. The regulation of T_H subsets by cytokines production from different cell types. Solid arrows indicate stimulatory effects (+); dashed arrows indicate inhibitory effects (–). Thin arrows indicate the cytokines of the T_H1 pathway; thick arrows, the T_H2 pathway.

10.6 T CYTOTOXIC (T_C) CELLS (TCL)

These cells are responsible for elimination of host cells harboring viruses and other pathogens or tumor cells. The T_C cells are also known as $CD8^+$ cells. These are activated by a complex process that involves at least three signals. The T_C cell receptor bind to class I MHC-antigen complex on the surface of virus infected cells or tumor cells (target cells) and provides first signal (Fig. 10.7). A CD28 (T_C cell) and B7 (target cell) complex formation acts as costimulation signal. Finally the T_C cells require exposure to IL-2 from T_H 1 cells. The cells then proliferate and differentiate to form active cytotoxic T lymphocytes (CTL) from precursor CTLs (CTL-Ps), which can attack target cell. CTL and T_H 1 cells release cytokines such as gamma interferon, whereas both macrophages and T cells release tumor necrosis factor-α. These cytokines inhibits virus reproduction, meanwhile they also activate macrophages and other phagocytes that can kill the infected cell.

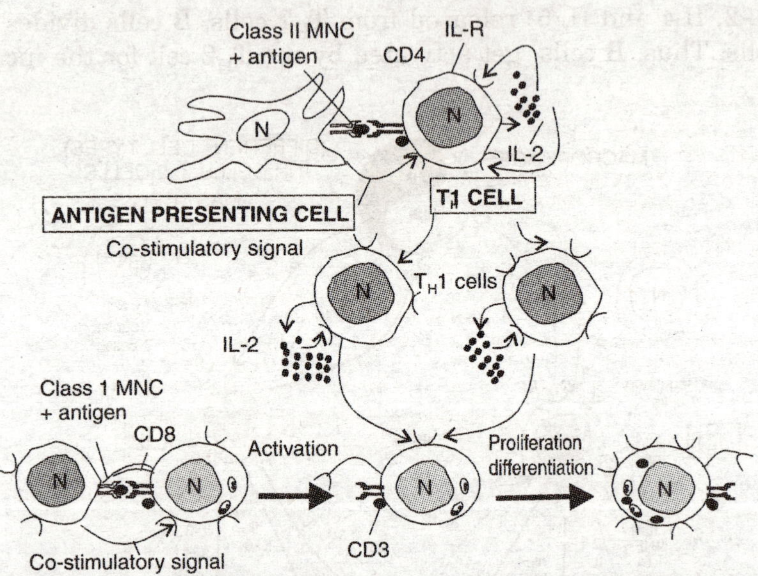

TARGET CELL	CYTOTOXIC T-LYMPHOCYTE PROGENITOR	Ag-ACTIVATED CTL-P	CTL
CELL STATE		ACTIVITY	
IL-2R expression	–	+	+
IL-2 expression	–	±	±
Proliferation	–	–	+
Effector cytotoxic function	–	–	+

Fig. 10.7. Model showing generation of effector cytotoxic T-lymphocytes (CTLs). After interaction with antigen—class I MHC complexes on appropriate target cells, CTL-Progenitors (CTL-Ps) begin to express IL-2 receptors (IL-2R) and lesser amounts of IL-2. Proliferation and differentiation of antigen-activated CTL-Ps generally require additional IL-2 secreted by T_H1 cells resulting from antigen activation and proliferation of $CD4^+$ T cells. In the subsequent effector phase, CTLs destroy specific target cells.

The cytotoxic T lymphocytes can destroy target cells in at least two ways. The CD95 pathway triggers apoptosis or programmed cell death. The perforin pathway is probably more often used by CTL through perforin and granzymes to induce both osmotic cytolysis and apoptosis.

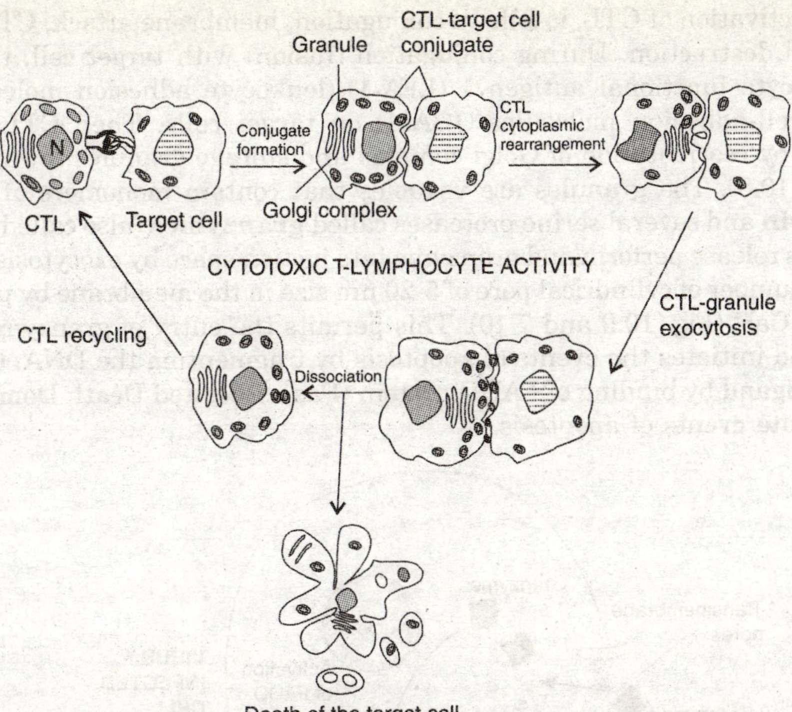

Fig. 10.8. The process of cytotoxic T. lymphocyte activity for destruction of target cell. T-cell receptors on a CTL interact with processed antigen—class I MHC complexes on an approximate target cell, leading to formation of a CTL/target-cell conjugate. The Golgi stacks and granules in the CTL reorient towards the point of contact with the target cell, and the granules' contents are released by exocytosis. After dissociation of the conjugate, the CTL is recycled and the target cell dies by apoptosis.

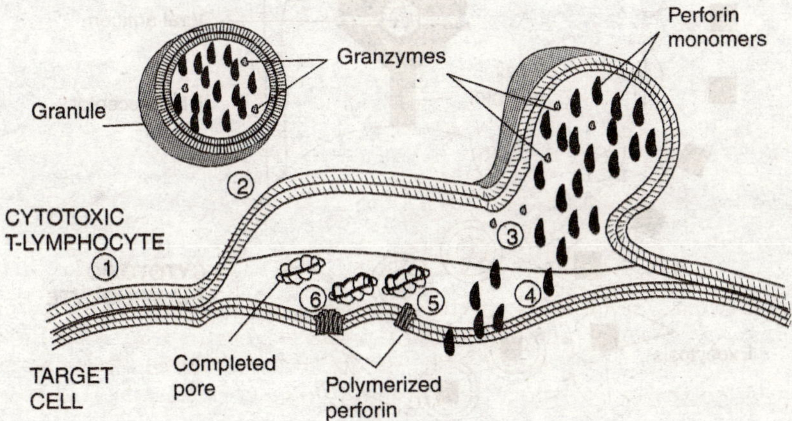

Fig. 10.9. CTL-mediated pore formation in target cell membrane. The rise in intracellular Ca^{2+} triggered CTL-target cell interaction, (1) induces exocytosis in which the granules fuse with the CTL cell membrane (2), and release monomeric perforin into the small space between the two cells (3) The released perforin monomers undergo a Ca^{2+}-induced conformational change that allows them to insert into the target-cell membrane (4). In the presence of Ca^{2+}, the monomers polymerise within the membrane (5), forming cylindrical transmembrane pores.

After the activation of CTL, it follows conjugation, membrane attack, CTL dissociation and target cell destruction. During conjugation (fusion) with target cell, the membrane integrin leukocyte functional antigen 1 (LFA-1) (leukocyte adhesion molecule) binds to intercellular cell-adhesion molecules (ICAM) on target cells. Then CTLs cytoplasmic changes occur by reorientation of Golgi complex and storage granules towards target cell junction (Fig. 10.8). The granules are vacuoles that contain monomers of pore forming protein **perforin** and several serine proteases called **granzymes** (also called fragmentins). These granules release perforin and granzyme into juction space by exocytosis. The perforin forms a large number of cylindrical pore of 5-20 nm size in the membrane by polymerization in presence of Ca^{2+} (Fig. 10.9 and 7.10). This permits the entry of granzymes into target cell interior and initiates the events of apoptosis by fragmenting the DNA. CTLs can also activates Fas ligand by binding of FADD protein (Fas associated Death Domain) on target cells that initiate events of apoptosis.

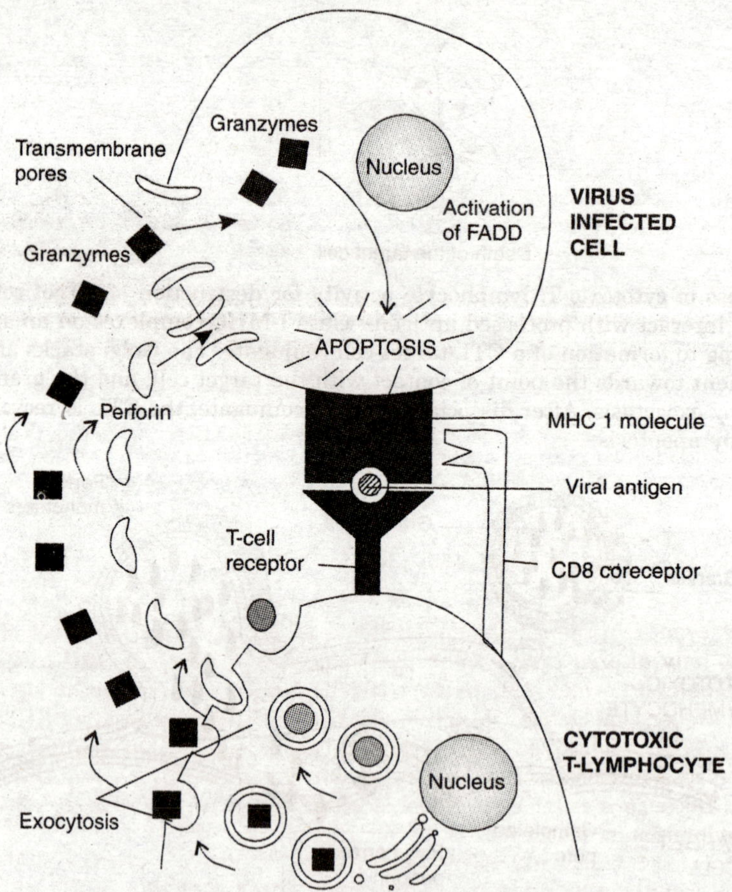

Fig. 10.10. How an effector cytotoxic T cell destroys a virus-infected target cell. T-cell cytotoxicity often involves the perforin pathway and leads to apoptosis and cytolysis. A cytotoxic T cell (down) contacts a target cell (top). The T cell secretes perforin that forms transmembrane pores in the target cells' plasma membrane. These pores allow the contents of the target cell to leak out (cytolysis). The granzyme activates the FADD (Fas associated death domain that starts events of apoptosis).

10.7 NATURAL KILLER (NK) CELLS

Natural Killer cells appear to kill tumor cells and virus infected cell by a process similar to that employed by CTLs. It is not having any immunological memory and not having any antigen specific receptors. The general carbohydrate binding protein activation receptor (AR) on NK cells interact with ligand on normal and altered self cells, inducing an activation of NK cell for killing. But, it does killing activity or not is based on the fact that it obtains negative signal from KIR (killer cells inhibitory receptor) and/or CD94/NKG2 (glycoprotein heterodimer) with class I MHC interaction. In normal cells class I MHC will be present that gives more negative signals and stops the killing activity (Fig. 10.11). Whereas in infected cells or altered cells class I MHC expression becomes decreased, thus killing signals becomes more leading to their destruction from the NK cells.

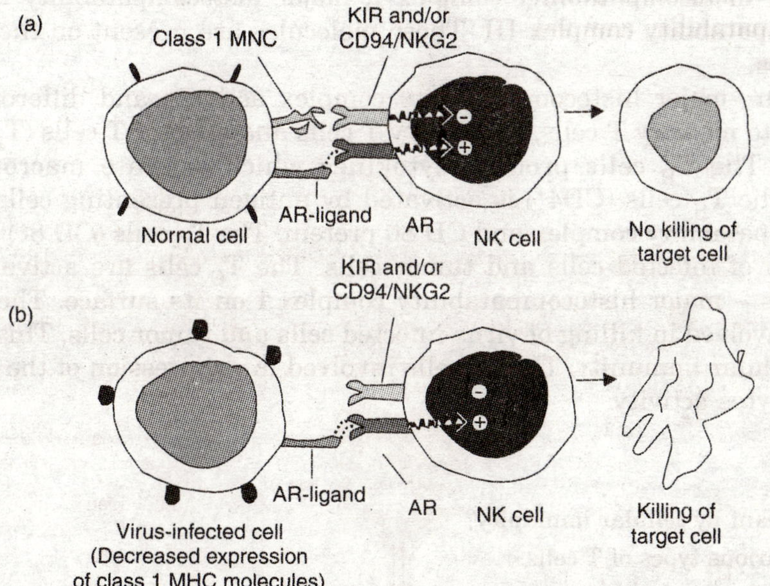

Fig. 10.11. Molecular events in opposing-signals model of how cytotoxic activity of NK cells is restricted to only altered target self-cells. An activation receptor (AR) on NK cells interacts with its ligand on normal and altered self-cells, inducing an activation signal that results in killing. However, engagement of inhibitory NK cell receptors such as KIR and CD94/NKG2 by class I MHC molecules delivers an inhibitory signal that counteracts the activation signal. Expression of class I molecules on normal cells thus prevents their destruction by NK cells. Because class I expression is often decreased on altered self-cells, the killing signal predominates, leading to their destruction.

T suppressor (T_S) cells are known to suppress the functions of B and T cells in immune response.

SUMMARY

Immunity mediated directly by immune cells is known as cellular immunity. It does not involve antibodies directly. Cell mediated immunity consists of different types of T lymphocytes. These include T helper cells (T_H cells), cytotoxic T cells (T_C cells), macrophages, natural killer cells antibody dependent killer cells. The cell mediated immune response involves the processing of antigens within the cells and presentation of processed antigens to T lymphcytes along with major histocompatability complex (MHC) molecules. Antigen processing is defined as the degradation of complex protein into antigenic peptide by protease enzyme series of events. Antigen presentation can be defined as delivering the processed antigen along with the MHC molecule by antigen presenting cells (APC) such as macrophages and dendritic cells.

A special group of proteins present on the cell surface are called major histocompatability complex (MHC) proteins. The genes encoding for MHC proteins are present on the chromosome 6 in humans and these surface antigens are Human leukocyte antigen (HLA) complex. The major histocompatability complex proteins are of three types and they are major histocompatability complex I, major histocompatability complex II and major histocompatability complex III. These molecules are present on the surface of different cell types.

The antigen—major histocompatability complex activate and differentiate naïve T lymphocytes into memory T cells, regulatory T cells and effector T cells (T_H cells, T_C cells and T_S cells). The T_H cells produce cytokines which activate macrophages and B lymphocytes. The T_H cells ($CD4^+$) is activated by antigen presenting cells with antigen-major histocompatability complex and CD 80 protein. The T_C cells (CD 8^+) are involved in the elimination of infected cells and tumor cells. The T_C cells are activated by antigen presenting cells – major histocompatability complex I on its surface. The natural killer cells are also involved in killing of virus-infected cells and tumor cells. This belongs to the non-specific cellular immunity. The T_S cells involved in suppression of the B lymphocytes and T lymphocytes activity.

EXERCISE

1. What is meant by cellular immunity? (BU Oct. 2002)
2. Describe various types of T cells.
3. Give an account on major histo-compatibility complex. (BU April 2001)
4. Explain antigen presentation.
5. Describe activation of T_H cells.
6. Illustrate the cytotoxic activity by T_C cells.
7. Write a note on Natural Killer cells. (BU April 2001)
8. Define cellular immunity. Briefly explain the cluster of differentiation. (BU Oct. 2002)
9. Give an account of 'T'-Immunocompetent cells. (BU Oct. 2001)
10. What is MHC? Explain its role in immunity. (BU April 2001, Oct. 2001)
11. Briefly explain cytosolic and endocytic pathway of antigen presentation.
12. Describe effect of cytokines produced by the T cells on different cells of immune system.
13. Name different cytokines produced by the T_H1 and T_H2 cells.
14. Explain molecular events involved in killing of target cells by NK cells.
15. Differentiate between Class I and II MHC molecules.

16. What is meant by perforin and granzyme?
17. List out the effector functions carried out by T_H cells subsets lymphokines.
18. Write a class of genes which codes for the MHC molecules in humans.
19. What do you mean by Class III MHC molecules in the immune system of human beings.
20. Enumerate the details of signal transduction processes in the APCs and TCR interactions that leads to the activation of T_H cells.

11

Effector Mechanisms

11.1 INTRODUCTION

Recognition of antigen by the immune system induces a variety of effector mechanisms associated with the humoral and cell-mediated responses. The cytokines are small proteins secreted by activated lymphocytes, macrophages and other immune cells that assist in regulating the development of immune effector cells and/or possess direct effector functions. The antibodies produced and complement proteins are other effectors of humoral immunity. Both specific cells (T_H and T_C cells) and non-specific cells (natural killer cells, macrophages, neutrophils, eosinophils) are effectors of cell-mediated immunity. In humoral immunity, system primarily functions to eliminate extracellular bacteria and bacterial products, whereas the cell-mediated immunity functions to eliminate altered self-cells (infected cells and tumor cells) and foreign grafts cells.

11.2 EFFECTOR MECHANISM IN HUMORAL IMMUNITY

Cytokines

Cytokines are low molecular weight regulatory proteins or glycoproteins secreted by white blood cells and various other cells in the body in response to a number of stimuli (Table 11.1). Cytokines help in regulating the development of immune effector cells and some cytokines act directly and function as effector molecules. Cytokines bind to receptors on the membrane of the same cell that secreted it (**autocrine** action) or other target cell (**paracrine** action), which are very near by or target cells which are far away placed in body (**endocrine** action). They regulate the intensity and duration of immune response by stimulating or inhibiting the activation, proliferation, and/or differentiation of various cells (Fig. 11.1). They also regulate secretion of antibodies and other cytokines.

Cytokines produced by lymphocytes were earlier known as **lymphokines**. Cytokines produced from monocytes and macrophages were formerly known as **monokines**. Cytokines secreted from some leukocytes and acting on other leukocytes are called **interleukins**. Cytokines that induce an antiviral condition in other cells are called **interferons**. Cytokines that help in the chemical attraction of cells (chemotaxis) are called **chemokines**.

Table 11.1: Cytokines sources and types from different leukocytes and platelets

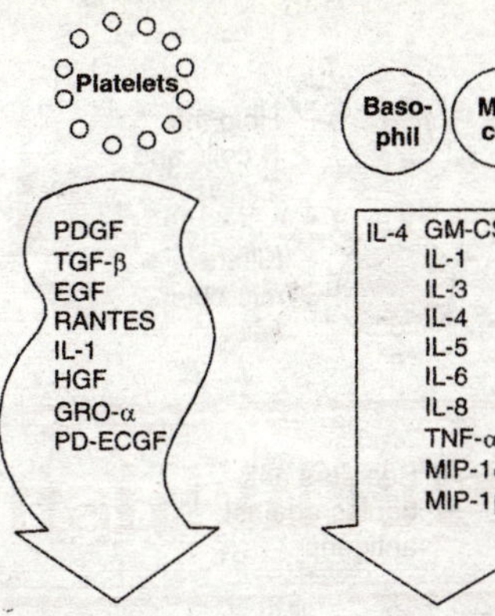

Platelets are involved in blood clotting and inflammation. Following endothelial injury they adhere to and aggregate at the surface of the endothelium, where they release the above cytokines, which activate various components of the complement system and stimulate the endothelium.

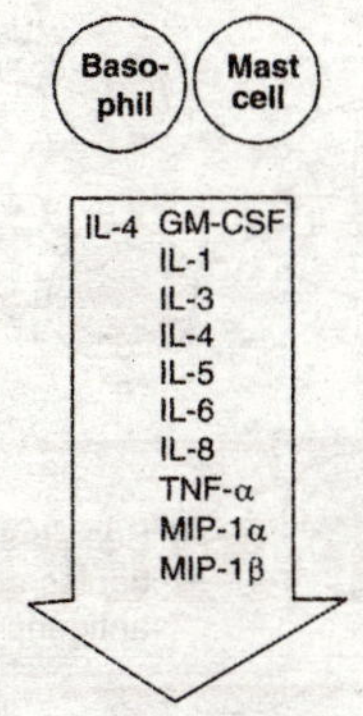

Basophils and mast cells are granulocytes that become activated when surface-bound IgE molecules bind to antigen. The physiological mediators (histamine) released following degranulation cause most of the adverse symptoms of allergy. However, these cells also play an important role in protection against protozoan and helminth parasites.

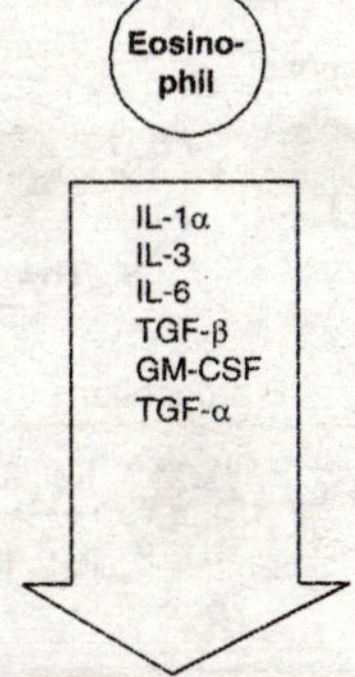

Eosinophils are poor phagocytes but undergo degranulation promptly in the presence of appropriate factors. Some eosinophil cytokines inhibit mast cell cytokine activitty/release and thuscan be regarded as having an antiinflammatory role. Also, they play an important role in protection against helminth parasites.

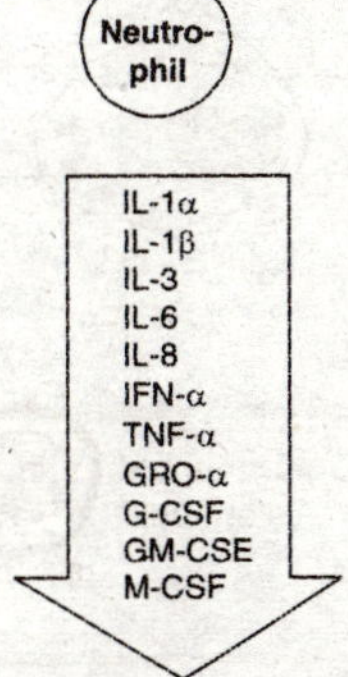

Neutrophils are normally stored in the bone marrow and released quickly into the circulation in response to various stimuli such as a bacterial infection. Their functions are directed toward killing and degrading bacteria.

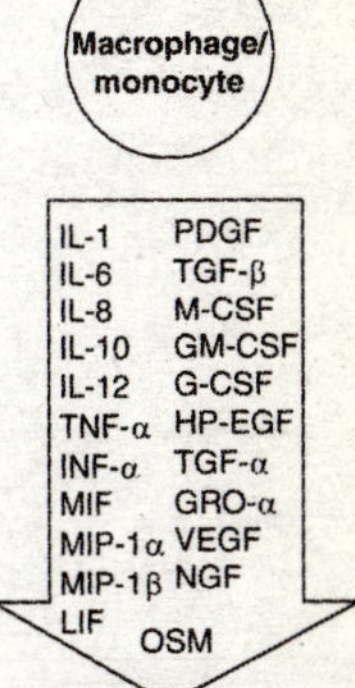

Macrophages/monocytes are highly phagocytic cells that accumulate at the site of an infection of tissue injury and secrete many cytokines, prostaglandins, and complement components. They also act as antigen-presenting cells.

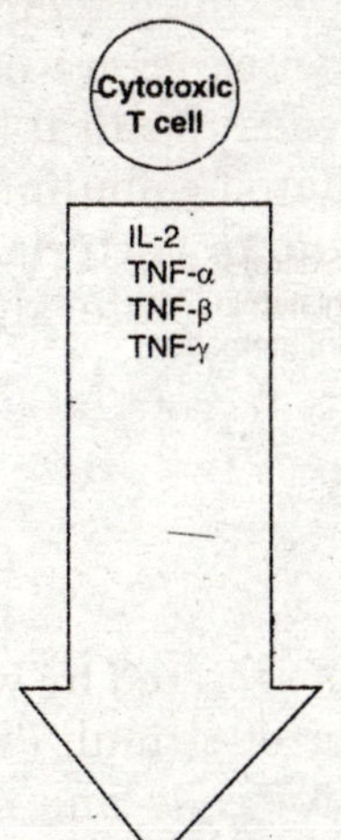

Cytotoxic T cells possess the ability to kill virus-infected cells or normal body cells expressing foreign determinants on their plasma membrane surface. They also secrete various cytokines that are important in the regulation of the immune response.

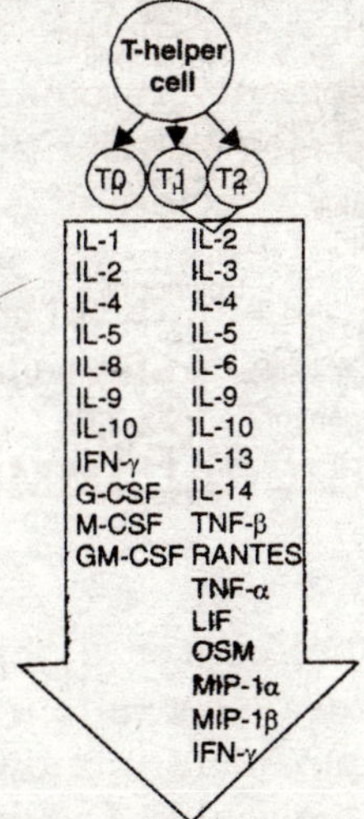

The T-helper cell is the most important cell in immunity. The various cytokines that it secretes are essential for the differentiation of immature B cells into antibody-secreting cells and also for the activation of cytotoxic T cells. Three subsets (T_H0, T_H1, T_H2) exist, although their specific activities are not fully understood. It is thought that T_H1 cells are involved in cellular immunity, whereas T_H2 cells play a role in humoral immunity. T_H0 cells represent a unique lineage of T-helper cells that produce their own cytokines.

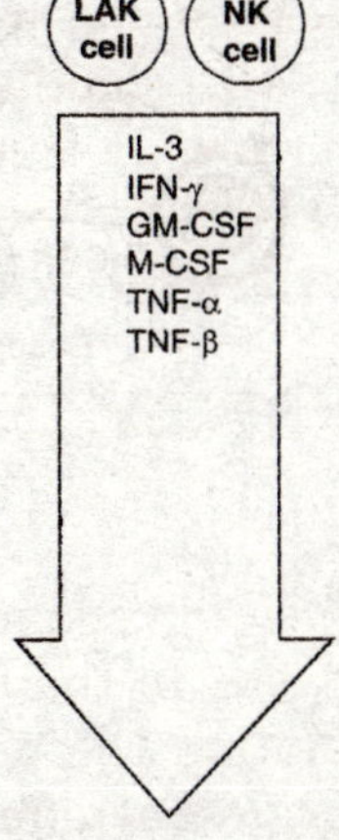

LAK and NK cells are nonphagocytic granulocytes that act against a variety of targets without the MHC restriction of T cells. They also kill a wide range of IgG-coated targets. LAK cells possibly represent a population of activated NK cells or T-helper cells rather than a specific lineage.

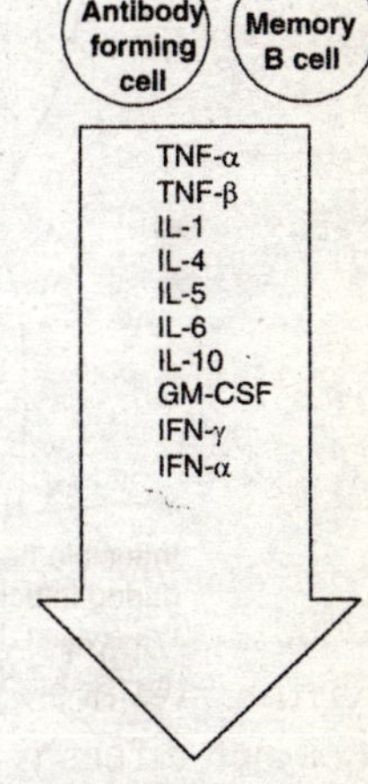

Antibody-forming B cells and memory B cells develop from antigenically stimulated B lymphocytes. Antibody-forming B cells lack surface Ig expression but synthesize large quantities of immunoglobulins that are released as free antibody. Memory cells are the resting cells derived from activated B cells, which upon reexposure to a specific antigen generate a rapid and vigorous secondary antibody response.

Cytokines, toxic to tumor cells, promote inflammation (a localized tissue response to injury characterized by pain, heat, redness and swelling resulted due to changed blood

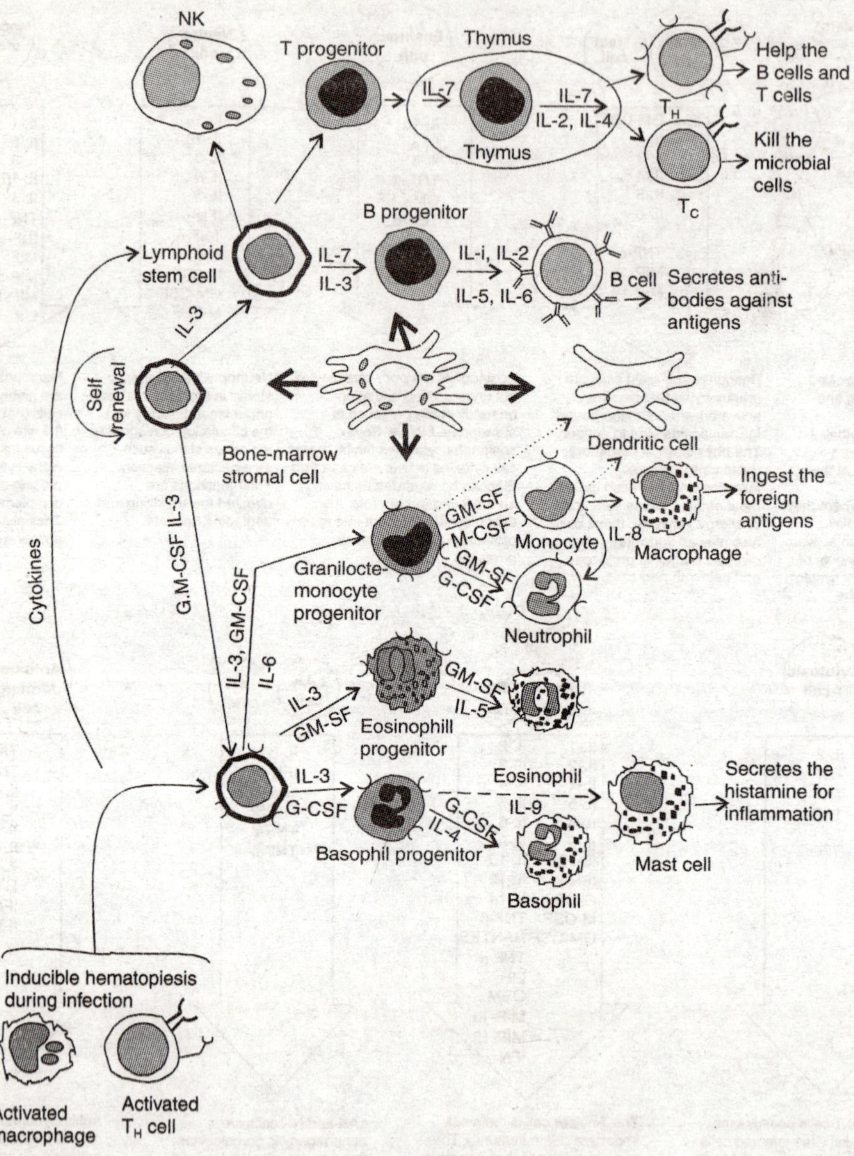

Fig. 11.1. Regulation of hematopoiesis in immune system by cytokines that stimulate the proliferation and/or differentiation of various hematopoietic cells. In the presence of infection, cytokines (thick arrows) produced by activated macrophages and T_H cells induce additional hematopoietic activity, resulting in rapid expansion of the population of white blood cells that fight infection. In the absence of infection, bone marrow stromal cells are the major source of hematopoietic cytokines (thick arrows). IL = interleukin: CSF = colony-stimulating factor; NK = natural killer cell. Stem cells are shown in thick outline; progenitor and immature cells in dark shades; stromal cell in without any shade; and mature differentiated cells in thin outline. Dashed lines indicate likely, but still hypothetical pathways.

flow, movement of phagocytes or other immune cells) or programmed cell death is known **as tumor necrosis factor** (TNF). Cytokines help in blood cells growth and differentiation (hematopoiesis) are called **hematopoietins**.

All these cytokines function as intercellular messenger molecules that induce particular biological activities after binding to a receptor on target cells. Monocytes, macrophages, T helper cells, B cells, dendritic cells, mast cells, natural killer cells, bone marrow cells, endothelial cells and other leukocytes are some of the cells that can produce cytokines. But T_H cells and macrophages are major cells, which produce cytokines. Their production is induced by viral, bacterial or parasitic infection, cancer, inflammation, T cell activation by antigen, or chemokines. Cytokines released from the parent cell can activate all the cells involved in the immune system to function in a defined way. Out of numerous physiological functions mediated by cytokines the major ones include development of cellular and humoral immune responses, induction of inflammation, regulation of hematopoiesis, control of cell multiplication and differentiation, fever, cell death and inhibition of replication.

11.3 THE COMPLEMENT SYSTEM

Complements are serum proteins or glycoproteins that play a major role in animal defense immune response. Originally Jules Bordet at Pasteur Institute discovered serum component necessary for the cell lysis and named it **alexin**. But later Paul Ehrlich coined the word **complement** for the alexin, because blood serum lytic activity was complemented along with antibodies. The complement proteins can lyse antibody-coated eukaryotic cells or bacterial cells (cytolysis). Liver hepatocytes, Monocytes, tissue macrophages and epithelial cells of gastrointestinal or genitourinary tract synthesize these proteins. Now more than 30 complement proteins have been identified. They include C1 to C9, factor D, H, I, C4b binding protein, C1 INH complex, S protein and properdin. The complement system acts in a cascade fashion (continuous series of reaction), the activation of one component resulting in the activation of the next. Usually within plasma and other body fluids complement proteins are in an inactive state.

There are three pathways of complement activation: the classical, alternative and lectin pathways. Each pathway has three stages, namely, initiation, amplification and membrane attack complex. They generally employ similar mechanisms, but specific proteins involved are unique to the first part (initiation) of each pathway (Fig. 11.2a).

These stages are explained in the figures. The bar in the Fig. 11.2b indicates the complements when associated with others, forms the enzymatic activity. Finally they form membrane attack complex made up of several complement proteins. This complex displaces the membrane phospholipids and forms the pore. Through these pores contents of the target cells are lost and thus the viability (Fig. 11.2b).

The classical pathway initiated when antigen (soluble or cell surface antigen) binding to the IgM or IgG, initiates C1, C4, C2 and C3 components. Each complement (C) generally breaks into smaller "a" fragment and larger "b" fragment except in C2 where C2a was larger and C2b was smaller. The activated C1 complex consists of C1qr2s2, which activates both C4 and C2 and forms complex C4a2b (C3 convertase) shows enzymatic cleavage activity, which can split C3 into C3a, C3b. The C4a2b3b complex (C5 convertase) can

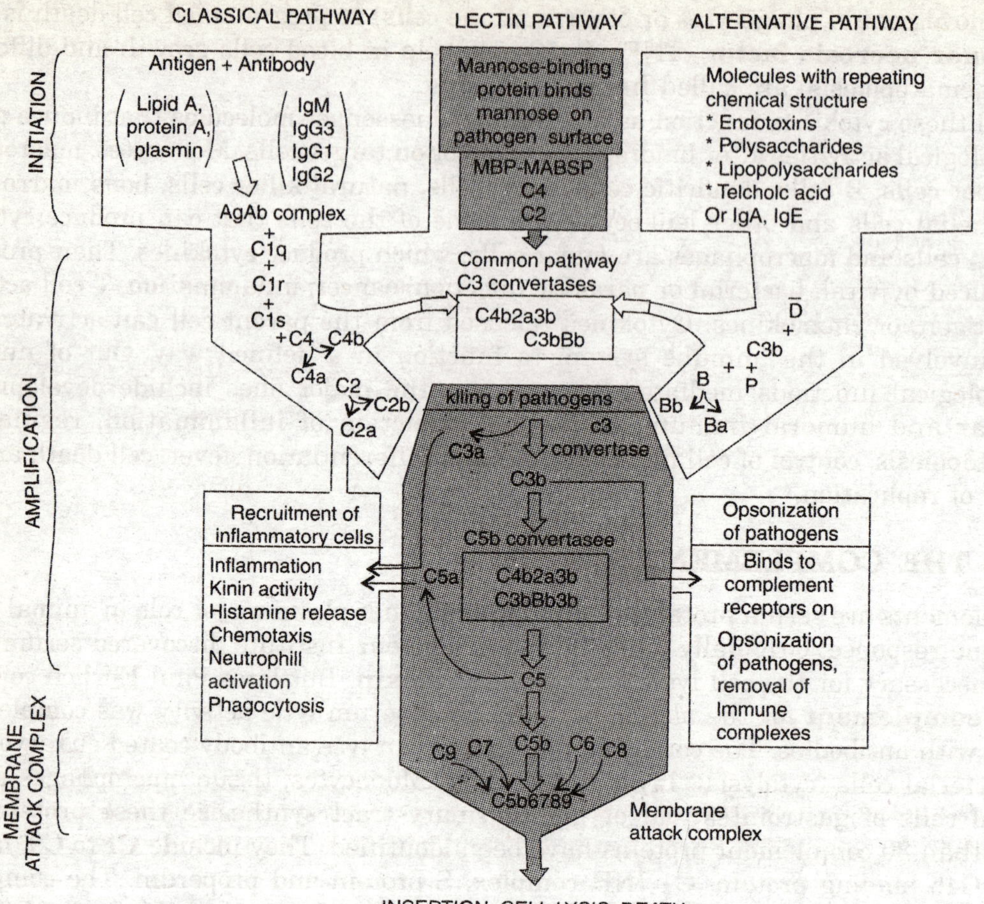

Fig. 11.2a. Complement activation. Within each pathway the components are arranged in order of their activation and aligned opposite their functional and structural analogs in the opposite pathway. There are three pathways of complement activation: (1) the classical pathway (left), which is mediated by antibody; (2) the lectin-mediated pathway (middle), which is triggered by a serum protein that binds to mannose residues of bacteria; and (3) the alternative pathway (right), which is triggered directly on pathogen surfaces. All three generate a crucial enzymatic activity (common pathway of C3 covertases) that in turn generates the effector activity of complement. The three main consequences of complement activation are recruitment of inflammatory cells (left box), direct killing of pathogens by the membrane attack complex (middle box), and opsonization of pathogens (right box).

split C5 to C5a and C5b. The C5b bound to the membrane and activates sequentially C6, C7, C8 and C9 to form membrane attack complex (MAC) (Fig. 11.3).

The alternate pathway can initiate by surface antigens of microorganisms (bacteria, fungi, some viruses and some parasites) or immunoglobulin (IgG-, IgA-, and IgE-) and antigen complexes. These activate C3 then factor B, factor D and properdin get activated and form the complex C3bBb (C3 convertase) and which then forms complex C3bBb3b (C5 convertase). This further forms C5a and C5b from the C5. The bound C5b forms membrane attack complex similar to classical pathway.

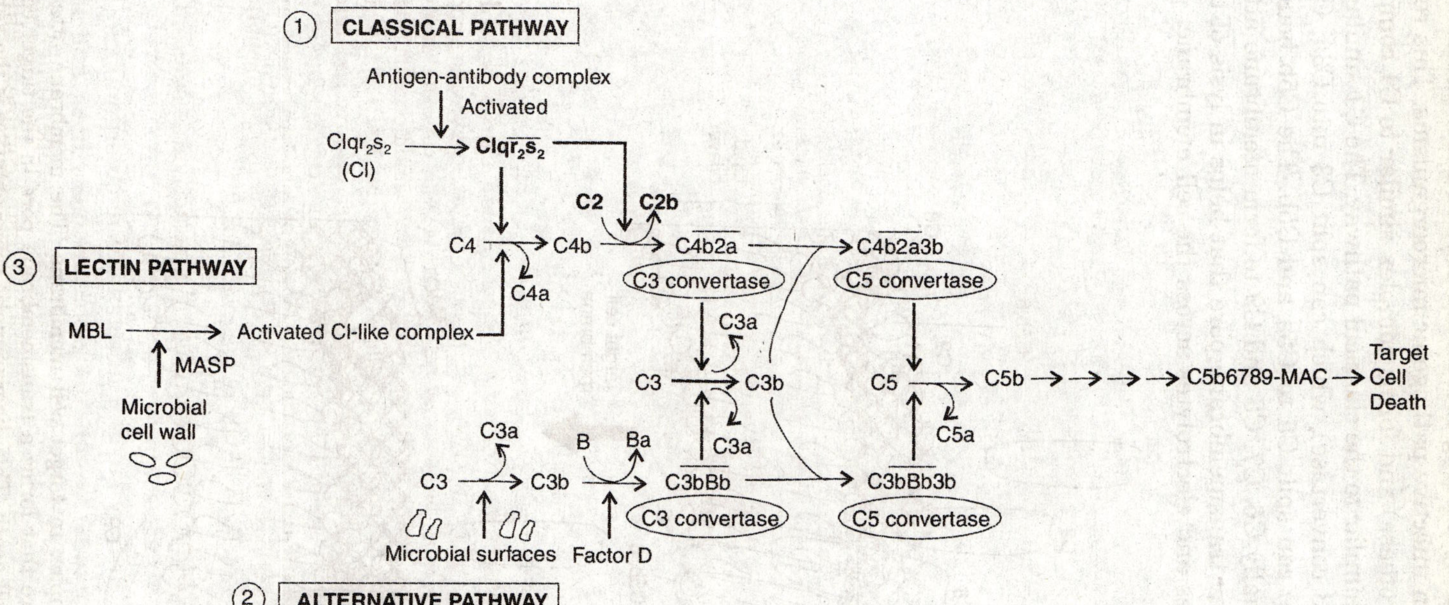

Fig. 11.2b. Model of the complement activation pathways in immune system. The classical pathway is initiated by binding of C1 complex to antigen-antibody complexes. The alternative pathway is initiated by binding of C3b to activating surfaces such as infective microbial cell walls. The lectin pathway is initiated by binding of the serum protein Mannose binding lectin (MBL) to the surface of a pathogen. The alternative pathway utilises some components (or factors) that are unique, whereas the classical and the lectin pathways have all components in common after C1 formation. All three pathways generate C3 and C5 convertases and bound C5b, which is converted into a membrane-attack complex (MAC) by a common sequence of terminal reactions. Breakdown of C3 is the major amplification step in all pathways, generating large amounts of C3b, which forms part of C5 convertase. C3b also has capacity to diffuse away from the activating surface and bind to immune complexes or foreign cell surfaces, where it functions as an opsonin. Thin arrows indicate reaction steps; thick arrows indicate enzymatic or activation activity.

The lectin pathway can initiate by binding of a serum protein MBL (mannose binding lectin) to carbohydrate residues on infective pathogenic microorganisms. This reacts with MASP (MBL associated serine protease) and forms complex similar to C1 complex. This complex activates C3, C4 and C2 similar to the classical pathway. The C1 similar complex gives raise to complex C4a2b (C3 convertase), which can split C3 into C3a, C3b. Then C4a2b3b complex (C5 convertase) can split C5 to C5a and C5b. The C5b bound to the membrane and activates sequentially C6, C7, C8 and C9 to form membrane attack complex (MAC). The MAC is a tubular transmembrane pore that helps in lysis of the target cells by out flow of cellular solutes and electrolytes across the cell membrane.

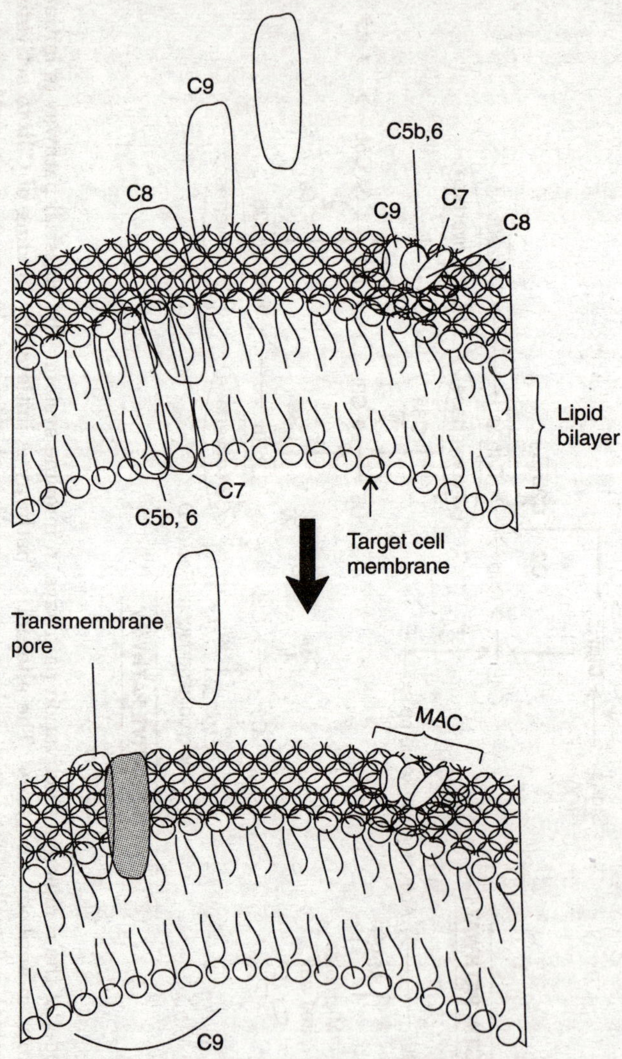

Fig. 11.3. The membrane attack complex in target cell membrane. The membrane attack complex (MAC) is a tubular structure that forms a transmembrane pore in the target cells' plasma membrane that results in cell death. This diagrammatic representation is the subunit architecture of the membrane attack complex. The transmembrane channel is formed by a C5b768 complex and 10 to 16 polymerised molecules of C9.

11.4 EFFECTOR MECHANISM IN CELLULAR IMMUNITY

The effector T cells produce various effector molecules, which may be membrane bound or soluble compounds. The membrane bound molecules include Fas ligand (FAS) on CD8$^+$ cells (CTLs), TNF-β on T_H1 cells and CD40 ligand on T_H2 cells. Each of these T cell also produces soluble effector molecules. The CTLs secrete cytotoxins (perforins and granzymes) as well as two cytokines IFN-α and TNF-β. T_H1 cells produce IL-2, IL-3, TNF-β, IFN-γ,

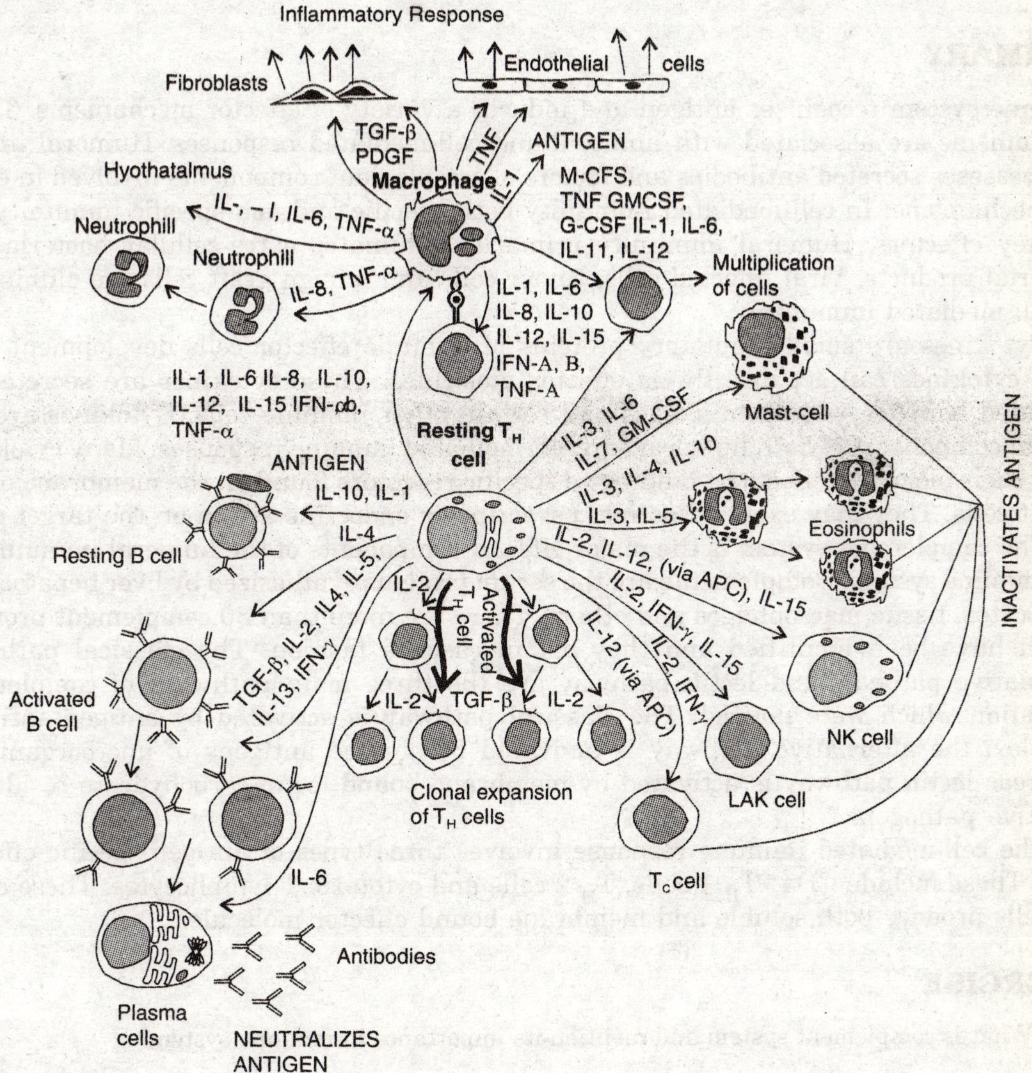

Fig. 11.4. Effector mechanism in cellular immunity. Interaction of foreign antigen with macrophages and the subsequent activation of resting T_H cells leads to release of numerous cytokines (thin arrows), generating a complex network of interacting cells in the immune response. A small subset of T_H cells secretes TGF-β, a cytokine that can inhibit the proliferation of T cells (see thick arrows). All the resulting cells work in their own way to fight against the foreign antigen.

high GM-CSF and T_H2 cells produce IL-3, IL-4, IL-5, IL-6, IL-10, IL-13, IFN-γ, low GM-CSF. Each of these membrane-bound and secreted molecules plays an important role in various T-cell effector functions (Fig. 11.4). The Fas ligand, perforins and granzyme, for example, mediate target cell destruction by the CTL; membrane-bound TNF-β and soluble IFN-γ and GM-CSF promotes macrophage activation by the T_H1 cells; and membrane-bound CD40 ligand and soluble IL-4, IL-5, IL-6 play a role in B cell activation by the T_H2 cells.

SUMMARY

Immune system recognizes antigen and induces a variety of effector mechanisms. These mechanisms are associated with humoral and cell-mediated responses. Humoral immunity posseses secreted antibodies and factors of complement components involved in effector mechanisms. In cell-mediated immunity both specific and non-specific immunity are the key effectors. Humoral immunity primarily eliminates extra-cellular bacteria and bacterial products. Viral infected cells, tumor cells and foreign graft cells are eliminated by cell-mediated immunity.

Cytokines are small regulatory proteins of immune effector cells development. But some cytokines can act directly as effector molecules. These cytokines are secreted by activated lymphocytes, macrophages and certain other immune cells. Cytokines are the critical components of both humoral and cell-mediated immune responses. Many cytokines exert their biological effects by binding to specific receptors found on the membrane of the target cells. They may exert autocrine, paracrine or endocrine action on the target cells.

The complement system is the major effector components of the humoral immunity of the immune system. Complements are the serum proteins synthesized by liver hepatocytes, monocytes, tissue macrophages and others. There are more than 30 complement proteins which have been identified and they act in cascade fashion. The classical pathway, alternative pathway and lectin pathway are the three main pathways of complement activation which were noticed. The classical pathway is activated by antigen-antibody complex; the alternative pathway is activated by surface antigens of microorganisms. Whereas lectin pathway is activated by membrane bound lectin-carbohydrate residue of infective pathogens.

The cell-mediated immune response involves three types of antigen-specific effector cells. These include $CD4^+$ T_H 1 cells, T_H 2 cells and cytotoxic T lymphocytes. These effector cells produce both soluble and membrane bound effector molecules.

EXERCISE

1. What is complement system and mention its importance in immune system? (BU April 2001)
2. Define cytokines.
3. Mention types of action of cytokines.
4. List various types of cytokines.
5. Elucidate in detail complement activation.
6. Describe formation of membrane attack complex.
7. Give the outlines of classical pathway of complement activation.

8. Differentiate between interleukins and interferons.
9. What is complement system? Explain briefly complement-mediated Ag-Ab reactions. (BU Oct. 2002)
10. Define hematopoietins.
11. Differentiate between MBL and MASP.
12. Name different units involved in membrane attack complex.
13. What do you mean by autocrine, paracrine and endocrine actions of cytokines?
14. Distinguish between lymphokines and monokines.
15. List out the lymphokines produced by the B cells or B lymphocytes.
16. Illustrate the different population of leucocytes and their cytokines.
17. Name major cytokines or monokines produced by the menophage.
18. What do you mean by effector mechanism in immune system?
19. Give an account on major components of effector mechanism and their functions.
20. How different pathways of complement activation are interlinked and ultimately ends in the membrane attack complex formation in the target cell.

12

Immunity to Infectious Diseases

12.1 INTRODUCTION

The term **host** means any organism that is used by another organism for shelter and food. Any organisms that obtain their food and shelter from the host organism are known as **parasites**. The parasite or parasitic organisms take entry into the host organisms by crossing all the barriers of the host and the process of entry is known as **infection**. Infection may or may not result in disease. When the host is visibly or sensibly injured by the parasite, **disease** will occur. The parasite that causes the disease is known as **pathogen** (*pathos* means sadness or pain). For example the mosquito can be called parasite but not pathogen, whereas *Plasmodium vivox* can be called pathogen. The capacity to cause disease is known as **virulent** pathogen i.e. those organisms able to produce serious disease or able to initiate an infection with very few cells (as low as 1 to 10 cells). An organism that cannot infect or cause disease is known as **avirulent** (**attenuated**).

For a particular disease to occur pathogenic organisms have to be virulent. The virulent pathogen, to induce disease in the host, have to follow three stages; infectivity, invasiveness and pathogenicity. **Infectivity** means ability to cause infection and establishing inside the host. **Invasiveness** means capacity of the pathogen to leave from initial site of infection and grow in other tissue of the body. **Pathogenicity** is the ability of the pathogen to injure the host to the maximum extent and production of disease is known as pathogenesis. All the three phenomena include complex events in the host organisms. The varieties of infection processes may occur in host organisms but infection occurs normally by portal of entry (means by which the pathogen enters). These include wound on skin surface, mucous membranes (eye, nose, throat, lungs, genitourinary tract or gastrointestinal tract) (*Clostridium tetani, Shigella, Neisseria gonorrhoeae*). Some enters through intact skin surfaces in the area of sweat glands or hair follicle (*Staphylococcus aureus*). Invasiveness occurs by multiplication of pathogen and then they gain entry to blood which moves in the entire body. Pathogenicity occurs by the increased pathogen number or by the production of poisons called toxins.

The animals are in continuous contact with the various virus, bacteria, fungi and parasitic organisms. If an organism to be pathogenic it should establish infection in a susceptible host and a series of host defense mechanisms has to be overcome by the pathogen. The host immunity includes both nonspecific (innate) and specific (adaptive) defenses. The host innate immunity acts as first line of defense and consists of physical, chemical and biological barriers. This innate or nonspecific immunity lacks immunological memory, that is, these responses occur to the same extent each time a pathogen or antigen is encountered. The specific immunity mechanisms are detailed in the earlier chapters with respect to cellular immunity, humoral immunity and effector mechanisms. In this chapter the emphasis is given to the non-specific or innate immune mechanisms.

12.2 PHYSICAL AND CHEMICAL BARRIERS IN NON-SPECIFIC IMMUNITY

Physical or mechanical barriers, along with the host's secretions (flushing mechanisms) act as first line of resistance against the infectious microorganisms. The intact skin, and mucous membranes along with chemicals greatly help in nonspecific host defense.

Skin consists of thick keratin rich cells that are resistant to pathogenic degradation, mild acidic condition (pH 5-6), dryness, sebum (oil) from sebaceous glands and shedding (of outer epithelial cells) activity are known to act as a barrier to the pathogenic invasion or entry. Sometimes pathogens gain access to the tissue under the skin. Here they encounter defense by specialized cells called Langerhan's cell (specialized dendritic cell that can phagocytose antigens/pathogens). These cells are associated with skin associated lymphoid tissue (SALT) or cutaneous associated lymphoid tissue (CALT) and act as antigen presenting cells which activate the T cells for destruction of pathogen. Some of the intraepidermal lymphocytes (cells of a SALT) can directly act as T cells to destroy the pathogen.

The mucous membranes of the eye (conjunctiva), respiratory, digestive and urinogenital systems consist of the stratified squamous epithelium that resists the penetration and also traps the pathogens for removal. These epithelial systems also known to contain several types of mucosal associated lymphoid tissue (MALT) that includes gut associated lymphoid tissue (GALT) functions in digestive tract and tonsils, and bronchial associated lymphoid tissue (BALT) that functions in respiratory system. They are having secretory antibodies and macrophages for the protection.

The chemical barrier includes antimicrobial substances produced by the animals especially lysozyme (muramidase), lactoferrin, gastric juices, salivary glycoproteins, sebeum oleic acid on the skin, urea and low pH in urine are known to have microbicidal effects when they come in contact with these compounds. In addition, blood, lymph and other body fluids contain many defensive chemical compounds such as bacteriocins, beta-lysin and other polypeptides.

Lysozyme is an enzyme present in sweat, tears and mucous that degrades the cell wall peptidoglycan material of bacterial organisms. As a result bacteria loses viability leads to the death. **Lactoferrin** is an iron chelating compound present in plasma thus makes iron unavailable for the pathogenic multiplication inside. The gastric juices pH will be around 2-3 in which most of the organisms engulfed are killed. **Sebum**, fluids secreted by the oil glands contain triglycerides degraded to form free unsaturated fatty acids (oleic acid) that are antimicrobial in nature. **Bacteriocins** are toxic proteins produced by normal microbiota on the mucous surfaces. These toxic proteins are able to cause the cell lysis of target bacterial organisms. **Beta-lysin** is a cationic polypeptide released from blood platelets, which can kill some gram-positive bacteria by disrupting their plasma membranes. Other examples for such polypeptides having lethal effect are leukins, plakins, cecropins and phagocytin.

The biological barriers include the saprophytic microbial population which generally associated with the healthy individuals is known as normal microbiota. These micro organisms by their activity provide an environment that is not suitable for the pathogens to survive in that environment. Some of the organisms such as *Bacillus, Bacteroids, Enterococcus, Lactobacillus, Propinobacterium, Prepto streptococcus, Pseudomonas, Staphylococcus* and *Streptococcus* are members of the normal microbiota of humans.

12.3 INFLAMMATION

Inflammation (*inflammation* means to set on fire) is an important nonspecific defense reaction to tissue injury caused by a pathogen or wound. It was characterized by cardinal signs or symptoms such as *calor* (increased temperature), *rubor* (redness), *tumor* (swelling), *dolor* (pain) and inflation of immune cells. The inflammation may be acute or chronic types.

The acute inflammatory response starts when injured tissue cells release chemical signals (inflammatory mediators-histamines, prostaglandins and leukotrienes) activates endothelial cells which lining the blood capillaries to produce P-selectins and E-selectins (cell adhesion molecules). These molecules attract and attach neutrophils to endothelial cells and neutrophils activated by inflammatory mediators. This activation occurs through integrins (cell adhesion receptor), which binds neutrophils tightly to endothelial cells. This results in neutrophils to change shape, squeeze through the endothelial wall (the process is known as **extravasation**) (Fig. 12.1) into the interstitial tissue fluid, then migrate to the site of injury and attack the pathogen or other cause of the tissue damage. The chemotactic factors released by pathogen or mast cells attract the neutrophils first, then lymphocytes, monocytes and macrophages depending on the severity. Increased blood flow to the injured sites results in swelling, redness and pain.

The chronic inflammation is a slow process and occurs when neutrophils and macrophages are unable to protect host from the tissue damage. The body attempt to separate injured site and isolate from the surrounding tissues by the formation of granuloma (*granulum* means a small particle and *oma* means to form). This results in formation of new connective tissue and causes permanent tissue damage.

12.4 INTERFERONS (IFNs)

Interferons are cytokines produced by many eukaryotic cells in response to inducers such as virus infection, double stranded RNA, endotoxins and pathogens. These are proteins or glycoproteins known to inhibit viral multiplication inside the cells.

Several classes of interferons are identified based on physicochemical and antigenic properties such as IFN-α, IFN-β and IFN-γ. The IFN-α and IFN-β and generally degrade the mRNA synthesis in the virally infected cells. They are also known to activate the macrophages, T cells, NK cells and neutrophils that are involved in specific immunity.

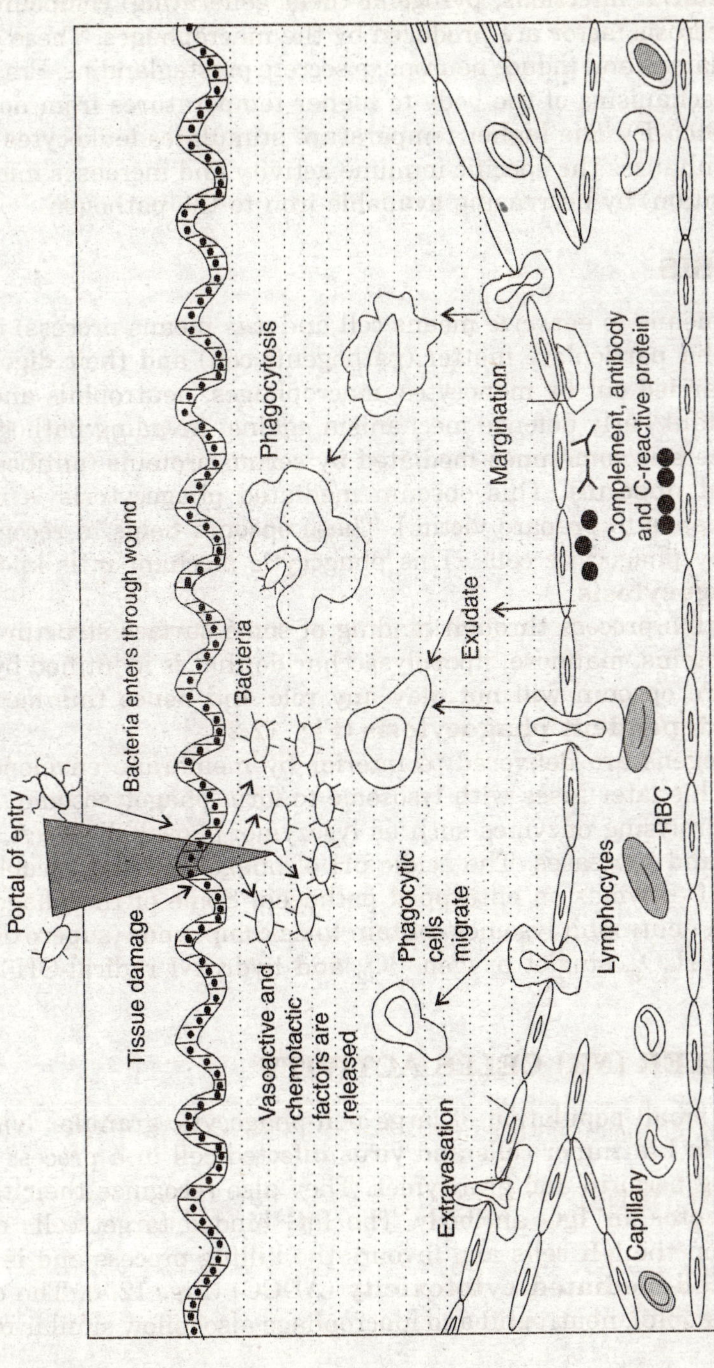

Fig. 12.1. Major events occurring in the inflammatory response. A bacterial infection causes tissue damage with production of various vasoactive and chemotactic factors. These factors induce increased blood flow to the damaged area, increased capillary permeability, and an influx of white blood cells, including phagocytes and lymphocytes, from the blood into the affected tissues. The serum proteins contained in the exudate have antibacterial properties, and the phagocytes begin to engulf the bacteria.

12.5 FEVER

During the viral or bacterial infections, pyrogenic (heat generating) compounds such as IL-1, IL-6 and tissue necrosis factor are produced by the macrophages. These compounds circulate to the hypothalamus and induce neurons to secrete prostaglandins. Prostaglandins reset the thermostat mechanisms of the body to higher temperatures from normal 37°C/ 98.6°F (rectum 37.5°C/99.5°F). The higher temperature stimulates leukocytes for the destruction of pathogen, enhances the specific immune activity and increases microbiostasis (pathogen growth inhibition) by decreasing available iron to the pathogen.

12.6 PHAGOCYTOSIS

Phagocytosis (*phagein* means to eat, *cyte* means cell and *osis* means process) is the process of engulfment of solid particulate matter (pathogenic cell) and their digestion. This process in the host is carried out by monocytes, macrophages, neutrophils and dendritic cells which is an important early defense mechanism against invading pathogens.

The phagocytic process is sometimes mediated by serum proteins (antibodies and/or complement C3b) called **opsonin**. This opsonin mediated phagocytosis is referred as **opsonization** (*opson* means to prepare victim). These opsonin helps in recognition and ingestion of pathogen by phagocytic cells. This phagocytic mechanism is known as **opsonin-dependent phagocytosis**.

Phagocytic cells do this process through binding of some surface structure of pathogens such as lectins, proteins, mannose, lipopolysaccharide that is identified by receptors on phagocytic cells. Here opsonin will not play any role and hence this mechanism is referred as **opsonin-independent phagocytosis** (Fig. 12.2).

Once ingested pathogens are delivered to interior by membrane enveloped vacuole known as phagosome. This later fuses with lysosome to form phagolysosome. Lysosomes contain a variety of hydrolysing enzymes such as lysozyme, phospholipase A, ribonuclease, deoxyribonuclease and proteases. The acidic pH of phagolysosome vacuoles favours the degrading activity of enzymes on entrapped pathogen. Some of the phagocytic cells such as macrophage and neutrophil lysomes contain toxic compounds (superoxide radical-O_2^-, hydrogen peroxide- H_2O_2, singlet oxygen- $^1O_2$ and hydroxyl radical-OH·) that kills the pathogens.

12.7 NATURAL KILLER (NK) CELLS ACTIVITY

Natural Killer cells are small population of large non-phagocytic granular lymphocytes. Their major function is to kill tumor cells and virus infected cell in a process similar to the CTLs that employing perforin and granzymes. They also recognize their target cells through F_C surface receptor for IgG antibody. The IgG binded target cells comes and binds to these receptor on the NK cells and favours the killing process and is known as **antibody-dependent cell-mediated cytotoxicity** (ADCC) (Fig. 12.3). The other non-specific cells such as eosinophil, neutrophil and macrophage also follow similar mechanism for target cell killing.

Nonspecific defenses such as skin, mucous membranes, and gastric secretions prevent many bacteria from entering host tissues. When bacteria do infect a host, certain immune responses such as antibodies, complement alters the invading bacteria so that they can be

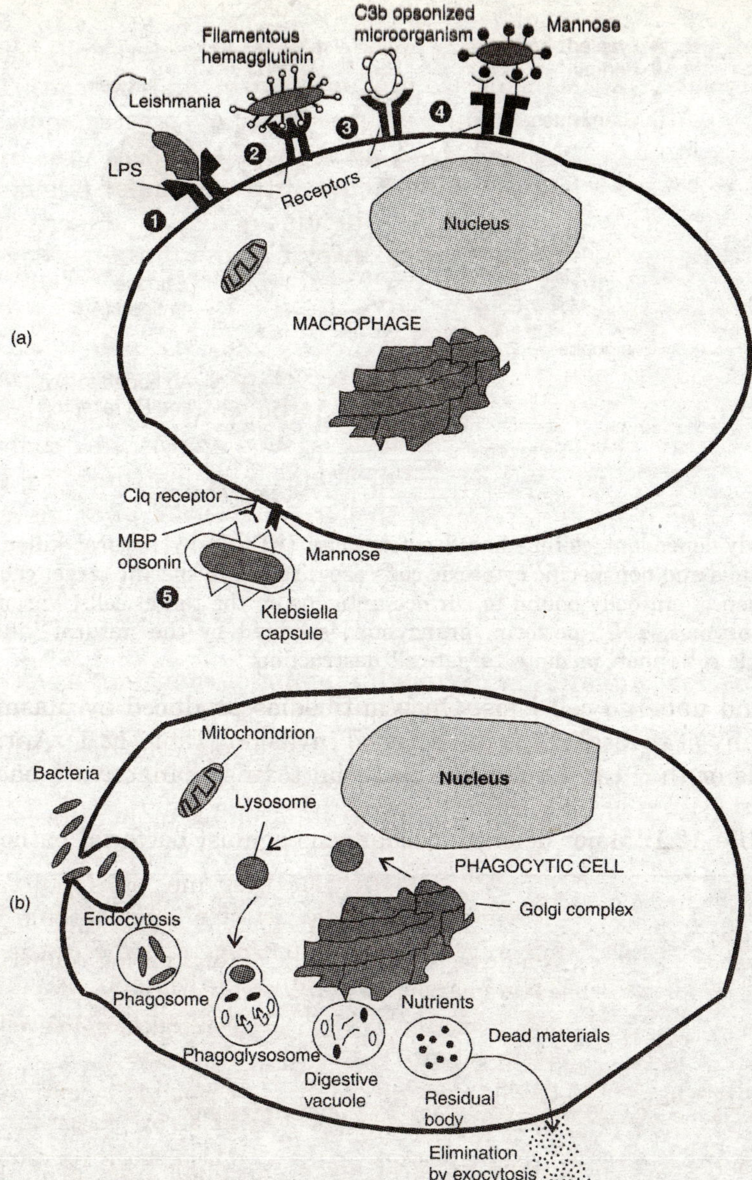

Fig. 12.2. Phagocytosis by macrophage. (a) Drawing shows receptors on a phagocytic cell, such as a macrophage, and the corresponding adhesions on foreign pathogenic microbial surfaces participating in phagocytosis. (1) The LPS binding site for lipophosphoglycan of *Leishmania* spp. (2) filamentous hemagglutinin of *Bordetella pertussis;* (3) binding to C3b on an opsonised cell; (4) the mannose-containing oligosaccharide side chain for lectin phagocytosis mediated by type 1 fimbriated; and (5) the participation of capsular polysaccharide of *Klebsiella pneumoniae* in nonpsonic binding mediated by the mannose receptor. (b) Drawing of phagocytosis, showing ingestion, intracellular digestion, and exocytosis.

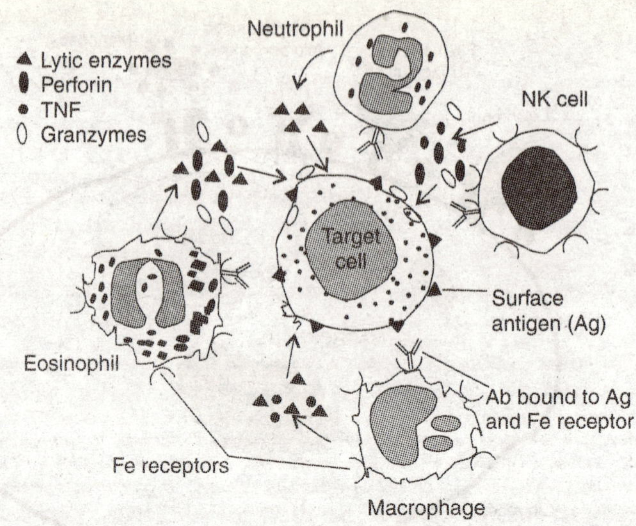

Fig. 12.3. Antibody-dependent cell-mediated cytotoxicity (ADCC) by natural killer cells. The natural killer cells and nonspecific cytotoxic cells are directed to specific target cells by binding to the Fc region of antibody bound to surface antigens on the target cells. Various substances (e.g., lytic enzymes, TNF, perforin, granzymes) secreted by the natural killer and nonspecific cytotoxic cells then mediate target-cell destruction.

phagocytized and undergo cell lysis. Once antibodies produced by plasma cells they can interfere with any of several steps in bacterial invasion (Table 12.1). Antibodies produced may act on cells or they can neutralize bacterial toxins or inactivate bacterial enzymes.

Table 12.1: Major defence mechanisms against bacterial pathogens

Pathogen	Pathogenesis	Major defence mechanism
Corynebacterium diphtheriae	Non-invasive Toxin-pharyngitis	Neutralizing Antibody
Vibrio cholerae	Non-invasive Toxin-enteritis	Neutralizing and adhesion-blocking Antibodies
Staphylococcus aureus	Locally invasive Toxic in skin	Optonized by antiody and complement killed by phagocytes
Mycobacterium tuberculosis	Invasive, evokes pathological events	Macrophage activation by cytokin from T cells
Mycobacterium leprae	Invasive, tissue injury	Macrophage activation by cytokin from T cells

Viruses first attack cells that line body passages. From this area they directly invade target organs such as lungs or liver or nervous system through blood. Interferons, secretory IgA and some IgG molecules act at mucous layers and prevent the entry. IgG and IgM act in the blood to neutralize viruses or to promote their destruction by complement. The fever and cellular responses such as cytotoxic T cells and NK cells are especially important in protecting the host from viral infection.

Fungal infections progress through a tissue as fungal cells invade and destroy one cell after another. Fungal skin infections probably are protected by IgA antibodies and helper T cells, which release certain cytokines that activate macrophages. The macrophages in turn, engulf and digest fungi.

Protozoa and helminths are quite dissimilar in size and complexity, but may use similar methods of invading the body. Many of them interact with their host in ways that do not endanger host survival and thereby ensure parasitic survival. Most of these are large and difficult to phagocytize, some may change the antigenic character or kills the host cells by toxins or cause allergy or develop cysts. The host combats these by cell-mediated immunity through production of IL-3, that activate macrophages or IL-5, that enhance the ability of eosinophils to fight worm infections.

SUMMARY

Infection and immunity involves interaction between the host and parasite. The parasite can be defined as any organisms that obtain food and shelter from the host organism. The host means any organisms that provide food and shelter to the parasite. Infection can be known as the colonization of the host organisms by a parasite. Any organisms which have the capacity to produce disease are called the pathogen. The virulence refers to the degree of pathogenicity by the pathogen. The diseases occur in three stages namely infectivity, invasiveness and pathogenicity. Both types of host immunity such as innate immunity and adaptive immunity protect the body against infectious diseases. The first line of defensive mechanism is innate immunity, which does not have the memory of the immunity. It consists of physical barriers, chemical barriers, biological barriers and general barriers.

The physical barriers include intact skin and mucous membrane. The chemical barriers include antimicrobial substances such as lysozyme, lactoferrin, gastric juices, low pH and others. The biological barriers include normal biota colonized in various organs of the body. The general barriers include inflammation, fever and phagocytosis.

Inflammation is one of the hosts' non-specific defensive reactions to a tissue injury that caused by a pathogen or wound. The symptoms of the inflammation include rise in temperature, redness, swelling, pain and others. It may be acute or chronic type. Cytokines produced by many eukaryotic cells in response to viral infection are called interferons. These bring about the destination of viral infected cells. Fever defined as an oral temperature above 98.6F (37°C) or rectal temperature above 99.5F (37.5°C). It takes place during infection and bring about the destruction of the pathogen. Phagocytosis involves the reorganization, ingestion, and destruction of parasitic organism by lysosomal enzymes. Large non-phagocytic granular lymphocytes are called natural killer cells. Their main infection is to kill tumor cells and virus infected cells.

EXERCISE

1. Define host and parasite.
2. What is meant by pathogen?
3. Mention different types of physical barriers for infection.
4. What do you mean by inflammation?
5. Explain phagocytosis and opsonization. (BU Oct. 2002)

6. **Describe in detail immunity to infectious diseases.**
7. Give an account of inflammation.
8. Write a short note on protective function of fever.
9. How do interferons protect cells by viral infection?
10. Briefly explain antibody-dependent cell-mediated cytotoxicity.
11. Define extravasation.
12. Differentiate between lysozyms and lactoferrin.
13. Name different microbiota of humans.
14. What do you mean by infectivity?
15. Give the difference between invasiveness and pathgenicity.
16. Briefly write about protection mechanism against pathogens.
17. How specific and non-specific immunity are interconnected?
18. Define pathogenicity and pathogenesis.
19. What is meant by virulent and avirulent pathogen?
20. What do you mean by SALT, GALT, MALT and BALT?

13

Vaccines

13.1 INTRODUCTION

Vaccines are the preparations from an infectious microorganisms that is injected to humans or other animals to induce protective immunity. The complete protection shown by the host towards infectious disease is called immunity. Vaccines may consist of a preparation of killed or living weakened (attenuated) microorganisms, inactivated bacterial toxins (toxoids), purified macromolecules, recombinant vectors or DNA vaccines.

The administration of vaccines to animals is known as **vaccination**. For the first time Edward Jenner and Louis Pasteur advocated vaccination for small pox and Rabies diseases respectively. Now vaccines are developed against most of the epidemic diseases (anthrax, cholera, diphtheria, German measles, hepatitis A and B, measles, mumps, polio and whooping cough). The immunity can be active immunity or passive immunity. The **active immunity** means the host organisms are active in the process of obtaining immunity. This can be obtained by natural infections or vaccines administration. The **passive immunity** means the host organisms obtaining immunity without being involved in the immunity process. This can be obtained from the prepared antibodies which comes from the different organisms such as child obtaining immunity from maternal antibodies or antitoxin injections given to snake bitten persons.

The successful vaccine must be able to stimulate immune responses that are similar to those resulting from natural exposure to the pathogenic microorganisms. Many efforts are being made to produce successful vaccines for the most dangerous such as AIDS, herpes, hepatitis C, malaria and tuberculosis, but not completely successful result is obtained.

13.2 VACCINES EFFECTIVENESS

Many physical, chemical and biological factors affect the vaccines effectiveness. An effective vaccine

a. must induce the right sort immunity
b. should be stable on storage

c. should be having sufficient immunogencity
d. should not be contaminated with unwanted proteins or toxins

13.3 TYPES OF VACCINES

There are many types of vaccines available for the various bacterial, viral, and Protozoan diseases. Based on the type of the materials used for the preparation vaccines are classified.

Whole-organism vaccines

The preparations consist of whole organisms that are killed (inactivated) or attenuated (avirulent live) are known as whole-organisms vaccines. Many of the common vaccines currently used for the bacterial or viral diseases come under this type. The anthrax vaccine, BCG vaccine, cholera vaccine, hepatitis A vaccine, measles vaccine, polio vaccine and yellow fever vaccine are example for whole-organism vaccines. The inactivated vaccines are effective, but require several booster doses and they do not stimulate cell-mediate immunity and secretory IgA production properly. Usually attenuated vaccines work effectively in a single dose and stimulate both humoral and cell-mediated immunity.

Purified macromolecular vaccines

The whole-organisms vaccines has some disadvantages such as some fail to give immunity, reverts to virulent form by mutation, causes unknown effect or acts as allergenic agents in the host. To overcome this purified specific macromolecules derived from the pathogenic microorganisms are used in the preparation of vaccines. The pure capsular polysaccharide (*Streptococcus* pneumonia vaccine), denatured toxins called toxoid (tetanus vaccine) and recombinantly produced surface antigen (Hepatitis B vaccine) are used for the vaccine preparation.

Recombinant-vector vaccines

The genes that encode major antigens of dangerous pathogenic organisms are introduced into non-pathogenic or attenuated microorganisms (bacteria or virus), which act as vector. This attenuated microorganisms are introduced into the host, where these vector microorganisms expressing antigenic proteins during their growth. These antigenic compound induce both cellular and humoral immunity in the host organisms. Adenovirus, canarypox virus, vaccinia virus, attenuated polio virus, attenuated *Salmonella* and *Mycobacterium bovis* have been used as vector vaccines.

DNA vaccines

The recombinant plasmid DNA encoding antigenic proteins are injected directly to the muscle cells of the recipient. Here muscle cells take up the DNA and express the antigenic proteins. This antigenic proteins stimulate both cellular and humoral immunity in the host. The DNA may integrate into the chromosomal DNA or maintained as episomal form in the cells. At present DNA vaccines against AIDS, hepatitis B, herpes, influenza and malaria are under human trials.

Synthetic peptide vaccines

As various pathogenic antigen sequence is known efforts are made to synthesize small peptides and introduced into the recipient organisms. Still lot of procedural progress has to be made in this category.

Multivalent subunit vaccines

The synthetic peptide vaccines are known to induce only humoral immunity and less of cellular immunity, thus antigenic synthetic peptide compounds which induce both humoral immunity and cellular immunity are attached to monoclonal antibodies on a solid matrix. This matrix approach is known as solid matrix-antibody-antigen (SMAA) complexes. The SMAA complex helps in making particulate matter, which facilitates phagocytosis and thus induce both humoral and cellular immunity. The other method includes incorporation of antigenic peptides into protein micelles or incorporation into lipid membrane vesicles to form liposomes.

13.4 VACCINES USED IN PRESENT TIMES

Large number of vaccines are currently used against infectious diseases. The Table 13.1 lists vaccines in general use and Table 13.2 contains some vaccines that are reserved for certain category of people. These practices are successful in completely eradication of the disease or reduction.

Table 13.1: Vaccines that are in general use

Disease	Vaccine	Remarks
Tetanus	Toxoid	Given together between 2 and 6 months in 3 doses (DTP); Diphtheria and Tetanus boosted every 10 years
Diphtheria	Toxoid	
Pertussis	Killed whole cells	
Polio	Killed (Salk) or Attenuated (Sabin)	
Measles	Attenuated	Given together (MMR) at 12-18 months
Mumps	Attenuated	
Rubella	Attenuated	
Haemophilus	Polysaccharide	Newly developed, can be used widely

SUMMARY

A preparation of either killed or attenuated micoorganism or inactivated bacterial toxin is called vaccine. Vaccination is the administration of vaccine to animal. It was first introduced by Edward Jenner for small pox. Later Louis Pasteur and other developed vaccines for rabies, anthrax, cholera etc. Vaccines induces active immunity in the same way as natural infection. Many types of vaccines are available for infectious disease. These include whole organism vaccines, purified macromolecular vaccines, recombinant vector vaccines, DNA vaccines, Synthetic peptide vaccine, and multivalent subunit vaccines. Vaccines serves best prophylactic agents for many infections diseases and saves many lives. Efforts are being made to develop vaccines against AIDS, malaria etc.

Table 13.2: Vaccines that are used only for certain people

Disease	Vaccine	Category of people
Tuberculosis	BCG	Tropics countries at birth UK: 10-14 years USA: At-risk only
Hepatitis B	Surface Antigen	At-risk (medical, nursing staff) Drug addicts Homosexuals Carriers contacting peoples
Rabies	Killed	At-risk (Animal workers) Bitten people
Hepatitis A	Killed/Attenuated	Travellers
Meningitis	Polysaccharides	Travellers
Yellow fever	Attenuated	Travellers
Typhoid	Killed/Mutant	Travellers
Cholera	Killed/Mutant	Travellers
Influenza	Killed	At-risk, elderly
Pneumococcal pneumonia	Polysaccharide	Elderly
Vasicella-Zoster	Attenuated	Leukaemic children

EXERCISE

1. What are vaccines?
2. Differentiate between active immunity and passive immunity. (BU April 2001, Oct. 2002)
3. Describe different types of vaccines. (BU Oct. 2001)
4. List disadvantages of whole-organisms vaccines.
5. Define vaccination.
6. What do you mean by DNA vaccines?
7. Give an account of types and properties of polio vaccine. (BU Oct. 2001)
8. Define immunity. Explain active and passive immunity. (BU Oct. 2001)
9. What is meant by SMAA complex?
10. What do you mean by multivalent subunit vaccine?
11. Write a note on pioneer work of Edward Jenner.
12. Define vaccination.
13. Explain work carried out by Louis Pasteur for advocating vaccination.

Part III
Plant Biotechnology

14

Plant Tissue Culture

14.1 INTRODUCTION

Since time immemorial man involved in the improvement of plants for his basic needs. Generally, crop improvement by conventional breeding methods are tedious and long time process. Now-a-days plant tissue culture procedures are used for fulfilling the objectives, which is not possible or impossible through the conventional methods. The purelines or inbreds production require six to seven generations of selfing. But two generations are sufficient for production of haploids through distant crosses or using pollen, anther or ovary culture, followed by chromosome doubling. This results in saving of four to six years. The production of plant derived valuable drugs such as vincristine, vinblastine, digoxin, taxol are produced at the concentration required for the growing market demand. The regular plant growth and their extraction from the plants not only takes long time but also is tedious one. The transfer of insecticidal gene, *cry* (crystal toxin protein) gene from *Bacillus thuringiensis,* into a plant cell and regeneration of whole plants containing and expressing this gene (transgenic plants) is method of improving the crop plant. Only by combination of tissue culture and genetic engineering it can be done. In conventional breeding methods it is impossible to produce such a recombinant plant having genes from prokaryotic microorganisms.

14.2 TERMS AND DEFINITIONS, USE OF GROWTH REGULATORS

The **tissue culture** is defined as *in vitro* culture of plant cells, tissues as well as organs on suitable medium. *In vitro* means in glass i.e. inside the glass vessel and *in vivo* means in life or living organisms i.e. inside the living organisms. The excised plant portion (bits of leaf, stem, root or any part of plant) (Fig. 14.1) used to start tissue culture is known as **explant.** Explant selection, the timing of excision and pretreatment are important factors which determines the success of culture. The process of dissection and removal to culture these tissues or organs is called explantation. The *in virto* culture of single or relatively small groups of plant cells is known as **cell culture** e.g. suspension cultures. Generally any plant cells or tissues cultured on suitable medium with proper environmental condition divides into group of unorganised mass of simple parenchymatous cells. This undifferentiated and unorganised mass of tissue is known as **callus.** Sometimes the term cell culture is also often used for callus cultures. The plant structures such as root tips, shoot tips, embryos etc. are cultured *in vitro* to obtain their development as organised structures, it is called **organ cultures.**

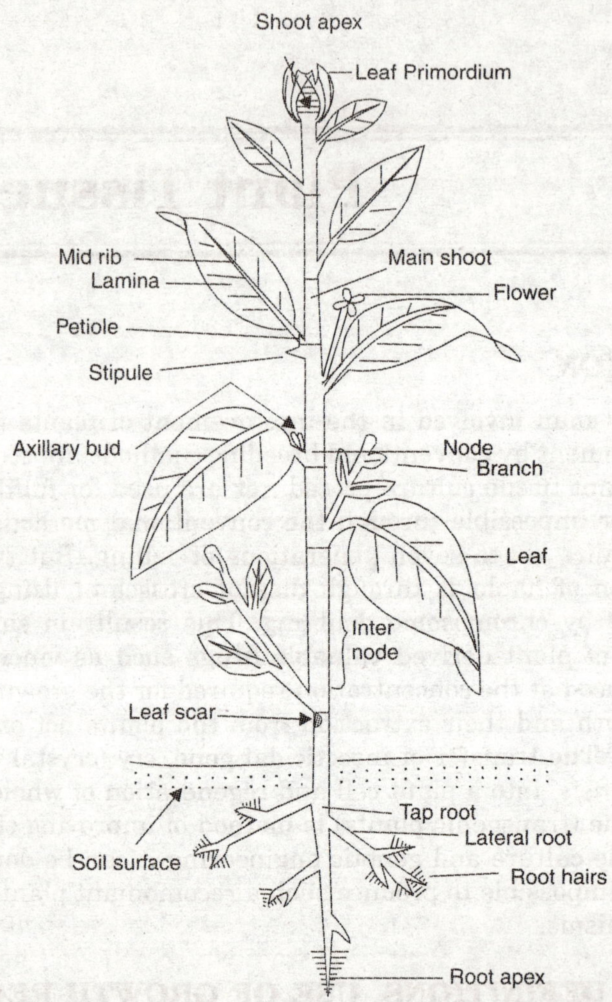

Fig. 14.1. Diagrammatic representation of the plant body showing different parts.

14.3 HISTORICAL DEVELOPMENT OF TISSUE CULTURE

The idea of growing isolated plant cells or tissue was started by Vochting (1878) and Rechinger (1893). **Totipotency** is the potential (inherent capacity) of a plant cell to develop into a complete plant. All cells of an organism having the ability to show the totipotency character. **Gottlieb Haberlandt** a German botanist in 1902, tried to culture isolated mesophyll cells from leaves of *Lamium purpureum* in Knop's salt solution enriched with asparagine, peptone and sucrose. The cells survived upto one month and increased in size, accumulated starch but failed to divide. He wanted to show the totipotency of the cells and not got the success, but able to predict the requirements for cell divisions *in vitro.* Thus, Haberlandt is considered **father of tissue culture.**

Robbins and Kotte in 1922 independently started working on root-tip culture and showed that meristematic tissues can be used for the success of culture system. Later White (1934) with tomato roots first showed indefinite culture of isolated roots. Efforts made to prove totipotency led to the development of techniques for cultivation of plant cells under defined conditions. This was made possible by the brilliant contributions from R.J. Gautheret in France and Phillip R. White in U.S.A. during the third and the fourth decades of 20th century. Most of the modern tissue culture media have been derived from the work of Skoog and coworkers during 1950s and 1960s.

Snow (1935) and Went and Thimann (1937) discovered hormonal characteristics of **auxin,** indole acetic acid which gave greater boost to maintain callus culture of plants indefinitely. First crude embryo culture was done by Hanning in 1940 and he cultured nearly mature embryos of certain Crucifers and grew them to maturity. The same method was utilised by Laibach (1925) to recover hybrid progeny from an interspecific cross in *Linum.* Steward's group at Cornell University (1948) made number of studies related to nutrition, technique of culture, quantitative analyses of culture growth and morphogenesis. They also made the contribution to identify the **coconut milk** as a major natural product which have the capacity to induce cell division in culture. Muir in 1953 clearly shown that plant cells can be grown in liquid (suspension) cultures similar to microorganisms.

Steward (1958) and Reinert (1959) discovered for the first time **somatic embryos production** (production of embryo-like structure from somatic cells) in carrot cultures independently. In 1960s Botany school at University of Delhi started working on *in vitro* culture of ovary, ovule and other reproductive organs of flowering plants. Kanta, Rangaswamy and Maheshwari (1962) first in India worked on *in vitro* pollination and fertilization of angiosperms.

First in 1964, Maheshwari and Guha by culturing anthers of *Datura* produced haploid plants from pollen grains (male gamete in plants). This made the beginning of **anther culture** or **pollen culture** for the production of haploid plants. This technique has been further developed by many workers including Nitch, J.P., Nitch, C. and coworkers. These workers showed that isolated microspores of tobacco produce complete plants.

Plant **protoplasts** are naked plant cells from which cell wall has been removed. In 1960, Cocking produced large quantities of protoplasts by using cell wall degrading enzymes. Protoplast production techniques have been considerably refined afterwards. It is now possible to regenerate whole plants from protoplasts and also to fuse protoplasts of different plant species. In 1972, Carlson *et al.,* produced for the first somatic hybrid plant by fusing the protoplasts of *Nicotiana gauca* and *N. infasoedii.* Since then many divergent somatic hybrids have been produced.

A successful establishment of callus cultures depended on the discovery during mid-thirties of IAA (indole-3-acetic acid), the endogenous auxin, and of the role of B vitamins in plant growth and in root cultures. First Skoog in 1944 reported continuously growing callus cultures were established from tobacco pith tissues cultured *in vitro.* This made possible by identifying the **kinetin** as major factor induced cell division. Later in 1957,

Skoog and Miller discovered root and shoot differentiation in culture system was regulated by auxin and cytokinin ratio. The first plant from a mature plant cell was regenerated by Brqun in 1959. Vasil and Hilderbrandt in 1965 clearly demonstrated a single isolated cell can divide and ultimately give rise to a whole plant (known as **regeneration**).

Ball in 1946 showed the greater potential of shoot apex for regeneration of plants. Later Morel and Wetmore regenerated whole plants from shoot apex measuring 100-250 mm length having two leaf primordia. This was successfully used for orchid propagation.

Within a brief period of time, the tissue culture techniques showed a tremendous progress. From the prime objective of proving the totipotency of differentiated plant cells, the technique now finds application in both basic and applied researches in a number of fields of enquiry.

14.4 TISSUE CULTURE METHODOLOGY (*In Vitro* Methods)

During culture of plant cells or organs mainly two basic problems have to be solved. First, to keep the plant cells and organs free from contaminating microbes (bacteria and fungi) and second, to ensure the desired development in the cells and organs by providing suitable concentration of nutrient and other environmental conditions. Contamination can be eliminated by using modern equipment and careful handling during various cultural operations. The second problem remains the area of active research and is likely to be so for quite some time in the future. It relies mainly on the manipulation of culture medium, especially growth regulators, and, to a much lesser extent, of other factors, including environmental conditions. The general technique of tissue, cell and organ culture is briefly outlined below in a rather simplified manner.

Laboratory Space

The tissue culture laboratory construction depends mainly on the nature and the type of activity. In general space for the following is very much needed: (a) washing, drying and storage of vessels, (b) preparation, sterilization and storage of media, (c) aseptic handling of explants and cultures, (d) maintenance of cultures and (e) observation of cultures. Now-a-days, provisions for washing and preparation can be made in a single room (media room), while the remaining work can be done in another room (culture room).

Tissue Culture Room

Tissue culture room should have the following facilities: (i) controlled temperature (usually at 25° ± 2° C with the help of air-conditioners and room heaters; higher or lower temperatures may be desirable in some cases), (ii) culture racks fitted with light (generally, 1000 lux or lower light regenerated by fluorescent tubes; automatic timers can be used to regulate photoperiod), and (iii) a shaker for agitation of liquid cultures. It is desirable to have a generator set for providing power to the culture room in the event of electricity cuts/failures which may often ruin the experiments.

Preparation of Culture Vessels

Glass culture vessels are used regularly as they are cheaper, reusable, clear and autoclavable. It is always better to use only borosilicate or pyrex glasswares as the ordinary soda glass may be toxic to some tissue types. Culture vessels made of clear plastic are available for a variety of purposes; these vessels are generally presterilized and disposable but certain types are autoclabale and, therefore, reusable. In general, plastic vessels are much costlier than glass vessels in the long run.

Plant tissues are generally cultured in culture tubes (rimless tubes of 25 × 150 mm or larger), flasks (long neck or even ordinary concial flasks) and Petri plates; but many specially designed dishes are also used. Wide mouth bottles are often employed especially for micropropagation work. Suspension cultures, of necessity, are maintained in long neck culture flasks. Culture tubes and flasks are usually stoppered with cotton plugs which are often wrapped in cheese cloth, but preparing such plugs on a large scale may be time consuming and inconvenient. Many workers use caps made of either polypropylene or a metal (aluminium/stainless steel) for this purpose; these caps have proved quite effective but their design may considerably affect their performance.

Culture vessels and other labwares are generally soaked in a suitable nontoxic detergent solution overnight, washed with a suitable brush, thoroughly rinsed clean with tap water followed by a rinse with distilled water and dried in a hot air oven (75-80° C). Culture vessels having contamination or some dried agar sticking to the wall are first autoclaved. Washed culture vessels should be stored preferable in a dust-proof cupboard or cabinet.

14.5 CULTURE MEDIUM

Cultured plant cells are similar to those required by whole plant. The nutrient mixture used in the medium depends on the type of material used for culture i.e. cells, tissues, organs and protoplast. Most of the nutrient medium consist of inorganic nutrients (macronutrients and micronutrients), carbon source(s), vitamins, plant growth regulators and other organic compounds (Table 14.1). If medium composed of chemically defined compounds, such medium is referred as a synthetic medium. If any medium consists of complex, chemically undefined compounds (fruit juices, plant extract, vegetable extract), such medium is known as natural medium.

Table 14.1: Nutritional composition of some plant tissue culture media commonly used in laboratory

Nutrients	White's	MS	B5	Nitsch's	N6	E1
	Amount (mg l^{-1})					
Macronutrients						
$MgSO_4 \cdot 7H_2O$	750	370	250	185	185	400
KH_2PO_4	–	170	–	68	400	250
$NaH_2PO_4 \cdot H_2O$	19	–	150	–	–	–
KNO_3	80	1900	2500	950	2830	2100
NH_4NO_3	–	1650	–	720	–	600
$CaCl_2 \cdot 2H_2O$	–	440	150	–	166	450
$(NH_4)_2 \cdot SO_4$	–	–	134	–	463	–
Micronutrients						
H_3BO_3	1.5	6.2	3	–	1.6	3

(Contd.)

Nutrients	White's	MS	B5	Nitsch's	N6	E1
	Amount (mg l^{-1})					
$MnSO_4 \cdot 4H_2O$	5	22.3	–	25	4.4	–
$MnSO_4 \cdot H_2O$	–	–	10	–	3.3	10
$ZnSO_4 \cdot 7H_2O$	3	8.6	2	10	1.5	2
$Na_2MoO_4 \cdot 2H_2$	–	0.25	0.25	0.25	–	0.25
$CuSO_4 \cdot 5H_2O$	0.01	0.025	0.025	0.025	–	0.025
$CoCl_2 \cdot 6H_2O$	–	0.025	0.025	0.025	–	0.025
KI	0.75	0.83	0.75	–	0.8	0.8
$FeSO_4 \cdot 7H_2O$	–	27.8	–	27.8	27.8	–
$Na_2EDTA \cdot 2H_2O$	–	37.3	–	37.3	37.3	–
EDTA Na ferric salt	–	–	43	–	–	43
Sucrose (g)	20	30	20	20	50	25
Organic supplements						
Vitamins						
Thiamine HCl	0.01	0.5	10	0.5	1	10
Pyridoxine HCl	0.01	0.5	1	0.5	0.5	1
Nicotinic acid	0.05	0.5	1	0.5	0.5	1
Myoinositol	–	100	100	100	–	250
Others						
Glycine	3	2	–	2	–	–
Folic acid	–	–	–	0.5	–	–
Biotin	–	–	–	0.05	–	–
pH	5.8	5.8	5.5	5.8	5.8	5.5

Agar 8 grams per liter
Growth regulators depending upon required concentration.

Large number of media composition are specified for a particular type of tissues and organs of plant. White's medium was designed for root culture. MS (Murashige and Skoog) medium and LS (Linsmaier and Skoog) medium was generally used for induction of organogenesis and regeneration of cultured plant tissues. Cell suspension and Callus cultures was carried out by B5 medium. Modified B5 medium known to be very useful for the protoplast culture and protoplast regeneration. The Nitsch's medium and N6 medium are much used for anther culture and other tissues. The MS medium is widely used in tissue culture work because of its success with large number of plant species and explant (Table 14.2).

14.6 PLANT GROWTH REGULATORS

Natural organic compounds produced by plants which influence (in low concentration) growth and development is known as **phytohormones.** Now synthetic compounds are also known to have such effects similar to natural ones. These are collectively called **plant growth regulators** (Table 14.3). There are four classes of growth regulators namely, auxin, cytokinins, gibberellins and abscisic acid. These compounds affect growth, differentiation and organogenesis of plant tissues in culture condition.

Table 14.2: Method for stock solution preparation of Murashige and Skoog (MS) medium in laboratory

Nutrient	Concentration in MS medium (mg/1)	Concentration in the stock solution (mg/1)	Volume to be taken/litre of final medium (ml)
Macronutrients (10 ×)			
NH_4NO_3	1650	16500	100
KNO_3	1900	19000	
$MgSO_4 \cdot 7H_2O$	370	3700	
$CaCl_2 \cdot 2H_2O$	440	4400	
KH_2PO_4	170	1700	
Micronutrients (100 ×)			
H_3BO_3	6.2	620	10
$MnSO_4 \cdot 4H_2O$	22.3	2230	
$ZnSO_4 \cdot 7H_2O$	8.6	860	
KI	0.83	83	
$Na_2MoO_4 \cdot 2H_2O$	0.25	25	
$CuSO_4 \cdot 5H_2O$	0.025	2.5	
$CoCl_2 \cdot 6H_2O$	0.025	2.5	
Iron source			
Fe EDTA-Na salt	40	Added fresh	
Vitamins			
Nicotinic acid	0.5	50 mg/100 ml	1
Thiamine HCl	0.1	50 mg/100 ml	0.2
Pyridoxine HCl	0.5	50 mg/100 ml	1
Others			
Myo-inositol	100	Added fresh	
Sucrose	30,000	Added fresh	
Agar	8000	Added fresh	
pH 5.8			

Auxins

The major function of auxin is induction of cell division and formation of callus in culture condition. It induces cell division, cell elongation and swelling of tissues and formation of adventitious roots in nature and *in vitro*. At low auxin concentrations, adventitious root formation occur, but at high concentrations root formation will not occur and callus formation takes place. The NAA (naphthalene acetic acid), IAA (Indole acetic acid), IBA (Indole butyric acid), 2,4,5-T (2,4,5-trichlorophenoxy acetic acid), 2-4-D (2,4-dichlorophenoxy acetic acid), *p*CPA (*p*-chlorophenoxy acetic acid), pictoram (4-amino-3,5,6-trichloro picolinic acid) are commonly utilized, but 2,4-D is most frequently used and highly effective.

Table 14.3: Characteristics of commonly used growth regulators in tissue culture

Growth regulator	Chemical formula	Molecular weight	Solubility
p-Chorophenoxy acetic acid (*p*CPA)	$C_8H_7O_3Cl$	186.6	Alcohol
2,4-Dichlorophenoxy acetic acid (2,4-D)	$C_8H_4O_3Cl_2$	221.0	Alcohol
Indole-3 acetic acid (IAA)	$C_{10}H_9NO_2$	175.2	1 N NaOH
Indole-3 butyric acid (IBA)	$C_{12}H_{13}NO_2$	203.2	1 N NaOH
α-Naphthalene acetic acid (NAA)	$C_{12}H_{10}O_2$	186.2	1 N NaOH
β-Naphthoxy acetic acid (NOA)	$C_{12}H_{10}O_3$	202.3	1 N NaOH
Adenine	$C_5H_5N_5 \cdot 3H_2O$	189.1	H_2O
Adenine sulphate (AS)	$(C_5H_5N_5) \cdot H_2SO_4 \cdot 2H_2O$	404.4	H_2O
Benzyl adenine (BA)	$C_{12}H_{11}N_5$	225.2	1 N NaOH
N-isopentenylamino purine (1PP)	$C_{10}H_{13}N_5$	203.3	1 N NaOH
Kinetin	$C_{10}H_9N_5O$	215.2	1 N NaOH
Zeatin	$C_{10}H_{13}N_5O$	219.2	1 N NaOH
Gibberellic acid (GA)	$C_{19}H_{22}O_6$	346.4	Alcohol
Abscisic acid (ABA)	$C_{15}H_{20}O_4$	264.3	1 N NaOH
Colchicine	$C_{22}H_{25}NO_6$	399.4	H_2O

Cytokinin

Cytokinin are adenine derivatives and helps in cell division, shoot induction and somatic embryo formation. The BAP (6-benzyl aminopurine), BA (6-benzyladenine), Kinetin (N-2-λ[furfurylamino]-1-H-purine-6-amine), Zeatin (6-[4-hydroxy-3-methyl-trans 2-butanyl amine and 2-iP (isopentyl adenine) are used frequently. Cytokinins with auxins promote cell division. The cytokinins at high concentration (1 to 10 mg/l) adventitious shoot formation is induced but root formation is inhibited. It also promotes axillary shoot formation by decreasing apical dominance. Cytokinins activates RNA synthesis and stimulates protein and enzyme activity in tissues.

Gibberellins and Abscisic acid

Gibberellins induce shoot elongation, enhance callus growth and growth of meristems or buds *in vitro*. GA_3 usually inhibit adventitious root and shoot formation.

Abscisic acid (ABA) either stimulates or inhibits callus growth depending upon species. It is also used in embryogenesis.

Sterilization Methods

All the materials, e.g., vessels, instruments, medium, plant material etc., used in culture work must be freed from contaminating microbes. This is achieved by one of the following approaches; (i) dry heat, (ii) incineration, (iii) autoclaving, (iv) filter sterilization, (v) wiping with 70% ethanol, and (vi) surface sterilization.

Dry Heat: The glasswares and Teflon plastic wares (empty vessels), and instruments may be sterilized by dry heat in an oven at 160-180° C for 3 hr. It is generally preferred to autoclave glass and plastic wares etc. and flame sterilize instruments like forceps etc.

Recently glass bead sterilizers (300° C) are being employed for the sterilization of forceps, scalpels etc.; these devices use dry heat.

Incineration: Instruments like forceps, scalpels, needles etc. are ordinarily flame sterilized by dipping them in 95% alcohol followed by flaming. This process of killing by burning is called incineration. It is customary to flame the mouths of culture vessels prior to inoculation/subculture.

Autoclaving (Moist heat): Culture vessels alone or with medium, small materials and equipment are generally sterilized by heating in an autoclave or a pressure cooker to 121°C at 15 p.s.i. (pounds per square inch; 1.06 kg/cm^2) for 15 (20-50 ml medium) to 40 (2 litre medium) minutes, the time being longer for larger medium volumes. Sterilization during autoclaving depends mainly on temperature. Certain types of plasticwares and some instruments, e.g., micropipettes etc. are also autoclavable. Care should be taken to properly stopper all the vessels and to open the autoclave only when its pressure gauge indicates zero pressure.

Sterilization by filtration: All heat liable compounds such as growth regulators, e.g., GA_3, zeatin, ABA (absicisic acid), urea, certain vitamins, and enzymes should not be subjected to higher temperatures. Such compounds are filter sterilized by passing their solution through a membrane filter of 0.45 μm or lower pore size. The membrane filter is held in a suitable assembly; the assembly together with the filter is sterilized by autoclaving before use. Filter sterilized heat labile compounds are added to autoclaved media after cooling. In case of agar medium they are added when the medium has cooled to about 40° C.

Laminar air-flow chambers are used to maintain aseptic condition in working space by blowing filter sterilized air through an enclosed (on all sides except one side) space. Air is first filtered through a coarse prefilter to remove large particles and then passed through a HEPA (high efficiency particulate air) filter which filters out all particles larger than 0.3 mm. This sterilized air blows through the cabinet at 1.8 km/hr which is sufficient to keep the enclosed working are in aspectic condition.

Wiping with 70% Ethanol: The surfaces that are not possible to be sterilized by any other methods e.g., platform of the laminar air-flow chamber, hands of the worker etc., are sterilized by wiping them thoroughly with 70% alcohol and then alcohol is allowed to evaporate.

Surface Sterilisation: All explants are treated with an appropriate sterilizing agent to kill contaminating microbes present on their surface; this is called surface sterilization. An explant is the excised piece of tissue or organ of plant used for culture. Surface sterilization procedure will depend mainly on the source and the type of tissue of the explant which will determine the contamination load and tolerance to the sterilizing agent.

The common sterilizing agents used for surface sterilization are calcium hypochlorite (9-10%), Sodium hypochlorite (2%), mercuric chloride (0.1-1%), silver nitrate (1%), bomine water (1-2%), H_2O_2 (10-12%) and antibiotics (4-5 mg/l) (Table 14.4). Among these calcium or sodium hypochlorite and $HgCl_2$ are the most commonly used with satisfactory results. The period of treatment varies from 5-30 min. As these agents are also toxic to plant tissues, the time and concentration used should be such that it causes minimum tissue damage. The rinsing of surface sterilized explant with pure sterile water after treatment should remove them as completely as possible. Otherwise they will not show any response in the culture condition.

Table 14.4: Disinfectant chemicals used for surface sterilization of explants

Disinfectant	Concentration	Exposure period
Benzalkonium chloride	0.01 - 0.1%	5-20 minutes
Bromine water	1 - 2%	2-10 minutes
Calcium hypochlorite	9 - 10%	3-30 minutes
Ethyl alcohol	75 - 95%	30-60 seconds
Hydrogenperoxide	3 - 12%	5-15 minutes
Mercuric chloride	0.1 - 1.0%	2-10 minutes
Silver nitrate	1%	5-30 minutes
Sodium hypochlorite	0.5 - 5%	5-30 minutes

In general procedure for disinfection of stem bit may be as follows: stem bits explants are selected for culture. Explants obtained from field should be first thoroughly washed under running tap water to remove the dust and epiphytotic microbes (microbes present on the plant surfaces) for 30 min. and may be first treated with 1% Cetavlon solution for upto 10-15 min to reduce bacterial contamination. These bits are treated with 0.2% $HgCl_2$ (acidified with a few drops of 1N HC1) for 10-15 min. This is followed by a rinse with 70% ethanol, and finally 4-6 rinses with sterilized distilled water (Fig. 14.2). The entire protocol is carried out in an aseptic area in a laminar air-flow cabinet. Most explant are large enough to be processed directly. But many explants, e.g., ovules, embryos, pollen grains etc., are rather small for efficient handling. In such cases, the organs containing them, e.g., ovaries for ovules, ovaries or developing seeds for embryos etc., are surface sterilized following which the desired explants are dissected out aseptically by using sterile materials.

14.7 TISSUE CULTURE RESEARCHES IN INDIA

The development of tissue culture methods throughout the world has initiated the interest in the Indian scientists. Professor P. Maheshwari of University of Delhi in 1950s has realized quite early the importance of *in vitro* culture technique and its application in the field of crop improvement. This made the beginning of tissue culture research in India. He considered complex tissues of plants such as ovary, ovule and control of fertilization for initiation of research in the field of tissue culture. This initial exploratory work on these organs formed the strong foundation in plant cell, tissue and organ culture in India.

This created lot of interest among the scientists in University of Delhi and other laboratories. Efforts were made to initiate the culture of pollinated ovaries of *Aerva tomentosa, Allium cepa, Althaea rosea, Anethum graveolens, Foeniculum vulgare, Hyoscyamus niger, Iberis amara, Linaria maroccana* and *Tropaeolum majus* between 1956 and 1966. Grate success with the culture of pollinated ovaries made it possible to culture ovaries and ovules at even earlier stage of growth. In case of *Zephyranthes* cultured ovaries from pollinated flowers formed fully developed seeds with completely differentiated embryos. Excised ovules containing zygote and primary endosperm nucleus also matured into normal-looking seeds but were without fully-differentiated embryos. Kappor (1959) noticed that these seeds were always smaller than the *in vivo* seeds. Whereas the *in vitro* formed fruits of *Anethum* were found to be larger than the naturally produced fruits. The occurrence of polyembryony due to cleavage and/or budding of zygotic embryos, on a culture medium was another important feature observed. This polembryony is important in the multiplication of plants which are known to form a limited number of seeds. Polembryony is of seeds. Polembryony is of most significant in umbellifers group because of low seed-set and germination.

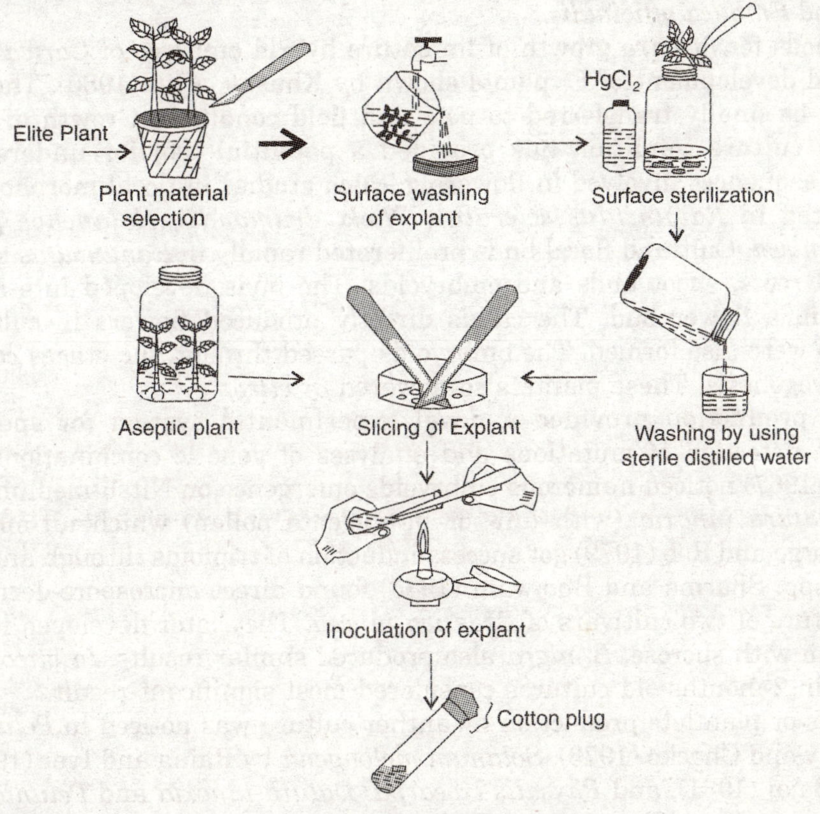

Fig. 14.2. Diagrammatic presentation of the steps followed for surface sterilization of plant material and inoculation of explant for culture initiation.

For the production of hybrid seeds, ovary and ovule culture is most suitable method. Joshi (1962) got success in growing *Gossypium hirsutum* ovules with 12-celled proembryo and 500 endosperm nuclei to maturity. The hybrid ovules of a cross between *G. arboreum* and *G. hirsutum* could be grown to maturity and viable seeds were obtained.

The adventive embryos from nucellus which is genetically uniform and reproduce the characters of maternal parent is very much important. Rangaswamy (1961) excised nucellus from micropylar-half of pollinated ovules of *Citrus microcarpa* (polyembryonic species) proliferated on a culture medium and the callus differentiated pseudobulbils which, eventually, developed into plantlets. Mitra and Chaturvedi (1972) reported embryogenesis from nucelli of unfertilized ovules of *Citrus sinensis* and *C. aurantifolia*. The embryos arose from nuclear cells or the callus. It was demonstrated that the pollination/fertilization stimulus is not essential.

Johri and Bhojwani (1965) shown organogenic potentiality of endosperm in *Exocarpus cupressiformis.* In *Scurrula* and *Taxillus,* shoot buds differentiated on a cytokinin medium, whereas auxin is also essential for *Dendrophthoe* and *Tolypanthus.* Continuously-growing tissue and organogenesis have also been observed in the mature endosperm of autotrophic taxa such as *Croton honplandianum, Jatropha panduraefolia, Putranjiva roxburghii, Codiaeum variegatum, Santalum album, Oryza sativa, Nigella damascena, Euphorbia geniculata* and *Emblica officinalis.*

The methods for *in vitro* growth of immature hybrid embryos of *Carica papaya* × *C. cauliflora* and development of F_1 plants shown by Khuspe *et al.* (1980). These seedlings obtained can be finally transferred to pots and field condition. Growth of young floral primordia in culture condition has provided a potential tool for understanding the morphogenic sequences involved in flowering. Such studies in floral morphogenesis have been conducted in *Ranunculus scleratus, Pholx drumondii, Kalanchoe pinnata* and *Browallia demissa.* Cultured floral buds proliferated rapidly in *Ranunculus* spp. and they differentiated roots, shoot buds and embryoids. The buds developed into shoots which finally ends in a flower bud. The callus directly produced flowers in culture. Rarely, mature seeds were also formed. The embryoids passed through the stages comparable to *in vivo* embryogenesis. These plants also flowered *in vitro.*

Haploids production provided a novel experimental system for speeding up of homozygosity, detection of mutations and analyses of genetic combinations. Guha and Maheshwari (1967) noticed numerous embryoids emergence on Nitsh-medium in cultured anthers of *Datura innoxia* (with uni- or bi-nucleate pollen) which ultimately formed plantlets. George and Rao (1979) got success induction of triploids through anther cultures of *Physalsis* spp. Sharma and Bhojwani (1985) found direct microspore-derived embryos in anther culture of two cultivars of *Brassica juncea.* They later developed into plantlets on B5 medium with sucrose. *B. nigra* also produced similar results. *In vitro* flowering of these plants in 2-months-old cultures considered most significant result.

Embryoids or plantlets production by anther culture was noticed in *Petunia axillaris* by Doreswamy and Chacko (1973), *Solanum melongena* by Raina and Iyer (1973), *Petunia obconica* by Bajaj (1981), and *Physalis ixocarpa, Datura innoxia* and *Petunia hybrida* by Chandra and coworkers (Chandra and Kothari, 1987). The anther culture technique leads to homozygous recombinants of *Hyoscyamus niger* with higher alkaloid content. Similar methods has great potential in the crop improvement. Haploids and used for one-step transfer of genotypes of cereals inbred lines and for the induction and characterization of mutants. For the isolation of protoplasts pollen grains are extensively used in number of plants.

14.8 OVARY CULTURE

Ovary is a female part of the flower which consists of either single or many ovules (egg containing part). La Rue (1942) made the first successful attempt to raise aseptic cultures of excised flowers and up to certain extent the limited growth of ovaries. Later Nistch in 1951 cultured pollinated ovaries of tomato on synthetic medium into mature fruits with viable seeds (Fig. 14.3). He also reported ovary culture of *Cucumis anguria, Nicotiana*

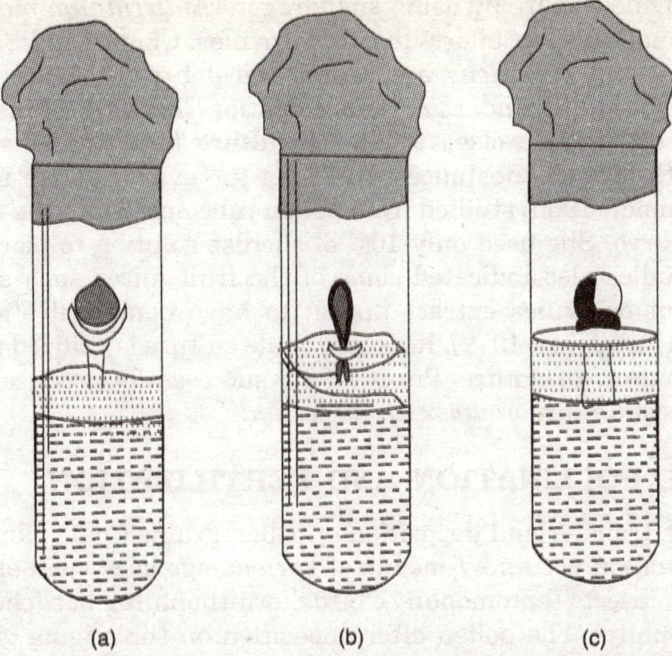

Fig. 14.3. Diagrammatic representation of development of ovary on synthetic medium a, b and c—three different stages of ovary culture; c—small fruit developed *in vitro* condition.

tabacum and *Phaseolus vulgaris* with reduced fruit size than fruits produced in nature. In India, similar efforts are made to culture ovaries of *Aerva tomentosa, Allium cepa, Althaea rosea, Anethum graveolens, Foeniculum vulgare, Hyoscyamus niger, Iberis amara, Linaria maroccana* and *Tropaeolum majus,* especially in Delhi University by using Nitsch's or White's medium supplemented with vitamins, hormones, amino acids and other growth inducing substances. Johri and Sehagal in 1966 by using Nitsch's medium amended with coconut milk produced larger size fruits in *in vitro* than naturally produced fruits.

From the studies it is discovered that auxins generally acts as pollination stimulus. Because production of seedless fruits is possible from the unpollinated ovaries in case of tomato (Nitsch, 1963). Such studies also indicated involvement of perianth or calyx (protective covering of flower) along with their protective role they also play a major role in the fruit development. Production of viable seeds from cultured ovaries is a big problem. In most of the ovary cultures, non-viable seeds are formed inside the fruits. However, in 1962 Sachar and Guha produced viable seeds in ovary culture of *Ranunculus sceleratus.*

Ovary culture is quite important method for horticulturists to obtain complete knowledge of fruit physiology for improving the quality of fruits. It is successfully used for the production of haploids and rare hybrids. Mess and Wagner (1976) produced parthenogenetic haploids of *Mimitus lutens* by pollinating their exposed ovules in ovary culture. It is important because haploid in this plant could not be produced by any other method. To study chemical effects on fruit development ovary culture can be used.

14.9 OVULE CULTURE

Ovules are zygote containing part of ovary, which later develops into seeds. White in

1932 first started ovule culture by using snapdragon (*Antirrhinum majus*). Later in 1958 Prof. Maheshwari made perfect efforts to culture ovules. Chopra and Sabharwal (1963) by using *Gynandropsis* and *Impatiens* and Guha and Johri (1966) by using *Allium cepa* successfully cultured ovules under *in vitro* condition. In 1961 Maheshwari and group produced viable seeds from excised ovules in culture for *Papaver somnifera* by using Nistch medium with growth substances. By using the ovary culture methods in orchids, Indian scientist Arnoldi (1960) studied from pollen tube entry into the ovule and up to the development of embryo. She used only 10% of sucrose solution to trace the details.

This type of studies also indicated some of the fruit juices such as cucumber, water melon, parent plant or ovules extract known to have beneficial effects on the ovules development (Joshi and Johri, 1972). Recently, ovule culture is utilized for the interspecific hybrids that fail to grow in nature. Punir (1967) successfully produced hybrids by using cross between *Gossypium arboreum* × *G. hirsutum*.

14.10 *IN VITRO* POLLINATION AND FERTILIZATION

Pollination means transfer and deposition of pollen grains (male gamete) on the stigma of the flower. Pollination occurs by means of various agencies such as air (anemophily), water (hydrophily), insects (entomophily), birds (ornithophily), bat (cheiropterophily) and snail and slug (zoophily). The pollen after deposition on the stigma grows and produces the pollen tube that contain male gametes. Pollen tube grows down the stigma through

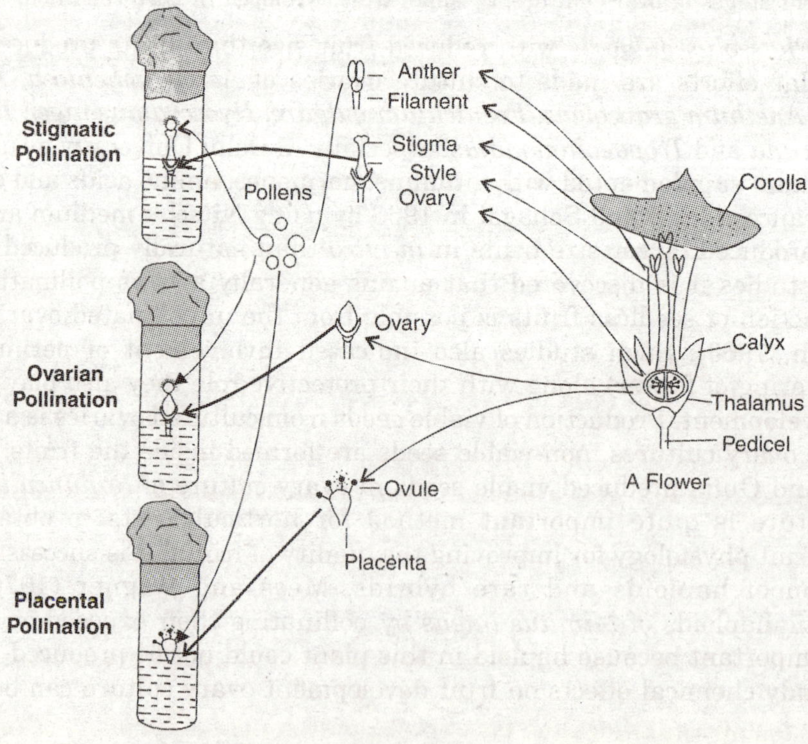

Fig. 14.4. Diagrammatic representation of various types of *in vitro* pollination.

style, ovule and last fuse with the female gamete. This fusion of male and female gametes is known as **fertilization**. These processes occurs in natural condition between the compatible species. Thus *in vitro* pollination and fertilization offer an opportunity for production of hybrid embryos among plants that cannot cross by conventional plant breeding methods or between incompatible species. This is due to barriers which blocks the growth of the pollen tube on the stigma or style. In such cases the pollen grains are deposited on stigma (stigmatic pollination) (Fig. 14.4), or applied on excised ovaries which are not having stigma or style (ovarian pollination), or applied on ovaries which are attached with placenta (placental pollination). These techniques are developed at University of Delhi to produce hybrids among species of the Solanaceae and Papaveraceae by Maheshwari and coworkers. Now *in vitro* pollination has been done in *Nicotiana tabacum, N. alata, Petunia hybrida* (Solanaceae), *Papaver somnifera, Eschscholtzia california, Argemone mexicana* (Papaveraceae) and *Melandrium album, M. rubrum, Agrostemma githago, Dianthus caryophyllus* (Caryophyllaceae).

All the materials such as pollen grains, ovary, ovule or entire pistil (ovary containing stigma and style) are surface sterilized and used with the suitable medium in the aseptic condition. Before collection of all these materials proper care has to be taken in selection of suitable materials for the *in vitro* culture techniques.

SUMMARY

Traditionally man was using breeding methods for crop improvement. Now a days plant tissue culture is being employed for crop improvement. This is because of the several advantages of plant tissue culture over conventional breeding methods.

Culture of plant cells, tissue as well as organs on suitable medium *in vitro* is called plant tissue culture. The plant part which is used for the regeneration of plant through tissue culture is called explant. The removal of explant is called explantation. Cell culture can be defined as in vitro culture of single or few plant cells. Culturing of plant cells or tissues on suitable medium under suitable environmental conditions gives simple, undifferentiated and unorganized mass of cells called callus. Organ culture can be defined as *in vitro* culturing of plant organs such as root tips, shoot tips, embryos etc. The potential of a plant cell to develop into a complete plant is called totipotency.

Plant tissue culture faces two major problems. First, to keep out contamination and second, to ensure the proper development of cells and organs by providing suitable medium and environmental conditions. Aseptic conditions can be maintained by using air flow chamber and careful handling of culture vessels. Either natural media or synthetic media can be used as plant tissue culture media. Some of the media used in plant tissue culture includes white, medium, MS medium and LS medium.

The organic compounds produced by plants or synthetic compounds affects the growth of plant are called plant growth regulators. These include auxin, cytokinin, gibberellins and abscisic acid. Auxin induces cell division and formation of callus in culture condition. Cytokinin induces cell division, shoot and somatic embryo formation. Gibberelllins induce shoot elongation; enhance callus growth and growth of meristem in vitro. Abscisic acid either stimulates or inhibits callus growth depending upon species.

The materials such as vessels, instruments, plant material etc., must be free from microbes. This can be achieved by sterilization process. Vessels can be sterilized by dry heat or autoclaving. Instruments can be sterilized by flaming. Media is sterilized by autoclaving, platform of the laminar air-flow chamber; hands of the worker are sterilized by using 70% alcohol and plant material by surface sterilization using 0.1% mercuric chloride or 10-12% H_2O_2 or any other surface sterilizers.

Ovum containing female plant organ is called Ovary. This can be used as explant, and such type of culture is called ovary culture. In ovule culture, ovule tissue used as explant. Now a days it is possible to produce hybrid embryos between sexually incompatible plants by *in vitro* pollination and fertilization.

EXERCISE

1. Define callus.
2. What is meant by tissue culture?
3. Give the meaning of totipotency and protoplast.
4. How tissue culture work started in India?
5. Explain ovary culture.
6. Write a short note on maintenance of aseptic condition. (BU April 2001)
7. Describe *in vitro* pollination and fertilization. Add a note on its significance. (BU April 2001)
8. Give an account of ovule culture. (BU April 2001)
9. List the contribution made by the different researchers towards development of tissue culture in India.
10. Explain preparation of explant.
11. In brief write the methods employed for sterilization purpose in tissue culture.
12. Write a note on components of the tissue culture medium.
13. Differentiate between phytohormones and plant growth regulators.
14. Mention effects of plant growth regulators in tissue culture.
15. Define incineration.
16. What do you mean by regeneration?
17. Give an account of applications of tissue culture in forestry and agriculture. (BU April 2003)
18. Write short notes on role of auxins in organogenesis. (BU April 2003)
19. Differentiate between cell culture and tissue culture.
20. What is meant by explant?

15

Callus Cultures, Suspension Cultures and Micropropagation

15.1 INTRODUCTION

The cultured explants on medium with suitable growth regulators combination undergo division and forms many cells. In culture condition certain mature and differentiated cells (parenchyma and often colenchyma) undergo many physiological changes to form meristematic (ability to divide) cells and is known as **dedifferentiation.** During dedifferentiation renewed and enhanced RNA and protein synthesis along with many chemical changes leading to the formation of new cellular components needed for meristematic activity. In the beginning cell divisions are confined to the cut ends, but later it covers the entire explant. The resulting cell mass is ordinarily unorganized, but it often consists of several cell types including fibers, and vascular elements. The tissues and cells cultured on an agar medium form an undifferentiated and unorganized mass of cells called **callus** (Fig. 15.1). For every 3-5 weeks callus cultures are subcultured because of cell growth, nutrient deficiency and medium drying. Therefore, calluses are easy to maintain and are the most widely used.

The tissues and cells cultured in a liquid medium produce a suspension of single cells and cells clumps of a few cells these are known as **suspension cultures.** To increase aeration and separation of cell clumps into individual cells, the liquid cultures are continuously agitated by gyratory shaker at 100-250 revolution per minute (rpm). Suspension cultures show faster growth than the callus cultures. Thus subculture should be done for very week.

15.2 CALLUS CULTURE

The callus tissue is basically a more or less organised mass of cells that usually arises on wounds of differentiated tissues and organs of the plant. Hence, in principle, **callus** can be defined as an unorganised and less differentiated parechymatous tissue. Generally callus culture is not homogeneous mass of cells, because it is made up of both differentiated (matured) and non-differentiated (immatured) tissues. The differentiated tissues of plant

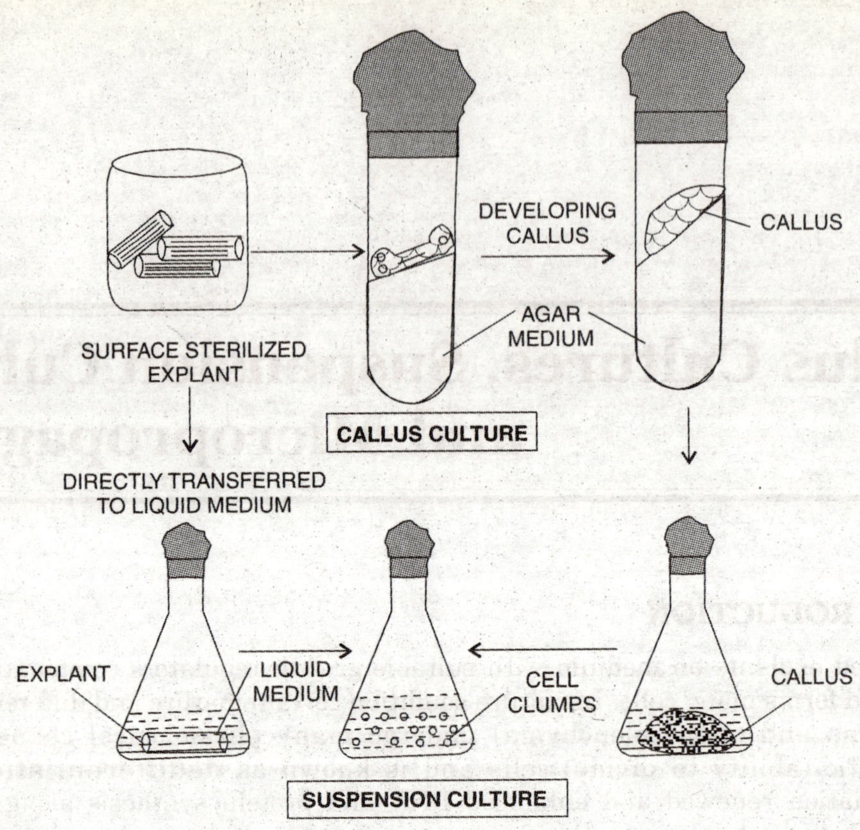

Fig. 15.1. Diagrammatic representation of callus and suspension cultures initiation.

(roots, stem, leaves, flowers, anthers etc.) can be used as a starting material (explant) for callus induction. These explant tissues generally show distinct planes of mitotic cell division, cell proliferation and organisation into specialised structures such as vascular systems. If only differentiated cells are present in an isolated explant, then **dedifferentiation** (i.e. cells undergo changes to become meristematic) must take place before cell division can occur. The mature parenchyma cells (and also collenchyma cells) present in explant usually undergo this process of dedifferentiation. If the explant already contains meristematic tissue when isolated, then this can divide immediately without dedifferentiation process. During dedifferentiation process, mature adult cells are temporarily able to revert from adult to the juvenile state. The rejuvenated cells have a greater growth and division potentiality. These cells under special circumstances are able to regenerate organs and/or embryos.

Exogenously supplied growth regulators present in the nutrient medium influence the callus formation from the explant. The type of growth regulator requirement and its concentration in the medium depends strongly on the genotype and endogenous hormone content of an explant. These requirements can be put into three categories:

1. Auxin alone (especially in monocotyledons)
2. Cytokinin alone

3. Both auxin and cytokinin

In cases where the callus is difficult to induce, or if young callus is needed, then immature embryos or seedlings or parts of these are used. It should be taken into account that the type of starting material (juvenile or adult) and the original position of the explant in the plant reflects the endogenous hormone level which have an important influence on processes such as cell division, organ and embryo formation.

The genotype, composition of the nutrient medium, physical growth factors (light, temperature, etc.) are also important for callus formation. To induce callus the Murashige and Skoog (1962) mineral medium, or modifications of this are used. Sucrose or glucose (2-4%) is usually employed as the sugar source. The effect of light on callus formation is dependent on the plant species; light may be required in some cases and darkness in other cases. The temperature of 22-28°C is normally employed for callus formation.

The different plant species may form different types of callus tissues that shows varied structure and growth habit; white or coloured, soft (watery) or hard, friable (easy to separate into cells) or compact. The callus growth within a plant species may also depend on factors such as the original position of the explant within the plant, age of the plant, physiology of the plant and the growth conditions.

Once the callus tissue formed from the explant, the callus is further transferred on to a new medium. This process of transfer is referred to as **subculturing**. When subcultured regularly on a agar medium, callus cultures will exhibit a S-shaped or sigmoid pattern of growth during each passage. There are five phases of callus growth:

1. Lag phase, where cells prepare to divide.
2. Exponential phase, where the rate of cell division is highest.
3. Linear phase, where cell division slows down but the rate of cells expansion continues to increase.
4. Deceleration phase, where the rate of cell division and elongation continues to decrease.
5. Stationary phase, where the number and size of cells remain constant.

Applications of callus cultures

Callus cultures provides an information on the regeneration potentiality of the plant.

It acts as a good source material for the protoplast isolation.

Callus can be used as inoculum for the initiation of the suspension cultures.

The callus can serve as a better source of cells for the genetic manipulation of the plant genomes.

Callus cultures can be maintained indefinitely in the *in vitro* condition for longer period of time.

15.3 SUSPENSION CULTURE

The individual cell or cell aggregates dispersed and growing in moving liquid media is known as **suspension culture**. In regular practice suspension culture is initiated by transferring pieces of undifferentiated and friable calli to a liquid medium. Which is continuously agitated by using rotary shaker. Suspension cultures have also been started from sterile seedlings or imbibed embryos or leaves by mechanical method. Leaves or the other tissue can be gently grinded or soft tissues can be broken up in a hand operated glass homogenizer. This homogenate, containing intact living cells, dead cells and cell debris is cleared by filtration and centrifugation, which is then transferred to liquid (broth) medium. In an enzymatic method for isolation of single cells pectinases which digest the pectin cell wall was employed. Generally the single cell cultures used in basic and applied research is isolated from cultured tissues.

At first stage cultures are initiated by placing freshly cut sections of plant organs (root, stem, leaves) on a solidified nutrient medium containing suitable growth regulators. In this condition the explant are transferred to a fresh medium to build up reasonable amount of callus tissue. This callus is transferred to a liquid medium and agitated to raise fine suspension of cells. Successful establishment of the suspension culture depends upon the initial callus.

A wide variety of explants and media compositions have been used with success for callus induction viz. MS, B_5, LS (Linsmaier and Skoog) and Blaydes medium etc. For these media vitamins, inositol, sucrose and growth regulators, especially auxin (2, 4-D at~1-5μM) are incorporated for induction of cell division. The inclusion of 1μM kinetin is also found beneficial for the induction. The growth rate and friability of callus produced can vary widely due to explants. The culture medium for suspension culture should be such that it maintains good growth of callus for the longer period.

In the suspension culture system orbital shakers are widely employed for the initiation and serial propagation of plant cells. They should have a variable speed control (30-150 rpm) and the stroke should be in the range of 4-8 cm orbital motions. The shaking of medium on a shaker serves two purposes. First it exerts a mild pressure on cell aggregates, breaking them into smaller clumps and single cells. Secondly, agitation maintains uniform distribution of cell and cell clumps in the medium. Movement of the medium also provides good gaseous exchange between the culture medium and air. The shaker should be kept in the air-conditioned room with good thermostat.

The Erlenmeyer flasks with wide-mouth are commonly used as culture vessels. The volume of the culture medium should be appropriate to the size of the culture vessel e.g. 20 ml per 100 ml flask and 70 ml per 250 ml flask. The flasks are normally sealed with aluminium foil. Flask closure must maintain sterility, allow gas exchange and reduce evaporation. Cotton wool plugs may be used for sealing flasks during autoclaving but not for culturing cells. They are a common source of contamination in flasks that are sitting on a shaker for several weeks.

Suspension cultures are used for production of secondary metabolites from different plant cell culture (Table 15.1).

Table 15.1. Different secondary metabolites produced by plant species used for increased production through tissue culture techniques with their bioactivity.

Compound	Plant Species	Bioactivity
Pyrethrins	*Chrysanthemum cincerariefolium*	Insecticidal
	Tagetus erecta	Insecticidal

(Contd.)

Compound	Plant Species	Bioactivity
Nicotine	*Nicotiana tabacum*	Insecticidal
	Nicotiana rustica	Insecticidal
Rotenoids	*Derris elliptica*	Insecticidal
	Lonchocarpus utilis	Insecticidal
	Tephrosia vogaeli	Insecticidal
	Tephrosia purpurea	Insecticidal
Azadirachtin	*Azadirachita indica*	Insecticidal
Phytoecdysones	*Trianthema portulacastrum*	Insecticidal
Baccharine	*Baccharis megapotamica*	Antineoplastic
Bruceantine	*Brucea antidysenterica*	Antineoplastic
Cesaline	*Caesalpinia gillisessi*	Antineoplastic
3-Deoxycolchicine	*Colchicum speciosum*	Antineoplastic
Ellipticine, 9-methoxyellipticine	*Ochrosia moorei*	Antineoplastic
Fararonine	*Fagara zanthoxyloides*	Antineoplastic
Harringtonine	*Cephalotaxus harringtonia*	Antineoplastic
Indicine *N*-oxide	*Heiotropium indicum*	Antineoplastic
Maytansine	*Maylenus bucchananii*	Antineoplastic
Podophyllotoxin	*Podophyllum peltatum*	Antineoplastic
Taxol	*Taxus brevifolia*	Antineoplastic
Thalicarpine	*Thalictrum dasycarpum*	Antineoplastic
Tripdiolide, triotolide	*Tripterygium wilfordii*	Antineoplastic
Vinblastine, ajmalicine, vincristine	*Catharanthus roseus*	Antineoplastic
Quinine	*Cinchona officinalis*	Antimalarial
Digoxin, reserpine	*Digitalis lanata*	Cardiac tonic
Diosgenin	*Dioscorea deltoidea*	Antifertility
Morphine	*Papaver somniferum*	Analgesic
Thebaine	*Papaver bracteatum*	Source of codeine
Scopolamine	*Datura stramonium*	Antihypertension
Atropine	*Atropa belladonna*	Muscle relaxant
Codeine	*Papaver* spp.	Analgesic
Shikonine	*Lithospermum erythrorhizon*	Dye, pharmaceutical
Anthraquinones	*Morinda citrifolia*	Dye, laxative
Rosamarinic acid	*Coleus blumei*	Spice, antioxidant
Jasmine	*Jasmium* spp.	Perfume
Stevioside	*Stevia rebaudiana*	Sweetner
Crocin, picrocrocin	*Crocus sativus*	Saffron dye
Capsaicin	*Capsicum frutesces*	Chilli dye
Vanillin	*Vanilla* spp.	Vanilla flavour

TYPES OF SUSPENSION CULTURE

Batch Culture

The cell suspension culture grown in a fixed volume of nutrient culture medium is known as **batch culture**. This is also known as **closed system** because they are incubated in a closed culture vessel within a single batch of medium. During incubation cell suspension increases in biomass by cell division and cell growth, this continues until a factor in the culture environment (nutrient or oxygen availability) becomes limiting and then growth ceases. The cells in culture exhibit the following five phases of a growth cycle and are represented in the Fig. 15.2.

(i) Lag phase, where cells prepare to divide by synthesising various materials required for cell division.

(ii) Exponential (log) phase, where the rate of cell division is maximum and increases in log scale.

(iii) Linear phase, where cell division slows but the rate of cells expansion increases.

(iv) Deceleration phase, where the rates of cell division and cell elongation starts decreasing.

(v) Stationary phase, where the number and size of cells remain constant by stopping cell growth.

During the subculturing of cells into fresh medium there is a lag phase in the cells. It is the initial period of a batch culture when no cell division is apparent. It may also be used with reference to the synthesis of a specific metabolite or the rate of a physiological activity. Then follows a period of cell division (exponential phase). It is a finite period of time early in a batch culture during which the rate of increase of biomass per unit of biomass concentration (specific growth rate) is constant and measurable. Biomass is usually

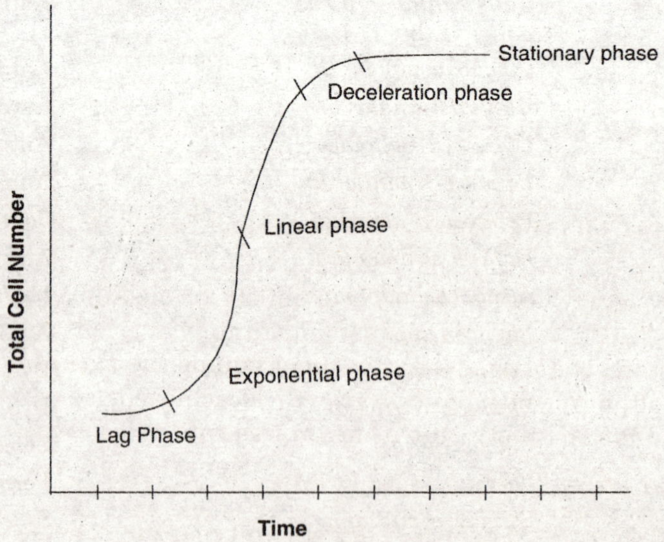

Fig. 15.2. Total number of cells per unit volume of culture to time in a bath suspension culture.

referred in terms of the number of cells per ml of culture. After 3 to 4 cell generations the growth declines. Finally, the cell population reaches a stationary phase during which cell dry weight is also lowered. It is the terminal phase of batch culture growth cycle where no net synthesis of biomass or increase in cell number is apparent. Cultures can be maintained continuously in exponential phase by frequent (every 2-3 days) subculture of the suspensions. Cells may remain viable for many days by utilising intracellular reserves or recycling metabolites released by lysed cells. The stability of cells in stationary phase depends upon the species and the nature of the growth limiting factor (cells brought to stationary phase by nitrogen limitation retain viability longer than when carbohydrate starved).

When such a stationary phase cell suspension is subcultured, the cells in succession pass through lag phase, a short period of exponential growth, a period of declining relative growth rate and then again enter stationary phase. Prior to the last phase, the cells expand in size. Traditionally such cultures at each subculture initiate from a relatively high cell density and the cells therefore accomplish only a limited number of divisions (cell number doublings) before entering stationary phase again. For example, a suspension culture of *Acer pseudoplatanus* initiated at 2×10^5 cells/ml will reach a final cell density of ca. 3×10^6 cells/ml corresponding to four successive doublings of the initial population. The cell generation time (doubling time) in suspension cultures varies from 24 to 48 hours in well established cell cultures. **Doubling time** (t_d) is the time required for the concentration of biomass of a population of cells to reach double.

For general purposes, the objective with cell suspension cultures is to achieve rapid growth rates, and uniform cells, with all cells being viable. Frequent subculturing in a suitable medium ensures these qualities. Cells should be subcultured at weekly intervals or less if they are to be used for experimental purposes. The size of the inoculum (dilution rate) will determine the lag period. Optimally suspension cultures passage length is 1-2 weeks, but the exact time and the dilution required must be determined for each cell line. Dilutions of 1:4 after 1 week or 1:10 after 2 weeks are commonly used. There is generally a minimum inoculum size, below that a culture will not recover.

Continuous Culture

In this method culture is continuously supplied with nutrients by the addition of fresh medium but the culture volume normally remains constant. Here cells are continuously in the exponential growth phase and are also known as **open system** of culture. A **continuous culture** is one in which inflow of fresh medium is balanced by outflow of corresponding volumes of culture including harvest of cells. Cells are constantly washed out with the out-flowing liquid. The rate of inflow of medium and culture harvest are adjusted so that the cultures are maintained indefinitely at a constant sub-maximal growth rate. In a steady state, the rate of cells washout equals the rate of formation of new cells in the system. A situation of balanced growth is achieved, i.e. majority of cells in the culture are in a similar metabolic state. The growth rate and cell density are held constant by a fixed rate of input of growth limiting nutrients and removal of cells and spent medium.

The continuous culture system may be of two types viz., chemostat and turbidostat.

Chemostat: In this system growth rate and cell density are held constant by a fixed rate of input of a growth limiting nutrient medium (nitrogen, phosphorus or glucose). In such a medium, all the constituents other than growth limiting nutrients are present at concentration higher than that required to maintain the desired rate of cell growth. The growth limiting substance is so adjusted that it increase or decrease in the growth rate of cells.

Turbidostat: In this fresh medium flows in response to increase in the turbidity so as to maintain the culture at a fixed optical density of suspension. A preselected biomass density is maintained by the washout of cells.

Applications of Suspension Cultures

- Suspension cultures are ideal for the production of secondary metabolites productions which are of therapeutic value.
- It also serves as a major source for obtaining individual isolated cells used for protoplast production.
- It also gives isolated cells which can be used for hybrids or cybrids production.
- Suspension cultures provide cells that are used for the immobilisation of plant cells.
- Suspension cultures are also used for the induction of mutation and genetic manipulation of plant cells.
- Cells obtained from suspension cultures are easily preserved by using cryopreservation technique than other tissues or organs of the plant.

15.4 MICROPROPAGATION (Clonal Multiplication of Elite Species)

Generally plants are propagated by sexual and asexual methods. In sexual methods seeds are produced by fertilization of egg by pollen grains and shows higher degree of heterogenicity. In asexual methods (vegetative propagation) new plants are obtained by mitotic division of earlier plant parts and they are genetical identical to parent. Multiplication of genetically identical copies of plant is known as **clonal propagation.** Individual plants obtained by this asexual methods is known as **clone**. Usually vegetative propagation is done by cuttings, budding and grafting. This is known as *in vivo* clonal propagation of plants and is often difficult, expensive and even unsuccessful in majority of time. Tissue culture is a better option for this purpose. Clonal propagation of plants through tissue culture is called **micropropagation.** This can be done in a short time and space. George Morel in 1960 used micropropagation methods for orchid plants production at commercial level.

In micropropagation shoot-tip, meristem and auxillary buds are the general explants used for the initiation of the culture. Shoot-tip and meristem consists of a group of actively dividing meristematic cells at the shoot apical portion (Fig. 15.3). Cutter in 1965 clearly defined shoot-tip and meristem. **Meristem** or **apical meristem** is the terminal portion of shoot contain active dividing meristematic cell mass and measures about 100-250 μm in length.

Shoot-tip or **shoot apex** is the terminal meristematic portion along with the one to three leaf primordia and measures about 100-500 μm in length. For *in vitro* clonal propagation shoot apex is enough as explant, whereas in case of meristem culture for the production of disease free plants clear excision of only meristem without surrounding tissue is very much necessary. **Auxillary buds** are the meristematic cells present in the axile portion of the node. These are either in dormant or active state depending upon the physiological state of the plant. Auxillary bud method was first used for carnation, but now this method is rarely used for carnation. This method is widely employed for the large scale micropropagation of strawberry and gurbera.

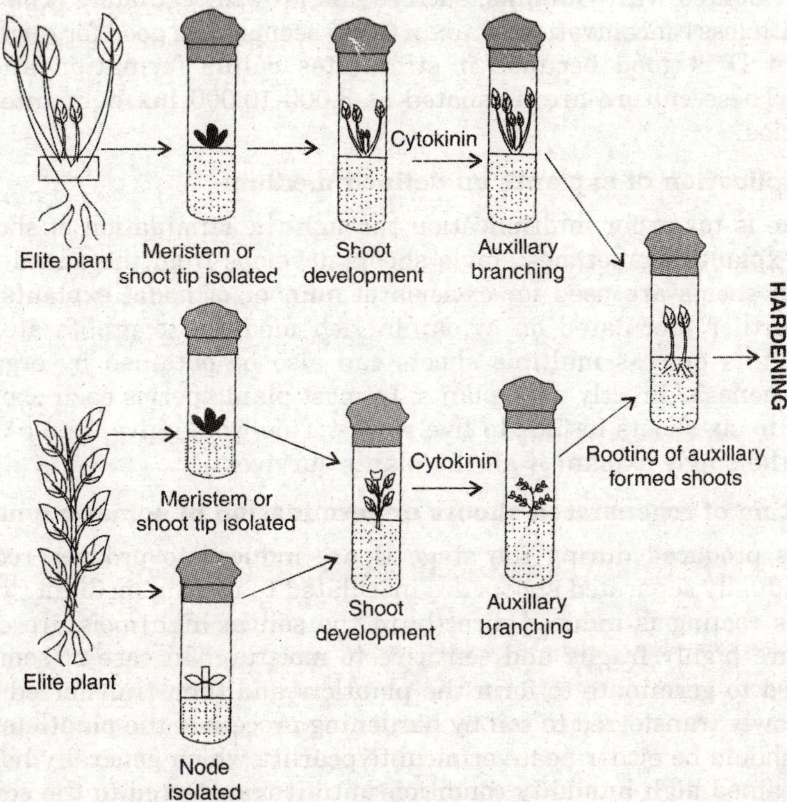

Fig. 15.3. Diagrammatic representation of *in vitro* clonal propagation (micropropagation) of plants by the auxillary bud method or shoot tip method. Top row: as applied to rosette plants; bottom row: applied to plants which show lengthwise growth.

Micropropagation or *in vitro* clonal propagation is a highly complicated process consisting of different steps or stages. Murasighe in 1978 recognized four stages for overall micropropagation of plants. Wherein stage I to III are followed under *in vitro* condition and stage IV is carried out in the green house condition. In 1981 Debergh and Maene introduced stage O for micro-propagation systems.

Stage 0: Selection of elite/stock plants

This is the first step of micropropagation, in this elite/stock plants (plants having a desired characters for the large scale multiplication) are selected and maintained in controlled environmental conditions for at least three months to initiate the culture. The elite plants are grown under low humidity, irrigation and without systematic microbial infections.

Stage I: Establishment of aseptic culture

In this stage selected explants are prepared for the inoculation and established on a suitable culture medium. The selected shoot-tip, meristem and auxillary bud explants are surface-sterilized by using the chemicals such as 0.1% $HgCl_2$ or 5% sodium hypochlorite or 70% alcohol either in combination or alone depending upon the surface contamination. These surface-sterilized explants are inoculated on to MS medium (or other suitable basal medium) supplemented with vitamins, sucrose and growth regulators. The cytokinins (1-3 mg/1 BAP) with less concentration of auxin (IAA) seems to be good for micropropagation. The auxin 2, 4 D is good because it stimulates callus formation and suppresses organogenesis. These culture are incubated at 3,000-10,000 lux light intensity with 16 hours photoperiod.

Stage II: Multiplication of explants on defined medium

Majority of time is taken for multiplication through the stimulation of shoots regeneration from the explants. Sometimes single shoots develops from the apical shoots and in such cases such shoots are used for excision of number of nodal explants. These nodal explants are further inoculated on cytokinin rich medium to proliferate the multiple shoots. During this process multiple shoots can also be obtained by organogenesis or somatic embryogenesis directly on explants. In most plant species each explant is known to produce five to six shoots in four to five weeks. This would give upto 5^{10}-6^{12} plants in one year from the single explant if all the plants survives.

Stage III: Rooting of regenarated shoots or germination of somatic embryo

Multiple shoots produced during the stage II are induced to produce rooting in fresh medium. Individually separated shoots are inoculated to rooting medium (with auxins) or in some species rooting is induced directly in the soil in high moisture condition. The rooted plants are highly fragile and sensitive to moisture. In case of somatic embryos, they are allowed to germinate to form the plantlets and then transferred to soil. These plantlets are slowly transferred to soil by hardening process of the plantlets. The medium for hardening should be either peat/vermiculite/pearlite which generally hold more moisture and maintained high humidity condition until it transferred to the soil.

Stage IV: Hardening

These plantlets are first prepared for soil condition from its *in vitro* conditions. This makes plants to become resistance to moisture stress and disease, making plantlets completely autotrophic from the heterotrophic nature in cultural conditions. The plantlets must be protected from the direct sunlight and the relative humidity should be gradually decreased over a period of time. During this acclimatization period plantlets develops well developed roots and also forms cuticular wax in the aerial tissues. Thus becomes suitable for complete transfer into field.

Some species or shoots formed *in vitro* condition appear glassy, brittle and water soaked and this is known as **vitrification**. Vitrification is due to poorly developed vascular bundles, abnormal functioning of stomata and abnormal wax quality and thus results in loss of plantlets. This can be overcome by addition of high concentration of agar (1%), use of growth retardants, bottom cooling of culture tubes and addition of hydrolysate compounds.

Meristem culture

Cultivation of shoot apical meristem *in vitro* is known as **meristem culture.** Meristem culture was developed by Morel and Martin in 1952 for elimination of rivers from *Dahlia*. Morel in 1965 employed meristem culture for micropropagation of orchid *Cymbidium*. Meristem culture involves the growth of an already existing shoot meristem and subsequent regeneration of adventitious roots from these shoots (Fig. 15.4).

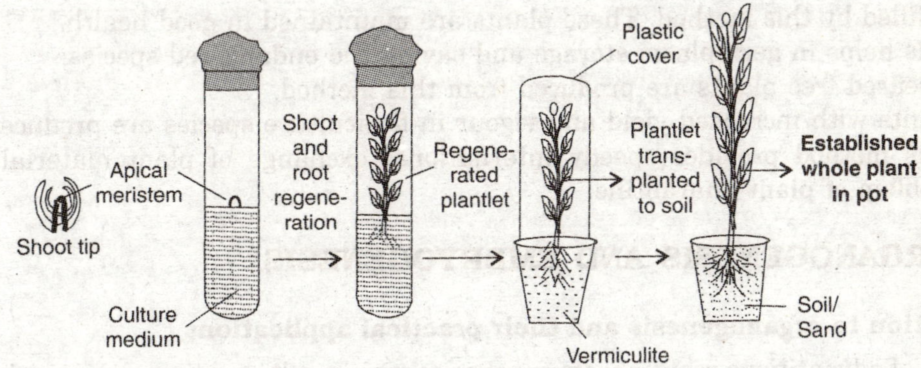

Fig. 15.4. Schematic representation showing steps in meristem-tip culture of plants.

Meristem lies in the 'shoot-tip' beyond the youngest leaf or first leaf primordium. It measures upto about 100 mm in diameter and 250 mm in length. Hence, a shoot tip of 100-500 mm would contain one to three leaf primordia in addition to the apical meristem. Normally shoot tips of upto 10 mm are used when the objective is virus elimination to obtain disease free plant. The shoot tip culture is followed for rapid clonal propagation for which much larger (5-10 mm) explants are used. Thus, most of meristem culture are essentially shoot-tip cultures. Nodal explants of various sizes are also employed for rapid clonal propagation.

If micropropagation is the main objective, the size of shoot tip used for culture is not important. But when the objective is to make stock free from a virus (or other pathogenic microorganisms) contamination, it is essential that the apical meristem should be excised with a minimum of the surrounding tissue. The shoot tip may be cut into fine pieces to obtain more than one plantlet from each shoot-tip. In case of some species such as cauliflower pieces of curd (the inflorescence) are used, while in plants having underground stems shoot tips or tissue pieces bearing buds of such stems may be used. Usually explants taken from actively growing plants at the beginning of growing season are the most suitable.

Meristem explant are then placed on MS medium which is known to be effective in majority of the species. For some species lower salt concentration is suitable. During the growth several fungicides (bavistin) or antibiotics (streptomycin/choramphenicol) can be added to the medium to completely remove the endophytic contamination. Similar to micropropagation, in meristem culture also follows similar steps namely (i) culture initiation, (ii) shoot multiplication, (iii) rooting of the developed shoots, (iv) transfer of plantlets to soil/pots.

Applications of micropropagation and tissue culture

- A small amount of plant tissue is sufficient for production of millions of clones in a year. Whereas it would take more years to get same number of plants.

- Micropropagation offers an alternative method of vegetative propagation for mass propagation. The large number of plants can be obtained in short period. Any particular varieties can be obtained in large quantity and time period to develop new varieties was reduced to 50%.
- Large number of plants can be maintained in less space and timely need can be fulfilled by this method. These plants are maintained in good health.
- This helps in germplasm storage and saving the endangered species.
- Diseased free plants are produced from this method.
- Plants with increased yield and vigour in floriculture species are produced.
- This method provides speedy international exchange of plant material without problem of plant quarantine.

15.5 ORGANOGENESIS AND EMBRYOGENESIS

Introduction to organogenesis and their practical applications

Induction of adventitious roots, shoots or other organs in cultured tissues is very important in the plant tissue culture methodology. The formation of shoots, roots, flowers can be initiated from the cultured plant cells or tissues. The induction of different adventitious organs of the plant from cultured tissues *in vitro* condition is known as **organogenesis** (Fig. 15.5).

Generally in tissue culture two types of organogenesis was noticed or induced. They are caulogenesis and rhizogenesis. Induction of adventitious shoot buds from the cultured tissues is known as **caulogenesis.** This was usually made after the formation of callus in the culture condition or directly from the explant which was placed on the medium. Whereas induction of adventitious roots from the cultured tissues is known as **rhizogenesis.** Rhizogenesis was induced only after the caulogenesis in the plant tissue

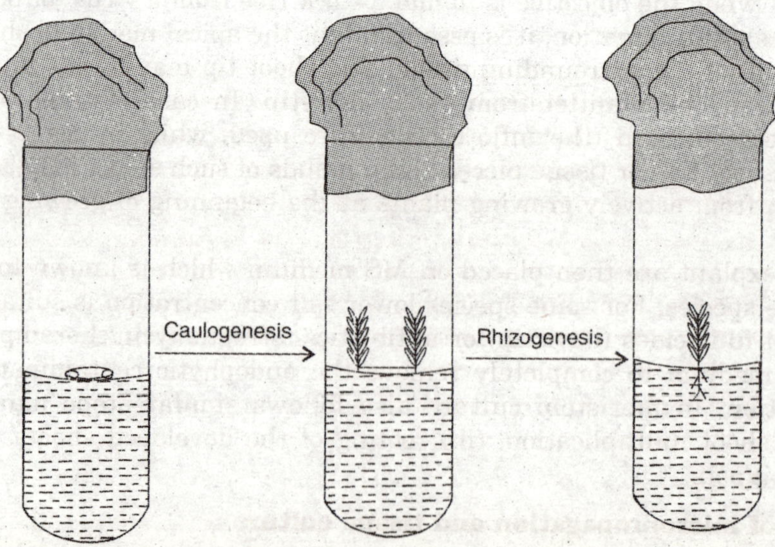

Fig. 15.5. Diagrammatic representation of organogenesis in plants.

culture methods. After both the type of organogenesis (shoot first and root next) the separate individual plant was transferred to the soil condition through the process of hardening.

Organogenesis has been obtained from many species of plants and from varieties of explants, namely, cotyledons, hypocotyl, stem, leaf, shoot apex, root, youn inflorescence, flower petals, petioles, embryos etc. which are cultured *in vitro* conditions. Explants from both immature and mature parts are made to induce to form callus. Then callus was regenerated to make complete plant.

Sometimes anomalous structures develops during the organogenesis and these structures are called organoids. The group of cells which give rise to the shoots and roots is known as meristemoids or nodule. Meristemoids are always developed in areas where the cells accumulates the high concentration of starch. These meristemoid generally forms roots from the inner cells thus they are endogenous in origin, while shoots formed from outer cells and are called exogenous in origin.

Organogenesis is due to dedifferentiation of the mature cells and followed by redifferentiation of cells. Dedifferentiation favours unorganised cell growth and the resultant developed callus has meristems randomly divided. Most of these meristems if provided proper *in vitro* conditions would redifferentiate shoot buds and roots. The whole plant regeneration from the cultured cells may occur either through shoot-bud differentiation or somatic embryogenesis.

Different types of tissue culture media are used for organogenesis such as MS, B5, White's and SH (Schenk and Hilderbrant, 1972) medium. The MS medium should contain high concentration of nitrogen as ammonium when compared to other media. It is observed difficulty in transfer of cells from B5, White's or SH medium to MS medium. The season of the year, donor plant conditions, age and physiological state of the parent plant have great impact on organogenetic potential of explant.

The organogenesis in case of plants is poorly understood and this involves various factors such as plant growth status, source of explant, culture medium, addition of growth regulators and environmental conditions. Skoog and Miller in 1957 confirmatively showed that growth regulators control the organogenesis in case of tobacco pith. They showed balance between auxin and cytokinin ratio helps in induction of different types of organogenesis. The high concentration of cytokinin known to induce shoot formation. Whereas high concentration of auxin induce the root formation. The inhibitory effect of one molecule of auxin on caulogenesis could be counteracted by two molecules of cytokinin. This is not always true with all the plant species.

In addition to auxin and cytokinin, other growth regulators is also known to have effect on organogenesis. Gebberellins commonly have inhibitory effect on shoot or root organogenesis in culture condition. Ethylene is shown inhibitory effect on shoot initiation, but promotes the bud growth once it is initiated. The physical condition of the medium, genotype of the plant materials, light, concentration of the sucrose and inorganic phosphate and abscissic acids are having profound influence on the organogenesis in the varieties of the plants.

Applications of organogenesis

- By organogenesis any plants are made to produce plantlets in very short period.
- It also provides a method for the production of large number of plantlets in case of tree species or any hard species where somatic embryogenesis is not possible.

- This helps in understanding the interplay of various chemical factors involved in morphogenetic change in plants.
- This acts as a simple bioreactor system where organ specific chemicals are produced.

15.6 SOMATIC EMBRYOGENESIS

Somatic embryogenesis means induction of somatic embryo other than the zygotic embryo from the culture tissues. The embryo developing from zygote is known as zygotic embryo or simply embryo. The somatic embryos are commonly called embryoids or nonzygotic embryos. Generally embryogenesis is not considered under organogenesis

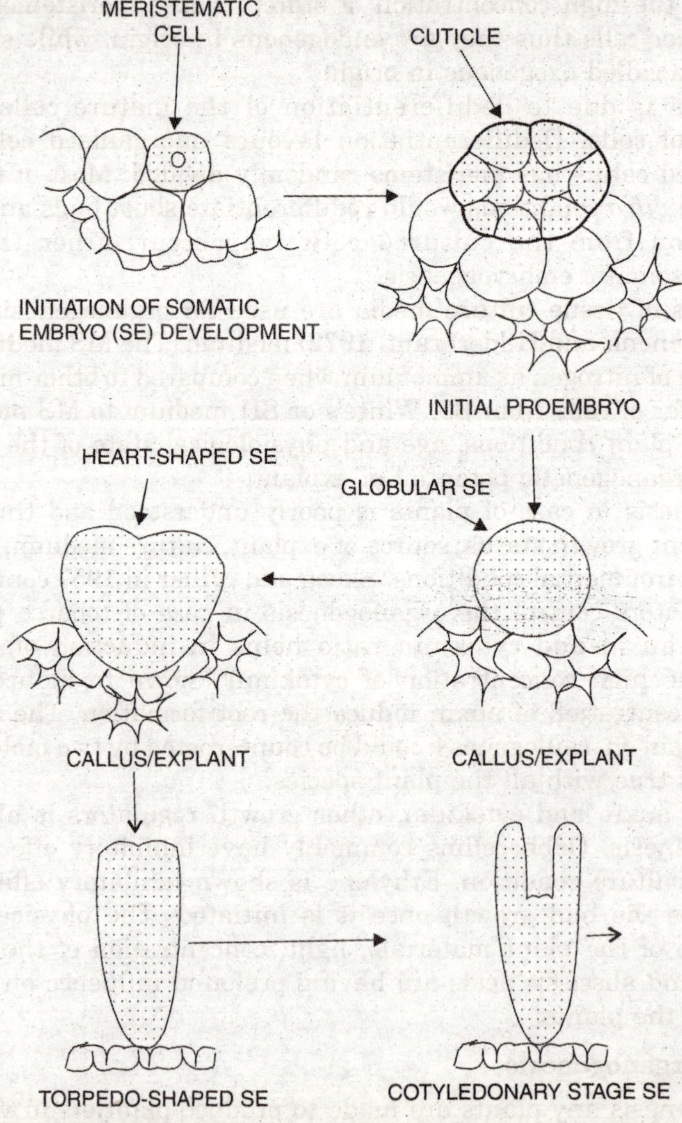

Fig. 15.6. Scheme showing development of a somatic embryo from a single superficial cell of the explant material.

because the embryo will not be having vascular connection with parent plant tissues and have independent existence. But other organs such as shoot, root or flower always having the vascular connection with parent plant tissues. Steward *et al.* (1958) and Reinert (1959) first observed first somatic embryogenesis in carrot suspension and agar cultures respectively. Sharp and others in 1980 concluded the somatic embryogenesis may be initiated directly from the explant or indirectly from the callus tissues.

Auxin is known to induce the somatic embryo formation. This always occurs in two phases viz., first induction phase and second maturation phase. Induction phase requires high auxin (0.5-40/60 mg/lt 2, 4 D) and reduced nitrogen (NH_4^+) in the induction medium and low auxin or growth regulator free medium (with high sucrose-6-40%) for maturation phase.

Initially somatic embryos derived from the single cells which later divide to form meristematic cell mass known as meristemoid. These cell mass contain dense cytoplasm, large starch grains, large nucleus and nucleolus with high dehydrogenase activity. They later become cutinized and losts cytoplasmic connection with the surrounding cells. This cell mass are called proembryo and develop in a definite sequence with characteristic morphology. First it forms globular structure and named as globular embryoid. This by further division forms heart-shaped embryoids and torpedo embryoid. Finally it forms cotyledonary somatic embryos. The cotyledonary embryoid is having a definite polarity i.e. a lower radicular end forms roots and plumular end forms shoot (Fig. 15.6). Thus they have full potential to develop into complete young plantlets.

During the development sometimes abnormalities may occur and forms embryogenic clumps or multicotyledons or irregular shaped embryos. This is due to high concentration of auxins and can be controlled by using abscissic acid with mannitol in the medium. In some species first formed somatic embryos (primary somatic embryo) further regenerate to produce second group of somatic embryos on them and these are called secondary somatic embryos, the process is known as secondary embryogenesis or recurrent embryogenesis (Fig. 15.7).

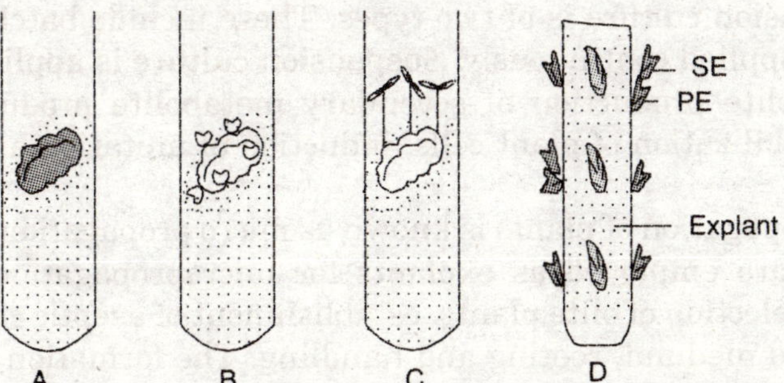

Fig. 15.7. Figures showing development of embryoid in tissue culture. A—callus, B—numerous embryoids in six week old cultures, C—plantlets developed from embryoids and D—an enlarged portion showing secondary embryoids (SE) on primary embryoids (PE) produced on explant.

Applications of somatic embryogenesis

- Somatic embryogenesis have tremendous advantage for mass production of plants in large numbers when compared to both regular clonal propagations (e.g. grafting, cuttings) and other tissue culture methods (e.g. micro-propagation).

- By somatic embryogenesis unlimited numbers of embryos can be produced from a single explant and it can be maintained for longer time.
- Somatic embryos are true to their parents and shows no or less tolerance to mutations.
- Like organogenesis, there is no need of shoot productions separately. Because they have complete potential of forming embryonic shoots, roots and leaves.
- This helps in scale-up of propagules with the automated culture systems because the somatic embryo can be produced in the liquid medium.
- The bioreactors can be used with somatic embryo production which helps in design of closed system for separation, sizing and distribution of propagules into culture vessel and even helps in automation of transplanting plantlets to soil mix in greenhouse tray at a rate of 8000 per hour. This cuts the cost, reduces contamination, saves space, time and labour.
- Somatic embryos serve as better source for protoplasts. Because cells of embryo have good potentiality of regeneration (in case of grass species, horticultural and forestry species protoplasts have less regenerative capacity.
- It also provide cells genetic transformations where callus induction is very difficult.

SUMMARY

The tissue culture medium with suitable growth regulators combination allows the growth and division of cells of explant. Cells of explant undergo maturation, differentiation and many physiological changes to form meristematic and it is known as dedifferentiation.

The unorganized and less differentiated parenchymatous tissues is called callus and culture is said to be callus culture. Once callus is formed from the explant and subcultured on to a new medium to maintain the viability of cells. Callus culture has several applications. These includes, it serves as a good source for the protoplant isolation, for suspension culture, for the genetic manipulation of the plant genomes, etc.

Suspension culture can be defined as growing of individual cell or cell aggregates in liquid media. Suspension culture is of two types. These include batch culture, continuous culture medium is supplied continuously. Suspension culture is applied for the production of secondary metabolite production of secondary metabolite production, for protoplast production, for immobilization of plant cells, induction of mutation and genetic manipulation of plant cells.

In vitro clonal propagation of plants is known as micro propagation. Shoot-tip, meristem and auxillary buds are employed as explants for micropropagation. This includes five major steps such as selection of elite plants, establishment of aseptic culture, multiplication of explants on defined medium, rooting and handling. The formation of different matured organs is called organogenesis under *in vitro* condition. The formation of embryo from somatic cells is referred to as somatic embryogenesis. Such embryo are called somatic embryo. These embryos are having varieties of application in improvement of plants.

EXERCISE

1. What is meant by micropropagation?
2. Differentiate between shoot-tip and meristem.

3. Explain different stages of micropropagation.
4. Describe meristem culture and add a note on its application. (BU April 2001)
5. Give the meaning of organogenesis and embryogenesis.
6. What is somatic embryogenesis? Give an account of its induction and importance. (BU Oct. 2001)
7. How somatic embryos are produced *in vitro*?
8. Define dedifferentiation.
9. Write a note on callus induction and its applications.
10. Critically give an account of suspension culture and its use in secondary metabolites production.
11. Mention the difference between batch culture and continuous culture.
12. Differentiate between primary embryoid and secondary embroid.
13. What do you mean by caulogenesis and rhizogenesis?
14. Define vitrification.
15. What is meant by recurrent embryogenesis?
16. Write a short note on role of auxins in organogenesis. (BU April 2003)
17. What is micropropagation? Describe various stages and applications of micropropagation. (BU April 2003)
18. Name different plants and their compounds having insecticidal properties used in tissue culture production.
19. Distinguish between chemostat and turbidostat.
20. What do you mean by clone?

16

Embryo Culture, Embryo Rescue and Their Application

16.1 INTRODUCTION

Embryo is the zygotic tissue obtained after fertilization of ovum by the male gamete. This embryo develops properly when nourishing tissue, the endosperm was present in the seed during the development. But in case of the cross between the two distant species result in the degeneration of endosperm tissue. Thus it hampers the embryo development and viable plant will not develop. To recover such hybrid embryo, culturing of embryo was followed. This culturing of embryo (hybrid) *in vitro* condition is known as **embryo rescue** and widely used for crop improvement. This method is also used to culture where the embryo is incompletely developed (orchids), to skip the seed dormancy for shortening the breeding cycle, determination of seed viability and microcloning of the source material.

16.2 CULTURING OF EMBRYO

Hanning in 1904 first cultured isolated mature embryo of cross between *Cochleria* × *Raphanus* and successfully obtained the plants. Hybrid seeds from *Triticum aestivum* × *Hordeum vulgare* are not obtained naturally. But hybrid seedlings have been obtained by embryo culture.

The embryo was dissected out carefully from the surface sterilized seeds (5-10% chlorox or 0.45% sodium hypochlorite for 5-10 minutes). Both immature and mature embryo can be used. Sometimes before placing embryo directly on to the medium they can be once again surface sterilized by using 70% ethanol for 30-60 seconds. Placing the embryo on endosperm tissue present on the medium was known to show good growth when compared to placing of embryo directly on the nutritive medium (embryo-nurse endosperm technique). Excision of the embryo along with suspensor (is an ephemeral structure found at radicular end of the proembryo) is always beneficial for the embryo development *in vitro.* The MS, White's and B5 medium with little modification is widely used for the embryo culture. The medium rich in sucrose/glucose, ammonium nitrate, vitamins, casein hydrolysate, natural plant extracts with growth regulators at pH range of 5 to 7.5 known to show good responses in the culture. The cultures are incubated at 25±2°C with light exposure of 4000 lux for 4 hours at the beginning. The callus obtained from the embryo tissue is generally having high regenerative capacity compared to those derived from the leaf, stem and root. The callus tissue is made to produce different organs (organogenesis) or embryoids (somatic embryos) and ultimately complete plants are produced.

16.3 APPLICATIONS OF EMBRYO CULTURE AND EMBRYO RESCUE

Viable embryo production from incompatible crosses

Interspecific and intergeneric crosses results in the non-viable seeds. Isolation and culturing of such hybrid embryo results in the production of rare viable plants.

Dormancy breaking

Seeds which are undergoing long periods of dormancy delay the breeding work in case of horticultural and crop plants. By embryo rescue this can be solved and breeding cycle can be shortened.

Embryo production from sterile seeds

Embryo will be immature stage in case of early ripening fruits as a result seeds will not germinate. In such cases by embryo culture methods seedlings can be produced from sterile seeds. E.g. plum, apricot and peach.

Haploid production

Interspecific crosses between *Hordeum vulgare* × *Hordeum bulbosum* results in the normal diploid after normal fertilization. But later during development chromosomes of *H. bulbosum* lost and result in the haploid embryo.

Micropropagation

As an embryo forms potential (young) organ for regeneration and thus can be used in micropropagation of hard species. This method was followed for the development of plantlets in case of *Pinus palustris*.

SUMMARY

The tissue obtained from zygote is called embryo. Embryo rescue widely used for crop improvement. It can be defined as culturing of the embryo *in vitro* condition. This method is caused to culture the incompletely developed embryo such as in orchids, to overcome the seed dormancy, production of seedlings from sterile seeds, production of viable plant from incompatible crosses, production of haploid plants and as source material for micro propagation.

EXERCISE

1. List the applications of embryo rescue.
2. Define embryo.

3. What do you mean by embryo culture? (BU April 2003)
4. Why is endosperm necessary for development of embryo?
5. How embryo culture is helpful for hybrid production?
6. What is meant by embryo rescue?
7. Name the plants where successful embryos are produced from hybrids.
8. Briefly explain production of seedling from sterile seeds.
9. Define embryo culture. Discuss the techniques and applied aspect of embryo culture. (BU April 2003)
10. How first cultured embryo of plants in tissue culture?

17

Haploids and Their Application

17.1 INTRODUCTION

Haploid plants (haploids) are the plants bearing only one complete set of chromosomes (n) whereas diploids contain two complete set of chromosomes (2n). Haploid plants was first discovered by Bergner, A.D. during 1921 in *Datura stramonium.* In natural condition haploids arises from parthenogenesis and they rarely produce character of male parent. The production of haploid individual from male cells/gametes is known as **androgenesis.** The haploid plants can be obtained *in vitro* from pollen grains by placing anthers or separated pollen grains on a suitable medium, this is known as **anther culture** or **pollen culture** respectively.

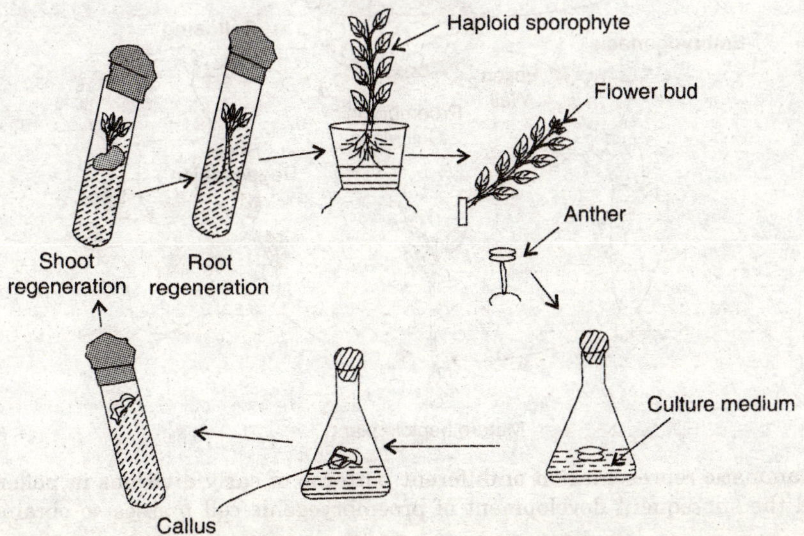

Fig. 17.1. Technique of anther culture.

17.2 PRODUCTION OF HAPLOIDS

Tulecke in 1953 first observed formation of haploid callus on medium in *Gingko biloba,* a gymnosperm. Later Guha and Maheshwari (1966) observed haploids in varieties of angiosperms. Anthers may be taken from the plants grown in open field or in pots which are preferably grown under controlled temperature, humidity and light.

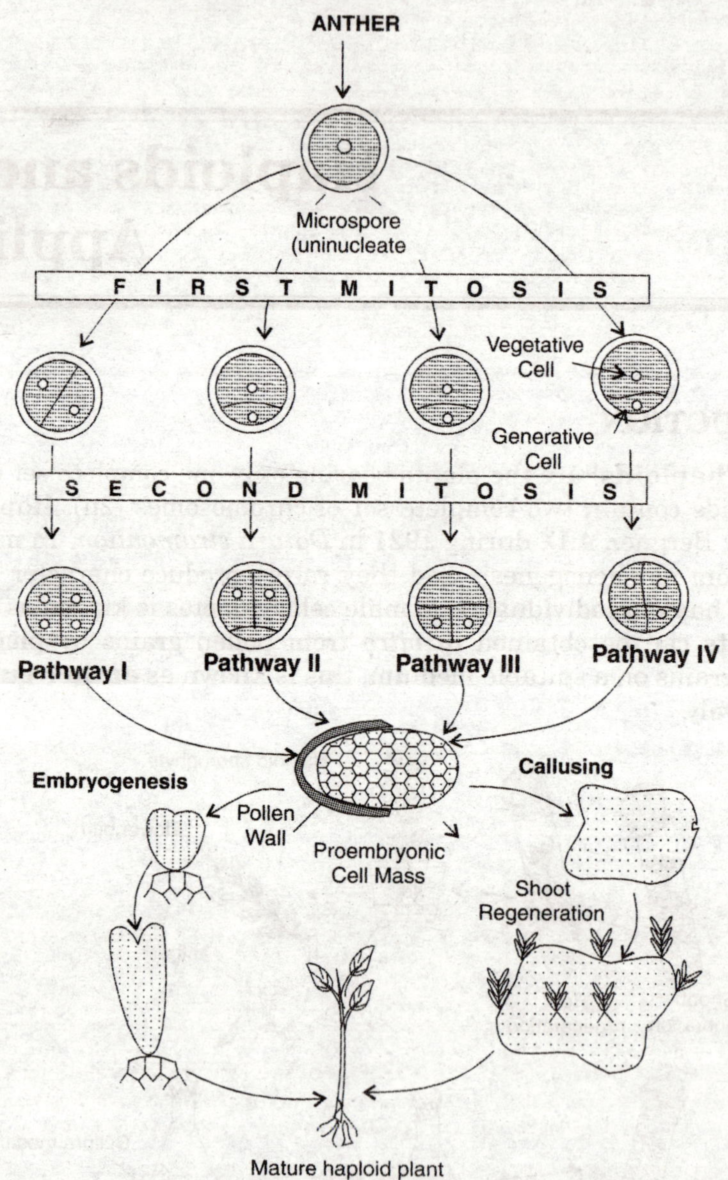

Fig. 17.2. Diagrammatic representation of different patterns of early divisions in pollen grains during androgenesis and the subsequent development of proembryogenic cell masses so obtained.

The particular stage of anthers at the time of culture is the very important factor which decides the success. Flower buds which are having premitotic microspores are collected, surface sterilized (Fig. 17.1). Then anthers are excised and placed horizontally on the culture medium. Adequate care should be taken to avoid any injury to anthers, otherwise it induces callus formation from the diploid, sporophytic anther wall cells. Sometimes pretreatment of anther such as exposure of excised flower buds to a low temperature (3-5°C for two days) or exposure to high temperature (35° for 24 hours), irradiation with gamma rays or X-rays, centrifugation are known to have increased haploids production in culture condition. In some cases pollen grains are taken out from the anther and directly cultured on a culture medium. Both solid or liquid medium can be used to culture anthers or pollen grains. The anther cultures are maintained in alternating periods of light (12-18 hr; 5000-10000 lux M^2) at 28°C and darkness (12-6 hr) at 22°C.

The cultured pollen grain follow mainly four (Fig. 17.2) pathways

i. the uninucleate pollen grain may divide equally to form two daughter cells of equal size. Later both the cells undergo further divisions e.g. *Dutra innoxia.*
ii. In few plants pollen grain divides unequally to give bigger vegetative cell and smaller generative cell. The vegetative cell divides number of times to form callus or embryo and generative cell degenerate after one or two divisions e.g. barley, wheat, chilly.
iii. In some other generative cell form callus or embryo and vegetative cell does not divide or forms only limited tissue e.g. *Hyoscyamus.*
iv. Lastly in some plants both unequal generative and vegetative cells involved in callus or embryo formation e.g. *Dutra* species.

After the maximum division the wall layers of anthers turn brown and after 3-8 weeks anther burst open due to excess pressure given by dividing cells. This forms fully formed callus or differentiate into globular embryo-like structure (somatic embryo) depending upon plant (Fig. 17.2). Later callus or somatic embryo is induced to form haploid plants by using appropriate concentration of growth regulators (auxins and cytokinins).

The production of haploid plants from *in vitro* culture of microspores or pollen grains is known as **androgenic haploids**. If haploid plants obtained from *in vitro* culture of unpollinated ovule or ovary is known as **gynogenic haploids.** This was first done by San Noeum in 1976.

17.3 APPLICATIONS OF HAPLOIDS

Haploid plants help in selection of desirable characteristics. The phenotype is the expression of single copy genetic information there by no masking of a character by gene dominance. The haploid plants act as good material for induction of mutations and various gene manipulations. Through doubling of haploid chromosomes homozygous condition can be obtained which helps in plant breeding and crop improvement. Haploids help in development of new varieties, selection of mutants which are resistant to diseases, developing asexual lines of trees or perennial plants, transfer of desirable genes.

SUMMARY

The plants possessing only one complete set of chromosomes is known as haploids. Under natural condition parthenogenesis gives haploids. Parthenogenesis rarely produces male haploids. However, such male haploids can be obtained *in vitro* from pollen grains or anthers, these culture techniques are referred as pollen culture and anther culture respectively. Haploids are of two types, these are androgenic haploids and gynogenic haploid. The haploids plants obtained from *in vitro* culture of pollen grains are called androgenic haploids and the haploids from *in vitro* culture of unfertilized ovule or ovary are called as gynogenic haploids. Haploid plants have several applications in plant improvement. These are used for the induction of mutations, for the production of homozygous diploid plants, for gene manipulations, for the selection of mutants which are resistant to diseases, to develop trees etc.

EXERCISE

1. What are haploids?
2. Define anther culture. (BU Oct. 2002)
3. Elucidate the different pathways followed by cultured pollen grain.
4. Give the applications of haploids. (BU Oct. 2001)
5. Who discovered haploids?
6. Name a few plants in which haploids are produced.
7. What are androgenic haploids? Discuss the technique of their production and application. (BU Oct. 2001)
8. Describe *in vitro* process of haploid production. Add a note on its application. (BU Oct. 2002)
9. Differentiate between anther culture and pollen culture.
10. Write a short note on androgenic haploids. (BU April 2003)

18

Somaclonal Variations

18.1 INTRODUCTION

Somaclonal variations are defined as the genetic variability present among cultured somatic cells. Plants derived from such cells or progeny of such plant is called **Somaclones**. The term somaclonal variation is first used by **Larkin** and **Scowcroft** in 1981. The term somaclonal variation is used for genetic variability present among all kinds of cells or plants obtained from cells cultured *in vitro*. Plants regenerated from tissue and cell cultures show heritable variation for both qualitative and quantitative traits. Somaclonal variation has been described in apple, sugarcane, potato, tomato etc. Some variants are obtained in homozygous conditions in the plants regenerated from cells cultured *in vitro* (R_0 regeneration) but most variants are recovered by progeny of tissue culture regenerated plants (R_1 generation). Chaleff in 1981 considered variants from tissue culture as R or R_0 plants and selfed progeny of R_0 plants as R_1 plants. Somaclonal variation considered a novel method of genetic variation and having much interest as a basic genetic process, as it contradicts the theory of clonal uniformity. But it is disadvantageous when the objective is micropropagation of considered genotypes.

18.2 PROCEDURE FOR ISOLATION OF SOMACLONAL VARIANTS

Mutants or variants for several traits can be far more easily isolated from cell cultures than from whole plant populations in the field condition. This is because a large number of cells, as high as 10^6 - 10^9, can be easily and effectively screened for mutant traits. Screening of as many plants would be very difficult to perform, ordinarily impossible. The mutants can be effectively selected for disease resistance, improvement of nutritional quality, adaptation of plants to stress conditions (e.g. Saline soils, low temperature), toxic metals (e.g., aluminium), resistance to herbicides and to increase the biosynthesis of plant products used for medicinal or industrial purposes (Fig. 18.1). The various approaches to the isolation of somaclonal variants can be grouped into two broad categories. They are screening and cell selection (Fig. 18.1) methods.

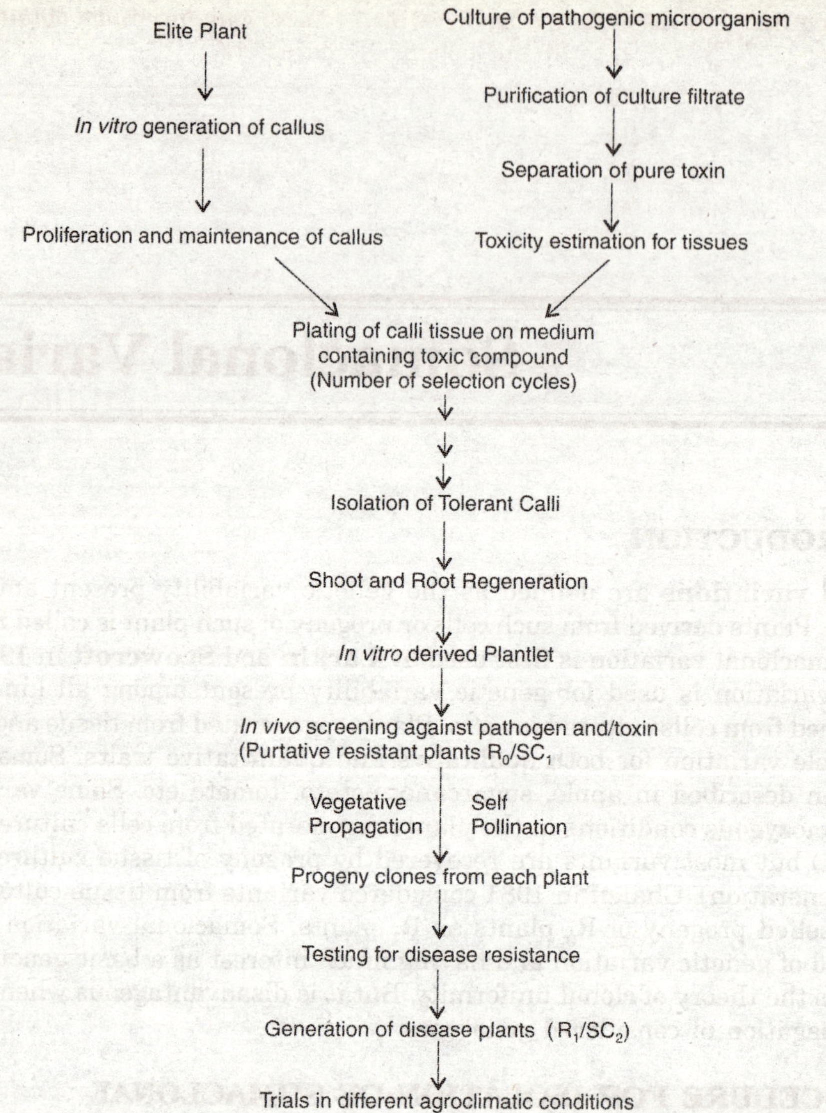

Fig. 18.1. Schematic representation of procedure followed for the isolation of disease resistant plants from culture.

Screening

It involves the observation of a large number of cells or regenerated plants from tissue culture for the detection of variant individuals. This approach is only feasible technique for the isolation of mutants for yield and yield traits, in general R_1 progeny (progeny of regenerated, R_0 plants) are scored for the identification of variant plants, and then R_2 progeny lines are evaluated for confirmation. This has been employed in number of crop plants (Table 18.1).

Table 18.1: List of plant species for which somaclonal variations obtained for different beneficial characters.

Plant	Characters
Barley	Grain yield, test weight, leaf shape, 1000 grain weight, height, lodging resistance, heading date, β-glucan content, ash
Brassica napus	Growth habit, glucosinolate content
Carrot	Higher carotene content
Datura	Solfonylurea, imidazolinone and triazolopyrimidine herbicide resistance
Fennel	Chlorotic foliage, altered stem height, reduced fruit production, varied oil composition
Maize	Glyphosate resistance
Oats	Avenin banding pattern variation, increased seed weight, seed number, grain yield
Potato	Yield, vine quality, maturity and morphology
Rice	Tiller no., particle size, flowering date, plant height, heading date, culm length, leaf shape, leaf colour, ear shape, sterility mutants
Sorghum	Improved acid soil stress tolerance
Soyabean	Varied height, lodging, maturity, seed protein and oil content
Sugarcane	Yield, sugar content, disease resistance
Tomato	Violet cotyledon colour, dwarf habit, modification of first floral node, early flowering, orange fruit colour
Wheat	Grain colour, height, tiller no., seed storage protein

The screening method has been profitably and widely employed for the isolation of cell clones that produce higher quantities of certain important biochemical agents; computer based automated cell sorting devices have also been used to screen as many as 1000-2000 cells/second from which desirable variant cells were automatically separated. The variant cells was further regenerated to produce complete plantlets.

Cell selection

In the cell selection method an appropriate selection pressure is applied which permits the preferential survival/growth of variant cells only during culture. Some examples of cell selection are, selection of cells resistant to various pathogen derived toxins, herbicides, acidity high salt concentration etc., when the selection pressure allows only the mutant cells to survive or divide, it is called **positive selection**.

On the other hand, in the case of **negative selection** method, the wild normal type cells divide normally and therefore are killed by the counter selection agent used in the medium, e.g., arsenate; the mutant cells are unable to divide as a result of which they are unaffected by the counter selection agent. These cells are subsequently rescued by removal of the counter selection agent from the medium. Negative selection approach is generally employed for the isolation of auxotrophic mutants.

The positive selection approach may be further sub-divided into four categories, namely, direct selection, rescued method, step-wise selection and double selection.

Direct selection

The cells resistant to the selection pressure survive and divide to form mature colonies; the selection agent used shows lethal effect on the wild type cells. The mutant cells are not affected by the selection agent and shows growth. This is the most common selection method employed for the isolation of cells resistant to toxins (produced by pathogens), herbicides, elevated salt concentration, antibiotics, amino acid analogues etc. in varieties of crop plants.

Rescue method

The selection agent used in the medium kills the wild type cells, whereas the variant cells remain alive but, usually, do not divide due to the unfavorable environment created by the selection agent. This selection agent is then removed to recover the variant cells. This approach has been widely used to recover low temperature and aluminium resistant variant cells in number of plants.

Stepwise selection

The selection pressure, selection agent e.g., salt concentration may be gradually increased from a relatively low level to the cytotoxic level in the medium; the resistant clones isolated at each stage are further subjected to the higher concentration of selection pressure. Such a selection approach is called stepwise selection. It may often favour gene amplification (which is unstable) in the cells or mutations in the DNA of organelles.

Double selection

In certain cases, it may be feasible to select for survival and/or growth on one hand and some other feature reflecting resistance to the selection pressure on the other hand, this is called Double Selection. An example, of such double selection is provided by the selection for resistance to the antibiotic streptomycin, which inhibits chlorophyll development in cultured cells. The selection was based on cell survival and colony formation in the presence of streptomycin (first selection trait) as well as for the development of chlorophyll in these colonies (second selection trait; only green colonies should be selected). This approach has been used for the selection of cells resistant to the herbicide amitrole, 2-4 D, and aluminium.

During *in vitro* selection procedure of cell variants by using selection agent was done at concentration that shows lethality upto 70-90% to wild type of cells. The concentrations that are too low should not be used, because cells fail to grow due to physiological reasons are wrongly considered as true variants. Whereas selection agent at too high concentrations also not advised because it may result in potential variant cells death. Cells should be grown in presence of selection agent for sufficient duration to ensure expression and growth of the resistant somaclones.

18.3 CHARACTERISATION OF SOMACLONAL VARIANTS

Majority of somaclonal variants, isolated through cell selections are often unstable, the frequency of stale variables may range from 8 - 62% perhaps depending on the species and selection agent employed in the test.

(a) Many selected somaclones are unable to exhibit their resistance during further screening or selection, obviously these clones are susceptible and are wrongly classified as **resistant somaclones (escapes)**.

(b) Several somaclones lose their resistance to the selection agent after a period of growth in the absence of selection pressure. Such clones are called **unstable variants** and may result from changes in gene expression and from gene amplification.

(c) Some variant phenotypes are quite stable during the cell culture phase, but not true, when plants are regenerated from the variant cultures, or when the regenerated plants reproduce sexually in case they are expressed in regenerated plants. Such changes are known as **epigenetic changes.**

Basis for somaclonal variation in plants

Number of factors are considered responsible for somaclonal variation in culture condition. They include

- Pre-existing chromosomal ploidy in the explant cells used for culture initiation.
- Gene mutations in the nuclear and cytoplasmic genomes that may be translocations, duplication, deletions and inversions.
- Nuclear fragmentations during first divisions at callus induction stages.
- Endomitosis or endoduplication during culture initiations.
- Mitotic crossing over during culture induction step.
- Activation of transposable elements in the cultured cells.
- Changes in gene expression and gene amplication.

18.4 SOMACLONAL VARIATIONS AND INDUCED MUTATIONS

Somaclonal variation are preferable to induced mutation for several reasons:

- **Chimaerism** is a major problem in induced mutations but not in somaclonal variation obtained in culture condition.
- Induced mutations are often associated with undesirable features like sterility, unusual character etc.
- *New alleles* and even *'new' mutations* have been recovered through somaclonal mutations which are generally beneficial in nature.
- The frequency of beneficial mutation is very high in the case of somaclonal variations.
- Highly effective selection can be employed *in vitro* for several economically important characters, which is virtually impossible in the case of mutation breeding.
- As astronomically large number of individuals can be effectively screened *in vitro*.

18.5 ADVANTAGES OF SOMACLONAL VARIATIONS

1. Somaclonal variation occurs in rather stable and high frequencies, which is a great advantage over conventional mutagenesis.
2. Some 'new' alleles or even 'novel' mutations may be isolated which were not available in the germplasm or through mutagenesis, e.g., jointless pedicel mutant in tomato.

3. This can be performed in all types of plants i.e., vegetatively or sexually or asexually propagated plants.
4. Use of somaclonal variation may reduce by two years the time required for the release of new variety as compared to mutation breeding. This is because somaclonal variations are usually free from undesirable features like sterility, while induced mutations are generally associated with such defects which necessitates one or two back crosses with parent variety.
5. A very effective selection can be practised at the cell level for several traits, e.g., disease resistant, salt tolerance and herbicide resistance.
6. This is the only approach for the isolation of biochemical mutants, especially auxotrophic mutants in plants, salt tolerance and herbicide resistance.

18.6 DISADVANTAGES OF SOMACLONAL VARIATIONS

1. The somaclones production is applicable only to those species cell cultures which regenerate complete plants.
2. Selected cell lines often show reduced or no regeneration potential.
3. Many selected clones show undesirable features such as reduced fertility, growth rate and even overall performance of the plant.
4. Many of the selected clones are unstable, that necessitates careful and detailed evaluation for identification of stable clones.
5. Most agronomic traits, e.g., yield quality etc., are not expressed at the cell level and hence cannot be selected for *in vitro.*
6. The variation observed is generally cultivar dependent.
7. The variation observed is not always stable and heritable.
8. The unpredictable and uncontrollable variations occur without much significance.

SUMMARY

The genetic variability present among cultured somatic cells is called somaclonal variations. Somaclones are the plants obtained from somatic cells showing variations. Somaclonal variation has been described in apple, sugarcane, potato, tomato etc. Cell culture serves best source for the isolation of somaclonal variations rather than whole plant populations. Somaclonal variants can be selected from many desirable characters such as, disease resistance, improvement of nutritional quality, adaptation of plants to stress conditions, resistance to herbicides and to increase the yield of pharmacological and industrial important plant products.

Somaclonal variants can be isolated by two basis methods. These include screening and cell selection. In screening either cells or regenerated plants are screened for the isolation of variants. In cell selection method variants are isolated by applying selection pressure. Such as using pathogen derived toxins, herbicides, high salt concentration etc. Cell selection methods are of two types, these are negative selection and positive selection, negative selection approach is generally employed for the isolation of auxotrophic mutants. The positive selection method is further divided into many types. These include direct selection, rescued method, step-wise selection and double-selection. Not all the somoclonal variants selected are stable. The frequency of variants may range from 8-62% that depends on the type of species and selection agent used.

The factors responsible for somoclonal variation are many. These include pre-existing chromosomal ploidy in the explant cells, gene mutations, nuclear fragmentation during first cell division at callus induction stage, endomitosis, mitotic crossing over, activation of transposable elements, changes in gene expression and gene amplification. Somoclonal variation is preferable. Somoclonal variations have both advantages and disadvantages over reduced mutations. In spite of having disadvantages, the somoclonal variants are preferred to induce mutants for obtaining the desired variant plants.

EXERCISE

1. What is meant by somaclonal variations? (BU Oct. 2001)
2. Give the definition for epigenetic changes.
3. Describe different selection methods of variants.
4. Give advantages of somaclonal variation.
5. Mention various limitations of somaclonal variation.
6. Explain how somaclonal variation is preferred to induced mutation.
7. What are somaclonal variations? Briefly describe their isolation, molecular basis and applications. (BU Oct. 2002)
8. Name the plant for which somaclonal variants are produced.
9. Discuss advantages and disadvantages of somaclonal variation.
10. Define positive selection and negative selection.
11. Write a note on stepwise selection and double selection.
12. Give the difference between direct selection and rescue method of selection.
13. Illustrate procedure followed for the isolation of disease resistant plants from culture.
14. List the plant species and their beneficial characters in which somaclonal variations are produced.
15. What are the factors responsible for somaclonal variation in culture condition?

19

Somatic Hybridization and Endosperm Culture

19.1 INTRODUCTION

The production of hybrid plants through fusion of two different plant protoplasts (wall-less naked cells) is known as **somatic hybridization** and such hybrids are called somatic hybrids. Carlson and others (1972) first produced somatic hybrids by using two different species of *Nicotiana (N. glauca × N. langsdorfii)*. Somatic hybridization involves the following four steps: (i) isolation of protoplasts, (ii) fusion of the protoplasts of desired species/varieties, (iii) selection of somatic hybrid cells, and (iv) culture of the hybrid cells and regeneration of hybrid plants from them. A brief, elementary consideration of these steps is presented below (Fig. 19.1).

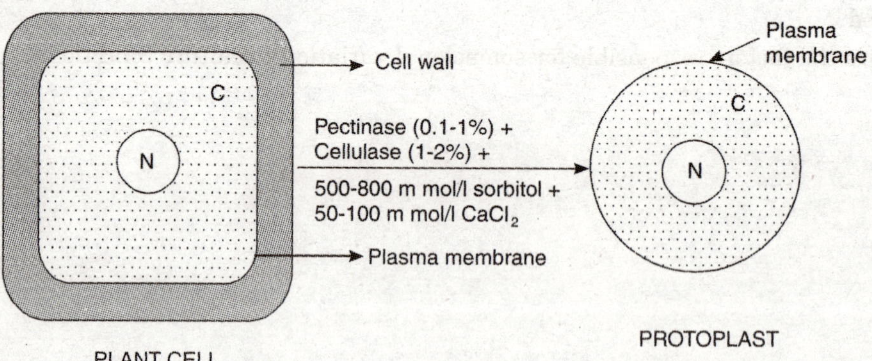

Fig. 19.1. Production of protoplasts by enzyme treatment (enzymes are depicted above the arrow). Osmoticum (indicated below the arrow) is added to stabilize the protoplasts and prevent them from bursting. N, Nucleus; C, Cytoplasm.

19.2 METHODS FOR ISOLATION OF PROTOPLAST

The word **Protoplast** was coined by Hanstein in 1880 for the living matter surrounded by the cell membrane. The isolated protoplast is highly fragile and outer plasma membrane is fully exposed. The plasma membrane is the only barrier between the interior of the living plant cell and the external environment. Isolation of protoplast can be done by three methods (i) mechanical, (ii) sequential enzymatic and (iii) mixed enzymatic methods.

Mechanical method of protoplast isolation was first done by Klercker (1982). This was done first by cutting the tissues which are earlier plasmolysed with a sharp knife into small pieces. Then these pieces were deplasmolysed by using dilute solution to release the protoplasts. But in this method very low concentration of protoplasts were obtained.

Sequential enzymatic method was first used by Takebe and others in 1968, wherein they used commercially available enzymes (Table 19.1) in two steps. The macerated tissue was first incubated in pectinase (degrade pectin cell wall) and then treated with cellulase (degrade cellulosic cell wall) for liberation of protoplasts.

Table 19.1: Various commercially available enzymes commonly used for isolation of protoplast

Enzyme	Source Microorganism
Cellulase onozuka R-10	*Trichoderma viride*
Cellulase onozuka RS	*Trichoderma viride*
Cellulysin	*Trichoderma viride*
Drisclease	*Irpex lacteus*
Hemicellulase	*Aspergillus niger*
Hemicellulase H-2125	*Rhizopus* species
Helicase	*Helix pomatia*
Macerase	*Rhizopus arrhizus*
Macero zyme R-10	*Rhizopus arrhizus*
Meicelase P-1	*Trichoderma viride*
Pectolyase	*Aspergillus japonicus*
Rhozyme HP 150	*Aspergillus niger*
Zymolyase	*Arthrobacter luteus*

In **mixed enzymatic method** both the enzymes were used together to reduce time and chances of contamination. Power and Cocking (1968) used this method for isolation of protoplasts from number of plant species and it gives very good yield as high as 2.5×10^6 protoplasts/gram of leaf tissue.

Protoplasts can be isolated by treating cells with a suitable mixture of cell wall degrading enzymes. The mixture of pectinase or macerozyme (0.1-1.9%) and cellulase (1-2%) is suitable for majority of plant parts. Additionally hemicellulase may be necessary for some tissues. Commercially available enzyme solution is adjusted to the pH value between 4.7 and 6.0 and is kept at temperature 25-30°C. The osmotic concentration of enzyme mixture and of subsequent media is increased by adding sorbitol/mannitol (500-800 m mol L^{-1}) to improve the plasma membrane stability. The enzyme mixture with cells and tissues are incubated for few to several hours (16-18). Then naked protoplasts devoid of cell wall are gradually released in the enzyme mixture (Fig. 19.1).

Protoplasts have been isolated from all plant parts, but leaf mesophyll is the most convenient tissue for this purpose. The fully expanded leaves are surface sterilized, their lower epidermis is peeled off with a pair of forceps and the peeled areas are cut into small (ca. 1 cm^2) pieces with a scalpel and suspended in the enzyme mixture. If epidermis cannot be peeled, leaf may be cut into ca. 1 mm^2 pieces and treated with the enzyme mixture vacuum infiltration may be used to increase the entry of enzymes into the tissues. After the period of incubation, protoplasts are washed with a suitable washing medium in order to remove the enzymes and the debris. The protoplasts may be cultured on a suitable medium in a variety of ways: (i) in agar medium (Bergmann's plating technique), (ii) in a thin layer of liquid medium or (iii) in small microdrops of 50-100 µ1. Protoplasts readily regenerated cell wall (within 2-4 days) and undergo mitosis to form macroscopic colonies which can be induced to regenerate whole plants.

Most of the time MS and B5 media, and their modifications are used for protoplast culture. The media are supplemented with a suitable osmoticum with an auxin and a cytokinin, their types and concentrations depending mainly on the plant species. After 7-10 days of culture, protoplasts regenerate cell wall and the osmolarity of medium is gradually reduced to that of normal medium. The macroscopic colonies are transferred onto normal tissue culture media (Fig. 19.2). Protoplasts are very sensitive to light; therefore, they are cultured in diffuse light or dark for the first 4-7 days.

19.3 TESTING THE VIABILITY OF ISOLATED PROTOPLASTS

The isolated protoplasts should be healthy and viable in order to undergo proper division and regeneration. This can be done by microscopic observation of untreated protoplasts or after staining the protoplasts with suitable chemicals to indicate active metabolism in the protoplasts.

(i) Phase contrast microscopy

Cytoplasmic streaming movement (cyclosis) and the presence of clear, healthy nucleus indicate that the protoplasts/cells are in viable state. For this phase contrast microscope is better because observation of unstained cells under bright field is highly difficult.

(ii) Tetrazolium reduction

In this test respiratory efficiency of cells is measured by reduction of 2, 3, 5-triphenyl tetrazolium chloride (TTC) to the red dye formazon. The formazon formed can be extracted and measured spectrophotometrically.

(iii) Fluorescein diacetate method

The 0.5% fluorescein diacetate (FDA) in acetone is prepared and stored at 0°C. This was added at 0.01% of final concentration to protoplasts suspension with osmatic stabilizer. After 5 min incubation the cells are observed under microscope with suitable filter. Esterase present in living viable cells cleaves non-fluorecing FDA into fluorescein (fluorescent compound) and accumulate inside the protoplast, which can be clearly visible as green colour protoplast under UV light, whereas in dead protoplast fluorescein does not accumulate and remains invisible.

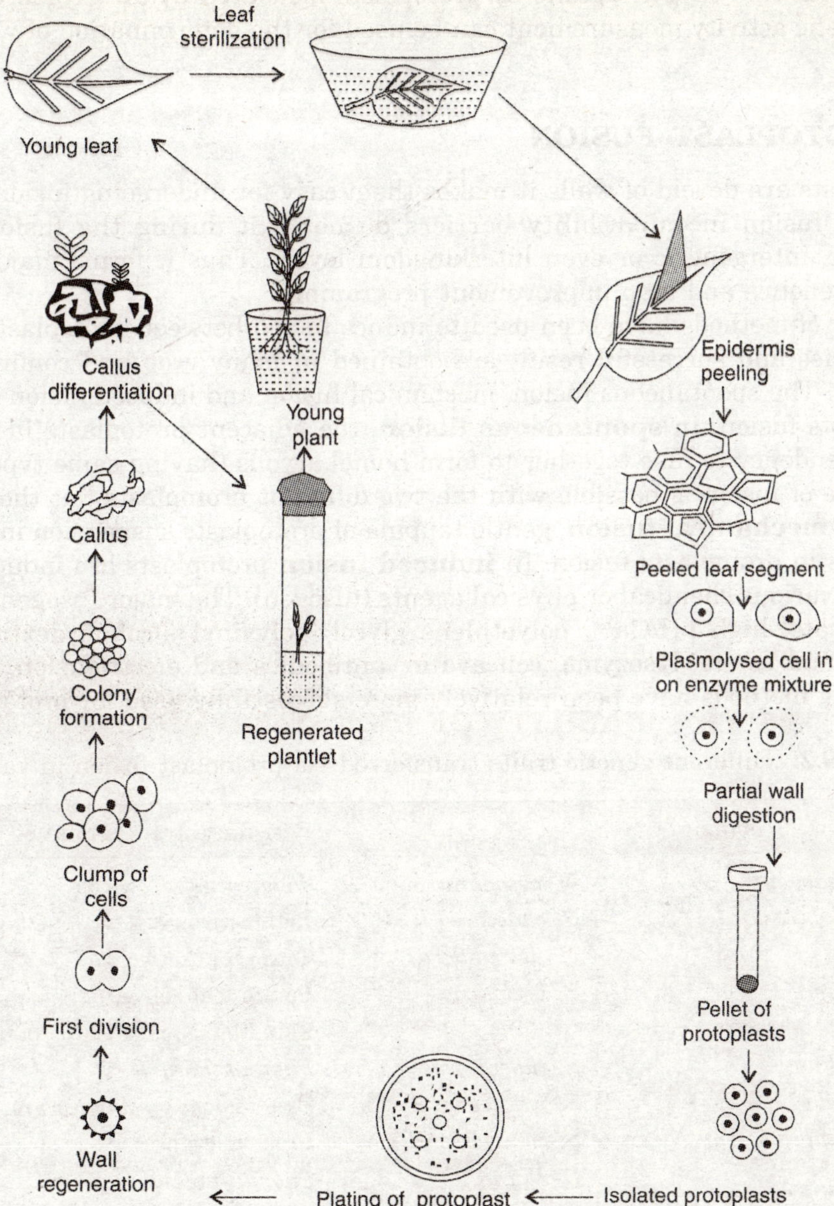

Fig. 19.2. Diagrammatic representation of the technique used for isolation, culture and regeneration of plants from leaf protoplasts.

(iv) Evans blue staining

The 0.025% of Evan's blue stain solution was used for staining the protoplasts. The stain gives colour to the dead protoplasts by becoming permeable to dead ones. Whereas viable protoplasts remains colourless due to impermeability of plasma membrane to the stain. This is the most simple method used for differentiation of protoplasts or cells.

Along with this oxygen uptake by protoplasts measured by an oxygen electrode and photosynthetic activity measurement can be used for the determination of viability of the protoplasts.

19.4 PROTOPLAST FUSION

As protoplasts are devoid of walls it makes them easy for undergoing fusions *in vitro*. In protoplasts fusion incompatibility barriers do not exit during the fusion process at interspecific, intergeneric or even interkingdom levels. Thus it forms major tool for somatic cell genetics and crop improvement programme.

Number of methods have been used to induce fusion between protoplasts of different strains/species and successful result are obtained in many crop and commercial plants (Table 19.2). The spontaneous fusion, mechanical fusion and induced fusion are the types of protoplasts fusion. In **spontaneous fusion**, the adjacent protoplasts in enzyme mixture have tendency to fuse together to form homokaryons (having same type of nucleus). Similar type of fusion is possible with the two different protoplasts but the frequency is very low. In **mechanical fusion**, gentle tapping of protoplasts suspension in a depression slide results in protoplasts fusion. In **induced fusion** protoplasts are induced to fuse in presence of various chemical or physical agents (fusogen). The major fusogen used include sodium nitrate, high pH/Ca^{2+}, polyethlene glycol, polyvinyl alcohol, dextran sulphate, fatty acids and esters, lysozyme, concavalin, antibodies and electrofusion. Out of these the following methods have been relatively more successfully used for protoplasts fusion.

Table 19.2: Different genetic traits transferred via protoplast fusion in various crop

Somatic hybrids obtained		Resistance trait plants
Nicotiana tabacum	+ *N. nesophila*	Tobacco mosaic virus
Solanum tuberosum	+ *S. chacense*	Potato virus X
N. tabacum	+ *N. nesophila*	Tobacco horn worm
S. tuberosum	+ *S. brevidens*	Potato leaf roll virus
		Late blight and PLRV Potato virus Y
S. circalifolium	+ *S. tuberosum*	*Phytophthora*
S. melongena	+ *S. saniteongsei*	*Pseudomonas solanacerum*
S. tuberosum	+ *S. commersonii*	Frost tolerance
B. napus	+ *B. nigra*	*Phoma lingam*
B. oleracea	+ *Sinapis alba*	*Alternaria brassiceae*
B. napus	+ *Sinapis alba*	*Alternaria brassiceae*
Citrus sinensis	+ *Poncirus trifoliata*	*Phytophthora*
Lycopersicon esculentum	+ *L. peruvianum*	TMV, spotted wilt virus, cold tolerance
S. lycopersicoides	+ *L. esculentum*	Cold tolerance
Solanum ochranthum	+ *L. esculentum*	Tomato diseases and insect pests

(Contd.)

Somatic hybrids obtained		Resistance trait plants
Brassica oleracea	+ *B. napus*	Black rot *(Xanthomonas campestris)*
B. oleracea spp. capitata	+ *B. oleracea* (Ogura CMS line)	Cold tolerance
S. tubersum	+ *S. cammersonii*	Frost tolerance
R. sativa (Japanese radish)	+ *B. oleracea var.*	Club rot disease *botrytis*
B. oleracea var. botrytis	+ *S. alba* + *B. carinata*	*Alternaria brassicola* and *Phoma lingam*
Citrullus lanatus	+ *Cucumis melo*	Club rot resistance
Hordeum vulgare	+ *Daucus carota*	Frost and salt tolerance
Solanum melongena	+ *S. sisymbrifolium*	Nematode
S. tuberosum	+ *S. bulbocastamum*	Nematode
S. tuberosum	+ *S. bulbocastamum*	Root knot nematode
Raphanus sativus	+ *Brassica napus*	Beet cyst nematode
Sinapis alba	+ *R. saticus* + *B. napus*	Beet cyst nematode
Quality characters		
N. rustica	+ *N. tabacum*	High nicotine content
B. napus	+ *Eruca sativa*	Low erucic acid
Agronomic characters (transferred through cybrid formation)		
N. tabacum	+ *N. syvestris*	Streptomycin resistance
N. tabacum spp.		Triazine resistance
S. nigrum	+ *S. tuberosum*	Triazine resistance
B. nigra	+ *B. napus*	Hygromycin resistance
B. napus spp.		Triazine resistance
N. tabacum	+ *N. sylvestris*	CMS
B. campestris	+ *B. napus*	CMS
N. tabacum	+ *N. sylvestris*	CMS
B. napus + B. campestris	+ *Raphanas sativa*	CMS and Triazine resistance
Oryza spp.		CMS
L. esculentum	+ *Solanum acaule*	CMS
B. napus	+ *B. tournefortii*	CMS

High pH/high Ca^{2+} treatment

Protoplasts of desired strains/species are mixed in almost equal proportion; generally they are mixed while still suspended in the enzyme mixture. The protoplast mixture is then subjected to a high pH at 10.5 and high Ca^{2+} concentration (50 m mol 1^{-1}) at 37 °C for about 30 min. This technique is quite suitable for some species, while for some others it may be toxic. This was first developed by Keller and Melchers (1973) and used for many intraspecific or interspecific somatic hybrids production.

Sodium nitrate treatment

The isolated protoplasts are transferred to 0.25 M sodium nitrate solution and then centrifugation results in protoplasts fusion. Power and others in 1970 developed this method.

Polyethylene glycol (PEG) treatment

Polyethylene glycol (PEG) induced fusion is the most commonly used as it induces frequency of fusion and low toxicity to most plant cells. The protoplast mixture is treated with 28-50% PEG (MW 1,500-6000) for 15-30 min followed by gradual washing of the protoplasts to remove PEG; protoplast fusion occurs during the washing. The washing medium may be alkaline (pH 9-10) and contain a high Ca^{2+} ion concentration (50 m mol 1^{-1}); this approach is a combination of PEG and high pH- high Ca^{2+} treatments, and is usually more effective than either treatment alone. PEG is negatively charged and may bind to cation like Ca^{2+} which, in turn, may link to the negatively charged molecules present in plasma lemma; they can also bind to cationic molecules of plasma membrane. During the washing process, PEG molecules may pull out the plasma lemma components bound to them. This would disturb plasma lemma organisation and may lead to the fusion of protoplasts located close to each other (Fig. 19.3).

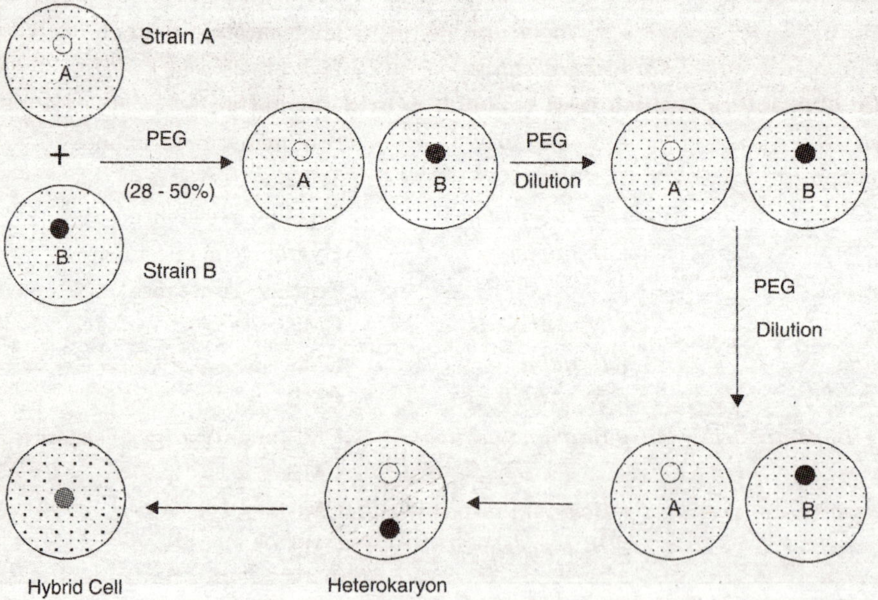

Fig. 19.3. Polyethylene glycol induced protoplast fusion. Protoplasts are first brought close together (aggregation) by PEG. Fusion occurs during PEG dilution due to disturbances created in plasma membrane.

Electrofusion technique

All the above fusion techniques are nonselective in that they induced fusion between two or more protoplasts. A more simple, selective and less drastic method is the electrofusion technique which utilizes low voltage electric current pulses to align the protoplasts in a single row like a pearl-chain. The aligned protoplasts are pushed, with a micromanipulator, at a gentle phase through the narrow gap between the two electrodes. When two protoplasts that are to be fused are appropriately oriented opposite the electrodes, a short pule of high voltage is released which induces the protoplasts to fuse. The high voltage creates transient disturbances in the organisation of plasma lemma, which leads to the fusion of neighbouring protoplasts. The entire operation is carried out manually in a specially designed equipment, called electroporator, under a microscope. Many workers feel that this fusion technique is more desirable than the others for a number of important reasons. This method was first followed by Senda *et al.* (1979) and later modified by Zimmermann and Scheurich (1981).

The techniques for protoplast fusion are pretty refined and highly effective for almost all the systems.

19.5 SELECTION OF HYBRID CELLS

During the process of protoplasts fusion there is a possibility of formation of homokaryons, heterokaryons and unfused parental protoplasts are present in the mixture. Hence proper selection of hybrid protoplasts or cells is of utmost important for the improvement. The biochemical, visual, albino or flow cytometric methods are employed for the selection process.

In biochemical methods some of the biochemical compounds are used in the medium (selection medium) which helps in the selection of parental and hybrid cells/protoplasts based on their difference in the expression of the characters. Some workers used antibiotics (drug sensitivity) or various salts (nitrate/chlorate-auxotrophic mutants) as compounds which are known to have effect on the growth of cultured protoplasts.

Drug sensitivity: In *Petunia* there are two species namely, *P. hybrida* protoplasts forms macroscopic callus on Murashige and Skoog medium and sensitive to (inhibited by) actinomycin D, whereas *P. parodii* forms small cells colonies and resistant to actinomycin D. The fusion protoplasts of these two species will be having character of both i.e. they form macroscopic colonies and are resistant to actinomycin D on the MS medium supplemented with actinomycin D, thus helps in selection. But parental protoplasts form either small colonies (*P. parodii*) or fails to divide and form the colonies (*P. hybrida*) because of the inhibitory effect of antibiotic.

Auxotrophic mutants: The original protoplasts have the capacity to grow in minimal medium is known as prototroph. The mutants of the prototroph which is not having the capacity to grow in the minimal medium is known as auxotroph. Various nitrate reductase deficient (nitrate non-utilising) and chlorate resistant mutants are known to occur in tobacco. The hybrid protoplasts are known to grow in the minimal medium and parental protoplasts are not able to grow in the minimal medium. It helps in the selection procedures.

In visual selection method the fused protoplasts are identified by fusing the cholorophyll rich parent with cholorophyll-deficient parent. The products of fusion are identified by using microscope because heterokaryons are bigger and green in colour, whereas parental protoplasts are either small and colourless. This is further differentiated by using suitable selective medium which supports good growth of only hybrid cells.

In case of albino mutants one parent protoplast was nuclear albino (chlorophyll-deficient) and the other one was cytoplasmic albino. Both of them are not able to grow into green somatic hybrids. But fusion product of both was green and develops into the well developed somatic hybrids. Here both cytoplasmic and nuclear components help together to develop chlorophyll, but alone is not sufficient for the chlorophyll development.

Flow cytometric method

Various research workers are using flow cytometry and fluorescent activated cell sorting for the viable protoplasts. These technique have also been applied for the sorting and selective enrichment of heterokaryons. The hybrid calli derived from this sorted material are known to regenerate hybrid plants.

19.6 REGENERATION OF PROTOPLAST

Various steps in regeneration of protoplasts are as follows:

1. The development of a cell wall around the membrane of protoplasts (**cell wall regeneration**): The development of a cell wall around the membrane of protoplasts is must for regeneration. Protoplasts produced should be protected from rupture of protoplasts (due to osmosis) until they regenerate strong cell wall. Cell wall formation of the protoplasts starts after a few hours of culturing and completed in two to several days. The more concentration of sucrose is added to the medium and less concentration of osmatic stabilizer is helpful in cell wall formation. Once the cell wall is formed the spherical shape of the protoplasts will be lost.
2. Induction of divisions in cells derived from the protoplasts to form callus tissue (**cell division**): Once the cell develops wall layer it was known to increase in its mass and directly undergoes cell divisions. There is a direct relationship between wall formation and its division. Poorly developed wall in protoplasts always shows improper cell divisions and after few divisions it forms small colonies of cells. Further divisions lead to formation of macroscopic callus after 2-3 weeks.
3. Induction of callus to regenerate whole plant (**plantlet formation**): The callus tissue is transferred onto osmotic free medium to develop good callus. Then it was induced to undergo differentiation to form shoot or embryo. This was further grown into well developed plants.

19.7 CYBRIDS (CYTOPLASMIC HYBRID)

Cybrids are cells or plants containing nucleus of one species and cytoplasm of both the parental species. These are generally produced during protoplast fusion in variable frequencies. Cybrid formation may result by fusion of normal protoplasts of one species with enucleated protoplasts (cytoplasm), elimination of the nucleus of one species from a normal heterokaryon, gradual elimination of the chromosomes of one species from a hybrid cell during the further mitotic divisions (Fig. 19.4). The cybrids can be produced in high frequencies by irradiation of one parental protoplast before fusion in order to inactivate the nuclei or by preparing enucleate protoplast of one species and fusing them with normal protoplast of other species.

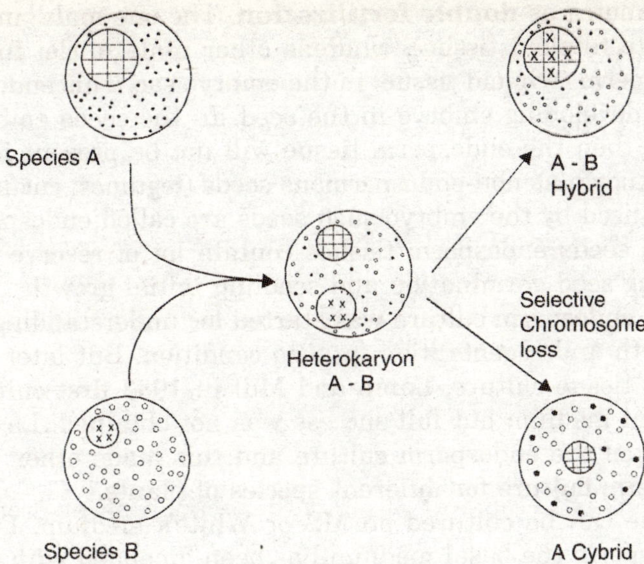

Fig. 19.4. Some fusion products resulting from protoplast culture. The fusion of protoplasts A and B results in a binucleate heterokaryon containing the cytoplasmic contents of the two original protoplasts. Fusion of the two nuclei results in a tetraploid hybrid cell or synkaryocyte. If one of the nuclei degenerates, a cybrid or heteroplast is produced.

19.8 PRACTICAL APPLICATION OF SOMATIC HYBRIDIZATION (HYBRID VS CYBRIDS)

Genetic recombination in asexual or sterile plants: Somatic hybridization is the only means of genetic recombination in plants that cannot reproduce sexually or which are completely sterile. The sterile plants are made to produce fertile diploids and polyploids by protoplast fusion of the parents.

Overcoming sexual incompatibility: Sexual crossing between two different species or two different genus fail to produce hybrids due to incompatibility barriers. This can be overcome by somatic cell fusion of two different species or genus.

Transfer of cytoplasmic characters: Cytoplasmic characters such as cytoplasmic male sterility, rate of photosynthesis, resistance to diseases or herbicides and low or high temperature tolerance are generally made to transfer to the other strains. To recover recombinants of mitochondrial or chloroplast DNA.

Mitochondria of one species can be combined with chloroplasts of another species. This may be very important in some cases and is not possible by sexual methods even between easily crossable species.

Recombinant organelle genomes, especially of mitochondria are generated in somatic hybrids and cybrids. Some of these recombinant genomes contain useful characters.

19.9 ENDOSPERM CULTURE AND PRODUCTION OF TRIPLOIDS

Endosperm is the nourishing tissue in seeds. During the process of fertilization in angiosperms the two male gamete/nuclei fuse separately with two different female nuclei and this process is known as **double fertilization**. The one male nuclei fuses with egg and forms the zygote (diploid tissue), whereas other male nuclei fuses with two polar nuclei to form endosperm (triploid tissue) in the embryo sac. This endosperm provides the nourishment to the developing embryo in the seed. In the seeds embryo completely utilises the endosperm then the endosperm tissue will not be present in the mature seeds and such seeds are known as non-endospermous seeds (legumes, cucurbits). If endosperm is not completely utilised by the embryo such seeds are called endospermous seeds (coconut, coffee). In such seeds endosperm tissues contain lot of reserve food materials and was consumed during seed germination and seedling initial growth.

In the beginning endosperm culture was started for understanding the potentiality of the endosperm growth and regeneration *in vitro* condition. But later its importance has been realised in the tissue culture. Lamp and Mill in 1933 first cultured endosperm of maize on the nutrient medium but full success was not obtained. La Rue in 1949 developed callus tissue from the endosperm culture and this made other workers to take up the work on endosperm culture for different species of plants.

Endosperm tissue can be cultured on MS or White's medium. For the induction of callus from an endosperm the basal medium has been amended with tomato juice, grape juice, maize juice and yeast extract. For some plants species asparagine and growth regulators was added in the basal medium. In some other cases addition of auxin, cytokinin, yeast extract or casein hydolysate and sucrose are known to produce good results.

Generally mature endosperm is considered not a good material for explant. Because endosperms undergo some changes after (12 days) pollination which makes it unsuitable for culturing. Thus immediately after a few (3-10) days of pollination endosperm was selected for this purpose. The entire seed or kernel was first surface sterilized and endosperm tissue was excised under sterile conditions. The excised endosperm was placed on nutrient medium. For the induction of callus from the endosperm culture should be incubated in darkness or diffuse light at 25 °C. But to induce shoot or other organ, the culture should be kept in bright light of intensity 2000-4000 lux. Sometimes association of embryo is found to be better for the response of endosperm in the cultural conditions.

19.10 APPLICATIONS OF TRIPLOIDS

Triploid plants are generally self-sterile and seedless. This character is very important in case of edible fruits such as grapes, mango, apple, banana, mul-berry, watermelon and others. In case of timber or oil yielding plants triploids are showing better performance than the diploid or tetraploids counterparts.

By breeding programme triploids are produced by crossing tetraploids and diploids plants. But in majority of the plants it is not possible and successful, thus endosperm culture was the better alternative and feasible method.

PRACTICAL APPLICATION OF TISSUE OR ORGAN CULTURE (SUMMARY)

In crop improvement methods transfer of gene and manipulation of genes in case of plants was done by conventional breeding. The importance was given to useful characters such as resistance to pathogenic diseases, insects and stress, high yielding varieties, nutritionally important varieties, long term storage capacity. This was a lengthy process and even results were not 100% successful. Now-a-days by enormous development in the field of chemistry, biochemistry, genetics, tissue culture, molecular biology and genetic engineering made it possible to genetic manipulation and transfer genes in plants is very easy.

- Plant cells which are in the culture condition is convenient to handle and use according to the requirement. Cells of plants grown in cultural conditions are similar to the plants grown in the natural conditions. Based on this plant cells which are known to be resistant to herbicides in culture are used for crop improvement of commercial varieties. This method is used for developing herbicide resistance in tomato, tobacco and carrot.
- Variation found in the cells or tissues in cultured condition can show heritable changes. Which accumulates in the somatic cells and expresses in the *in vitro* regenerated plants. This somaclonal variation found in plants was considered for development of disease resistance, high nutritional content, male sterility, quantity and quality of plants and pigmentation.
- Somatic cell hybridization helps in the fusion of desired cytoplasm and protoplasm. This is possible only by the cultured cells. This helps in the varied combination of cytoplasmic genes and nuclear genes. The fusion of interspecific or intergeneric protoplasts is possible in production of new characters in the desired plants.
- By culturing hybrid embryos it is possible to obtain the plants with the useful characters. This helps in solving the post-fertilization development which prevents normal seed development.
- Through ovule or ovary culture it is possible to overcome the problems involved in the embryo development.
- Culture system of plants provides the production of haploids by culturing anthers or ovary cells.
- The protoplasts offer good material for uptake of manipulated DNA directly or microinjection of desired genes by micropipettes or gene gun or bacterial or viral mediated gene transfer system.

Cultured cells are considered to be good for the infection by *Agrobacterium tumefaciens* and *A. rhizogenes* which successfully transfer the desired genes into the cells.

Culture of tissue also helps in the production of trioploids, polyploids in all the types of crops.

SUMMARY

Somatic hybridization can be defined as the process of production of hybrid plants through fusion of two different plant protoplasts. The hybrids formed by somatic hybridization are called somatic hybrids. Somatic hybridization involves the four steps. These include (i) isolation of protoplasts (ii) fusion of protoplasts of desired species (iii) selection of somatic hybrid cells (iv) culture of the hybrid cells and generation of hybrid plants from them.

Protoplasts can be defined as the cell without cell wall. It can be isolated by three methods. These include mechanical method, sequential enzymatic method and mixed enzymatic method. In mechanical method the cell wall is removed by means of mechanical force. In enzymatic method the cell wall is removed by pectinase and cellulase, either sequential are used together. Protoplasts used for culturing should be healthy and viable in order to undergo proper division and regeneration. This viability can be checked by microscopic observation of stained protoplasts or by using suitable chemicals to indicate active metabolism in the protoplasts (Tetrazolium reduction test).

After isolation of protoplasts from different species, they are subjected to fusion. The fusion of protoplasts can be done by using high pH or high Ca^{2+} treatment, sodium nitrate treatment, polyethylene glycol treatment and electrofusion techniques. Three types of protoplasts are formed after fusion process. These include homokaryons, heterokaryons and unfused protoplasts. The heterokaryons can be selected by biochemical (drug sensitivity) method, visual method, albino selection method and flow-cytometric method. Following selection of heterokaryons the generation of hybrid plant is done by three steps. These include cell wall regeneration, induction of cell division to form callus, and induction of callus to regenerate whole plant.

Plants containing nucelus of one species and cytoplasm of both the parental species are called cybrids. The cybrids may be formed by fusion of normal protoplasts of one species with another protoplast without nucleus.

Somatic hybridization has several applications. By somatic hybridization it is possible to do genetic recombination asexual plants, to overcome the sexual incompatibility and to transfer cytoplasmic characters such as cytoplasmic male sterility, rate of photosynthesis, resistance to diseases or herbicides etc.

The nourishing tissue to the embryo in seeds is called endosperm. It is formed by the fusion of one male nucleus with two polar nuclei. Hence, endosperm is triploid in nature. Triploid plants can be produced by the culture of endosperm tissue. Plants developed by endosperm culture are seedless and this character is very important in production of seedless edible fruits such as grapes, mango, apple, banana, watermelon etc.

Tissue culture or organ culture has several advantages over conventional breeding in the improvement of plants.

EXERCISE

1. Define somatic hybridization.
2. Explain different methods of isolation of protoplast. (BU Oct. 2001)
3. Describe various tests used for testing viability of protoplasts. (BU Oct. 2001)
4. Briefly explain protoplast fusion. (BU April 2001)
5. Give an account on hybrid cells selection. (BU April 2001)
6. What do you mean by cybrids?
7. Write a note on application of somatic hybridization. (BU Oct. 2001)
8. Explain how triploids are produced.
9. Give a detailed account on practical application of tissue or organ culture.
10. What is somatic embryogenesis? Give an account of its induction and importance. (BU Oct. 2001)
11. Define double fertilization.
12. Name different enzyme available for protoplast isolation.
13. Write a short note on fusogens. (BU April 2003)

14. What are the steps involved in regeneration of protoplast in culture conditions?
15. Explain polyethylene glycol assisted fusion of the protoplast.
16. List somatic hybrids produced in different plant species along with the resistant traits.
17. How spontaneous fusion occurs in the plant protoplasts and its conditions necessary for fusion?
18. What is meant by auxotrophic mutants?
19. Differentiate between hybrids and heterokaryons.
20. Give an account on electrofusion of protoplasts in plants.

20

Use of Plant Cell, Protoplasts and Tissue Culture for Genetic Manipulation of Plants

20.1 INTRODUCTION

Modern advances made in the gene technology made it possible to genetically manipulate the plants for the crop improvements. This genetic manipulation can be done by introduction of foreign genes (from both prokaryotic or eukaryotic organisms) into plant cells, protoplasts and tissues of both monocot and dicot plants. Today a large number of plants are genetically modified to produce transgenic plants in many important agricultural and commercial crops. Besides the selectable marker genes, a number of other foreign genes conferring insect resistance, herbicide resistance, viral, bacterial and fungal resistance, human genes and genes related to increase shelf life of plants or plant products have been transferred to plants from a wide range of plants, bacteria, viruses and humans. In these introduction of genes in plants resulted in expression and inherited in stable manner without any detrimental effect to the progenies. These transgenic plants also showed equal type of character in field condition similar to the laboratory condition.

20.2 *AGROBACTERIUM TUMEFACIENS*

In nature there are varieties of organisms especially microorganisms which have the natural ability to change the composition of the genome of their host organisms for their survival. This offers greater amount of chance for their survival in a new host environment. One such organism is *Agrobacterium tumefaciens,* which changes the genome of the host organisms after infection by integrating the DNA segment of its own. Hence this microorganism is known as **nature's genetic engineer**. *A. tumefaciens* is an aerobic, rod shaped, gram negative bacterium discovered by Smith and Townsend in 1907 as an infective agent of a widespread crown gall disease of plants. This bacterium forms overproduction of infected tissues to form tumor (mass of undifferentiated cells). They enter the plant cells at wound site generally around the crown of roots below the soil surface and hence this is known as crown gall tumor. *A. tumefaciens* contains an autoreplicative, extra-chromosomal DNA segment known as plasmid Zaena *et al.* (1974). This plasmid consists of genes responsible for transfer of host cells into tumor in plants called **T-DNA (Transferred DNA)**, which gets transferred from the bacterial plasmid to plant cells genome (natural genetic engineering). The plasmid is named as **tumor inducing plasmid (Ti-plasmid)** (Fig. 20.1) by van Larenbecke and others in 1975. This T-DNA expresses itself along with the normal host DNA.

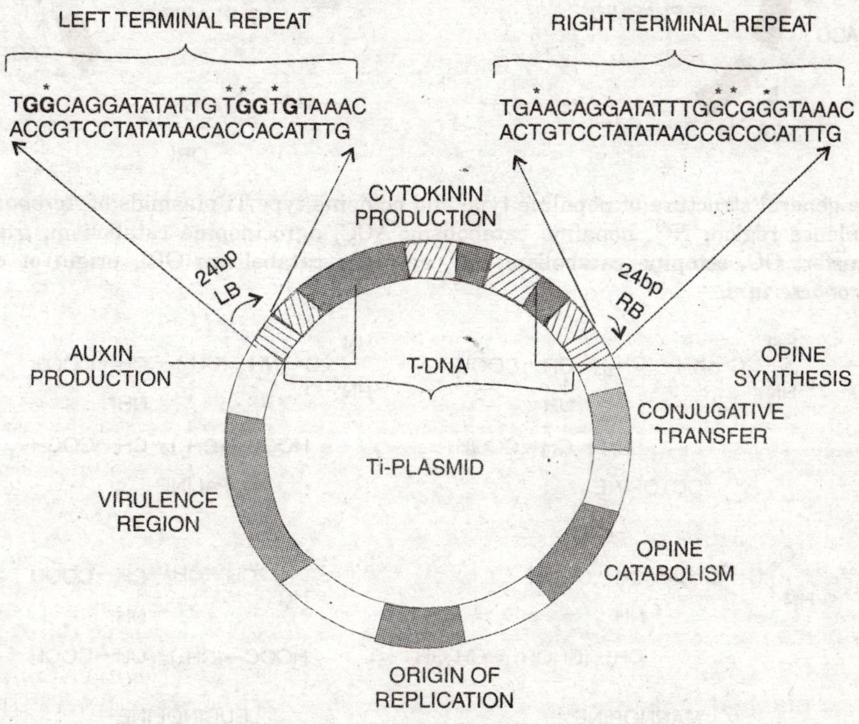

Fig. 20.1. Genetic map of a Ti-plasmid of *agrobacterium*

The asterisks above the sequences indicate the 4 bp that differ in the two border sequences.

Tumor inducing plasmid (Ti-plasmid) is a megaplasmid of about 200 kb and consists of T-DNA, *vir* region, *ori* sites, *tra* site (Fig. 20.2). Ti-plasmid also contains genes for the synthesis and catabolism of a special type of nitrogenous compounds known as opines (e.g. nopaline, octopine, leucinopine, mannopine, succinamopine, agropine and agrocinopine), auxins and cytokinins. Based on the type of opines synthesized and DNA homology Ti-plasmid can be of three major types namely, nopaline Ti-plasmid, octopine Ti-plasmid and agropine Ti-plasmid. The structural formulae of a number of opines are represented in Fig. 20.3. Opines are utilized by *Agrobacterium* present in the intercellular spaces of the tumor as their carbon and nitrogen sources. Normal plant cells do not synthesizes opines in the absence of the plasmid. Auxins and cytokinin synthesized by plasmid genes is responsible for tumorous growth of the plant cells. *Vir* region regulates the transfer of T-DNA into plant cells. *Ori* regions are responsible for the origin of replication for the plasmid multiplication. The *tra* part is helpful in conjugative transfer of the plasmid.

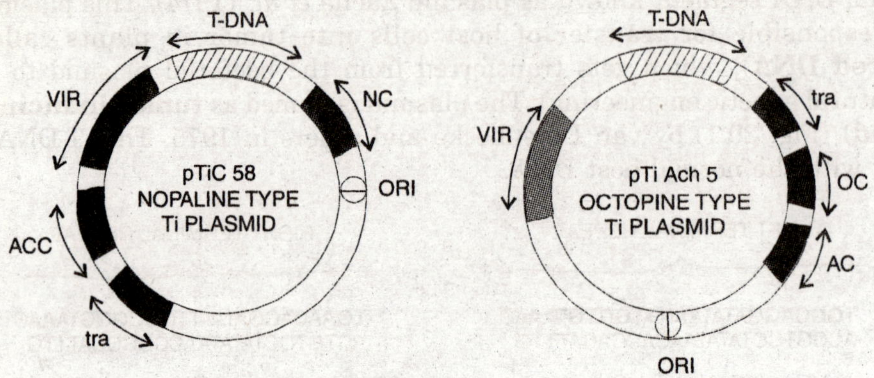

Fig. 20.2. The general structure of nopaline type and octopine type Ti plasmids of *Agrobacterium*. VIR, virulence region; NC, nopaline catabolism; ACC, agrocinopine catabolism; *tra*, conjugative transfer; OC, octopine catabolism; AC, agropine catabolism; ORI, origin of replication in *Agrobacterium*.

H_2N, HN=C–NH–$(CH_2)_3$–CH(–NH–CH(CH_3)–COOH)–COOH
OCTOPINE

H_2N, HN=C–NH–$(CH_2)_3$–CH(–NH–CH(COOH)–$(CH_2)_2$–COOH)–COOH
NOPALINE

O=C(H_2N)–$(CH_2)_3$–CH(–NH–CH_2–$(CHOH)_4$–CH_2OH)–COOH
MANNOPINE

$(H_3C)_2$CH–CH_2–CH(–NH–CH(COOH)–$(CH_2)_2$–COOH)–COOH
LEUCINOPINE

O=C(H_2N)–CH_2–CH(–NH–CH(COOH)–$(CH_2)_2$–COOH)–COOH
SUCCINAMOPINE

O=C(H_2N)–$(CH_2)_2$–CH(C=O)–O–CH–$(CHOH)_3$–CH_2OH, NH, CH_2 (ring)
AGROPINE

Fig. 20.3. Structural formulae of different types of opine

20.3 STRUCTURE OF TRANSFERRED DNA (T-DNA)

The transferred DNA (T-DNA) portion is approximately 23 kb in length and is transferred into the plant genome in tumor cells from the Ti-plasmid after infection of *Agrobacterium* (Fig. 20.4). It consists of 24 bp direct repeat sequence on the border known as TL (left

T-DNA) and TR (right T-DNA). In between these segments there are genes responsible for tumor formation (oncogenes-three genes) and opine synthesis. The *iaa M* and *iaa H* codes for typtophan-2-monooxygenase and indoleacetamide hydrolase respectively and helps together in synthesis of an auxin, indole acetic acid from the tryptophan precursor. The *ipt* gene codes for isopentenyl transferase helps in synthesis of a cytokinin, isopentenyl adenine. The *iaa M* and *iaa H* together constitute the *tms* (tumor with shoot) locus and *ipt* constitute *tmr* (tumor with root) locus. When mutation in *iaa M* and *iaa H* gene results in the absence of auxin and forms the shoots in tumor because of cytokinin. Whereas mutation in *ipt* gene results in lack to cytokinin and forms the roots because of auxin. One more locus tml (tumor large) mutation results in the production of large tumor. The opine biosynthesis gene are present near the right (TR) border of T-DNA.

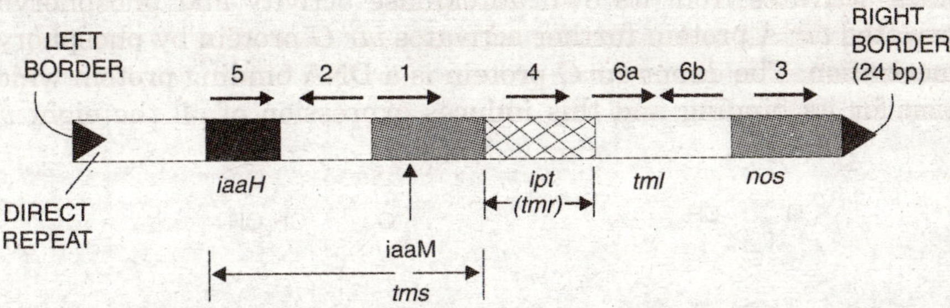

Fig. 20.4. Structural organisation of a nopaline type Ti plasmid T-DNA of *Agrobacterium*. Arrows represent the direction of transcription and numbers indicate the transcriptional units. *iaaM,* encodes tryptophan-2-monooxygesase; *iaaH,* encodes indoleacetamide hydrolase [both involved in auxin biosynthesis, together constitute the *tms* (shooty tumour) locus]; *ipt,* endodes isopentenyl transferase (cytokinin biosynthesis); *tml*, large tumour locus; *nos,* encodes nopaline synthase.

The virulence (*vir*) genes are involved in the transfer of T-DNA into plant genomes, but they will not be transferred themselves. These *vir* genes only mediate the T-DNA transfer and consists of eight operons (*vir A, vir B, vir C, vir D, vir E, vir F, vir G and vir H*) having about 40 kb length (Fig. 20.5). The eight operons consist of about 25 genes and

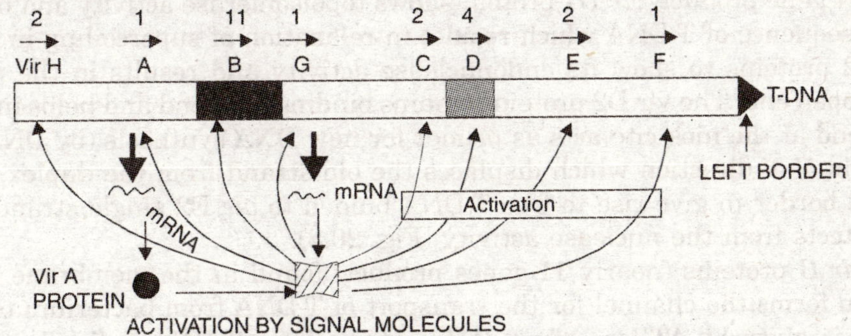

Fig. 20.5. Structural organisation of the *vir* region of an octopine type Ti plasmid of *Agrobacterium*. Operons *vir A* and *vir G* show low levels of constitutive expression. In response to plant phenolic signal molecules, *Vir A* protein phsophorylates itself and *Vir G* protein, which then induces all the 8 operons of *vir* regulon. The arrows, above the letters shows the operons, indicate the direction of transcription, and the numbers above the arrows represent the total number of genes in different operon.

of which *vir A* and *vir G* only code for only one protein each and other codes for many proteins. The *vir A* and *vir G* are constitutive genes and regulate the expression of all other genes in *vir* operons. The genes in T-DNA are not required for the transfer of the DNA.

The proper mechanism of T-DNA transfer is not completely understood. The transfer is regulated by the *vir* genes products especially *vir A* and *vir G* proteins. The *vir A* protein is activated by the phenicolic signal molecules acetosyringone (AS) and α-hydroxyacetosyringone (OH-AS) (Fig. 20.6) which are produced by the all dicots plant tissues during wounding. These signaling molecules were discovered by Stachel *et al.* in 1985. *Vir A* protein acts as receptor for phenolic signal molecules binding and work along with *Chv E* protein present in inner membrane of the bacterium. Binding of phenolic signal molecules activates from its own autokinase activity and phosphorylates itself. This phosphorylated *vir A* protein further activates *vir G* protein by phosphorylation and results in dimerization. The dimer *vir G* protein is a DNA binding protein which requires phosphorylation for its binding and this induces expression of all the eight *vir* operons (Fig. 20.7).

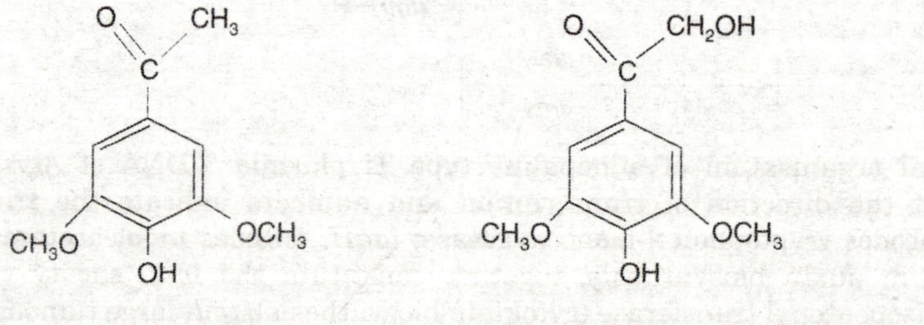

CH_3, O, C, CH_3O, OCH_3, OH
Acetosyringone (AS)

CH_2OH, O, C, CH_3O, OCH_3, OH
A-Hydroxy-Acetosyringone (OH-AS)

Fig. 20.6. Structures of phenolic signal compounds; acetosyringone and a-hydroxyacetosyringone. These signal molecules are produced by wounded plant tissues, and constitute the wound response. These compounds activate the *vir* operons whose function is essential for T-DNA transfer into plant cells chromosomes.

The *vir* D1 gene product, *vir* D1 protein shows topoisomerase activity and binds to the right border sequence of T-DNA which results in relaxation of supercoiling in DNA. This allows *vir* D2 proteins to show its endonuclease activity and results in the nick in the DNA with 3′ open end. The vir D2 protein remains binding at 5′ end and helps in transport. The open 3′ end at the nick site acts as primer for new DNA synthesis (by DNA polymerase) enzyme in 5′-3′ direction which displaces the old strand from the duplex. This later nicked at left border to give rise to free T-DNA binded to *vir* E2 single stranded protein and thus protects from the nuclease activity (Fig. 20.8).

Various *vir B* proteins (nearly 11 genes product) found in the membrane along with *vir* D4 protein forms the channel for the transport of T-DNA from bacterium to the plant cells. Energy is given by ATPase of *vir B11* protein. Then T-DNA is finally transported into the nucleus through nuclear pore. The single stranded T-DNA converted into double stranded DNA and integrates to chromosomal DNA at random sites by homologous recombination (Fig. 20.9). Then T-DNA starts expressing along with the plant genes and this remains as most stable part in the plant cells.

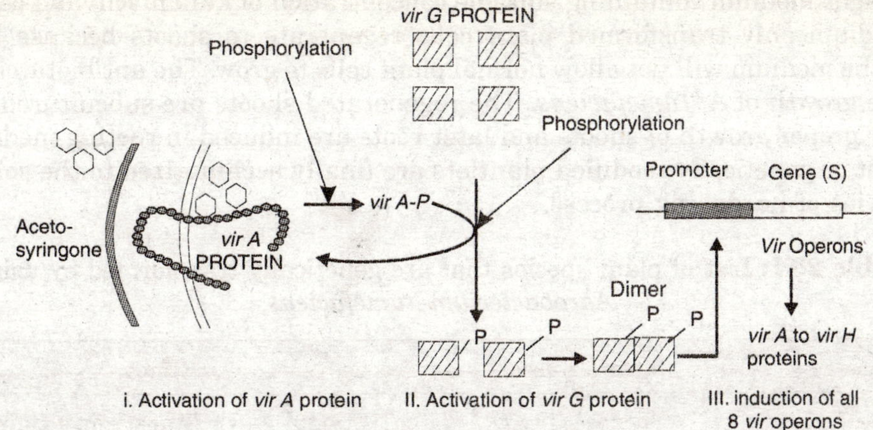

Fig. 20.7. Process of activation of *vir* operons by plant phenolic signal molecules acetosyringone/a-hydroxyacetosyringone that leads to expression of 8 operons. I. *Vir A* protein is activated by the phenolics to autophosphorylate (*Vir-P*); II. Phosphorylated *Vir A* protein phosphorylates *Vir G* protein, and *Vir G* forms dimers; III. Activated *Vir G* induces the 8 operons of *vir* regulon to yield proteins from *Vir A* to *Vir H* operons.

Agrobacterium nucleoid ('chromosomal') genes are also involved in the attachment of *Agrobacterium* to specific host plant cells. These genes include; the *chvA* and *chvB* are two linked loci that synthesize and excrete β-1,2-glucan, the *cel* locus synthesizes cellulose fibrils, the *pscA* (*exoC*) affects both cyclic glucan and acidic succinoglycan synthesis and the *att* locus affects cell surface proteins.

20.4 TUMOR FORMATION ON PLANTS USING *A. TUMEFACIENS* (MONOCOTS VS DICOTS)

Tumor can be induced in any dicotyledonous plants by making the wound on the surface sterilized stem or root or any other plant segments. The *A. tumefaciens* was made to contain recombinant gene in the Ti-plasmid before induction of tumor. This recombinant gene should be generally a beneficial gene which has to express in the plants continuously (Table 20.1). The removal of oncogenes involved in synthesis of auxin and cytokinin from T-DNA in Ti plasmid is called **disarming.** This disarming was not affecting transforming ability of the T-DNA from the bacterium. But cells did not show tumorous growth even with T-DNA in the cells. Now it was clear that only RT and LT is required for such transformation and not the entire T-DNA. This gave a idea of using Ti-plasmid as vector for the transferring any gene to the plant cells without any difficulty.

The required gene along with marker gene (antibiotic-kanamycin resistance) was inserted within T-DNA region of disarmed Ti-plasmid of *Agrobacterium tumefaciens.* This recombinant *A. tumefaciens* was cultured in suspension culture along with the wounded plant segments for two days. Generally circular leaf discs (a few mm diameter) are taken out from surface sterilized leaves. This forms the co-culture of *A. tumefaciens* with plant tissues (Fig. 20.10). At incubation period *A. tumefaciens* infects the leaf discs through wound done during disc preparation. These discs are transferred onto the fresh shoot induction agar medium containing suitable concentration of kanamycin and cerbenicillin. In this medium only transformed plant cells regenerate to shoots because kanamycin present in the medium will not allow normal plant cells to grow. The antibiotic carbenicillin inhibits the growth of *A. tumefaciens.* The regenerated shoots are subcultured onto fresh medium for proper growth of shoots and later roots are induced in rooting medium. These recombinant or genetically modified plantlets are finally acclimatized to the soil condition through series of hardening process.

Table 20.1: List of plant species that are genetically transformed by using *Agrobacterium tumefaciens*

Plant Species	Material used	Vector	Result
Alfalfa *(Medicago saliva)*	Stem segments	pTiBO542	Transformed callus and plants
Asparagus *(Asparagus officinalis)*	Stem segments	pTiC58	Transformed callus
Cotton *(Gossypium hirsutum)*	Hypocotyl segments	pCMC204	Transformed plants
Potato *(Solanum tuberosum)*	Stem or leaf segments	pTiBO542/ pGA 472	Transformed callus
Soybean *(Glycine max)*	Hypocotyl, embryonic axes	pNon 9749 pTV1100	Transformed plants
Tobacco *(Nicotiana tabacum)*	Leaf discs	pTiBO542/ pGA472	Transformed callus and plants
Tomato *(Lycopersicon esculentum)*	Leaf discs	pMON200SE	Transformed callus and plants
Walnut *(Juglans regia)*	Somatic embryo	pCEN594	Transformed plant

In the early 1980s number of workers in Europe and U.S.A. used this bacterium to insert foreign genes into plant cells or tissues, they were then able to recover whole plant from transformed cells. The resulting plants were found to incorporate foreign genes into their own genomes and thus had acquired new, stable and inherited characters. This led to the use of bacterial genes for antibiotic resistance and for the production of unusual amino acids. The Belgium biotechnology company, Plant Genetic Systems in 1987 using endotoxin gene from *Bacillus thuringiensis* developed transgenic tobacco plant expressing the toxic gene in the plant. These plants provide excellent protection against pest. Followed by this all over the world scientists reported the genetic manipulation of crop plants.

A. tumefaciens normally does not infect the monocotyledonous plants and does not form tumors. This is due to the fact that wound response is not present in such plants. The wound responses are necessary to initiate *A. tumerfaciens* through induction of *vir* genes.

Especially the signaling molecules are not synthesized which triggers the infection process. However in few cases they are known to case infection with tumor formation e.g. *Asparagus officinalis, Chlorophylum, Narcissus, Dioscorea bulbifera* and *Oryza sativa.*

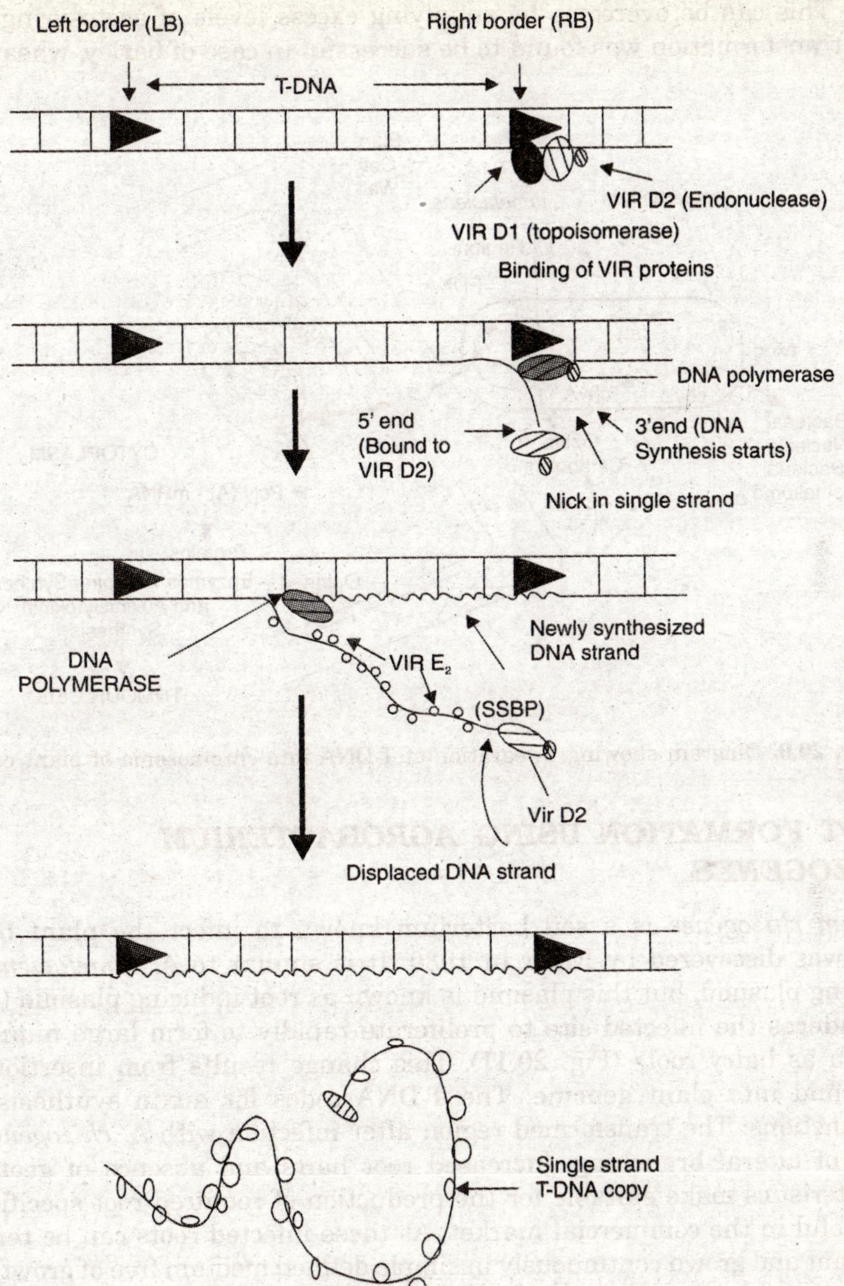

Fig. 20.8. Diagrammatic representation of the generation of T-DNA copy (single-stranded) necessary for transfer into plant cells genome. LB and RB, left and right borders of T-DNA; SSBP, single-strand binding proteins.

This infection process can be increased with effective transformation by adding external acetosyringone or wound exudates during the co-culture of plant cells with *A. tumefaciens*. Some monocot plants produce compounds which suppress the *vir* gene induction mechanism. This can be overcome by supplying excess levels of acetosyringone. By this mechanism transformation was found to be successful in case of barley, wheat, maize and rice.

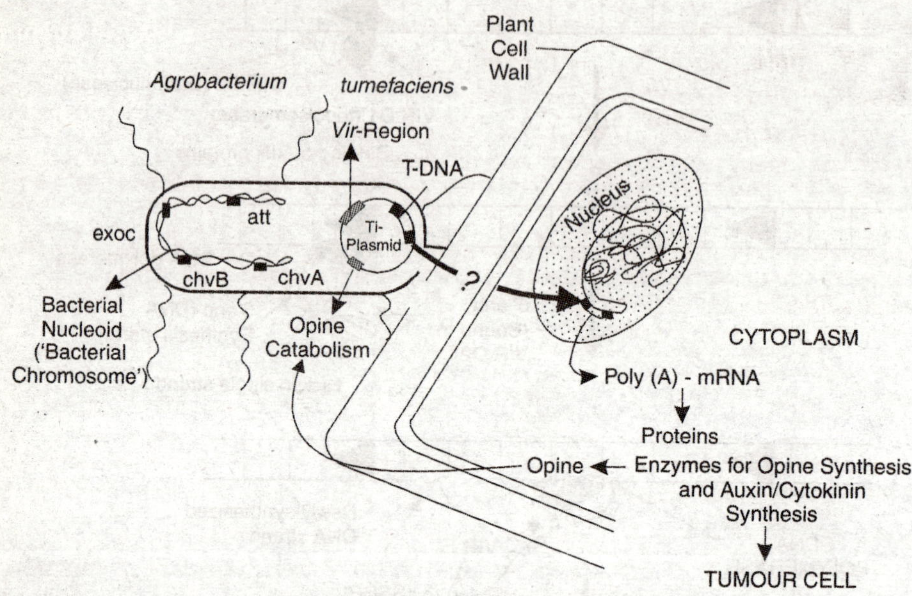

Fig. 20.9. Diagram showing integration of T-DNA into chromosome of plant cell.

20.5 ROOT FORMATION USING *AGROBACTERIUM RHIZOGENES*

Agrobacterium rhizogenes is a soil bacterium known to infect the plant belonging to dicots. This was discovered by Riker in 1930. It is similar to *A. tumefaciens* in having tumor inducing plasmid, but this plasmid is known as root inducing plasmid (Ri-plasmid) because it induces the infected site to proliferate rapidly to form large number of roots and is known as hairy roots (Fig. 20.11). This change results from insertion of T-DNA from Ri-plasmid into plant genome. The T-DNA codes for auxin synthesis and other rhizogenic functions. The transformed region after infection with *A. rhizogenes* produces high degree of lateral branching, increased root hairs and absence of geotropism. All these characteristics make suitable for the production of required root specific chemicals which are useful in the commercial market. As these infected roots can be removed from the parent plant and grown continuously in simple defined medium free of growth hormones (auxins). But normal root cells requires external supply of auxins and growth hormones. These hairy root cultures are having great potentiality in the production of root-derived secondary metabolites from dicot plant.

Number of root derived compounds such as tropane alkaloids, atropine, catharanthin, betalanin, nicotine, anatabine, hyscyamine, shikonin and sterioidal precursors, namely, solasidine having significant commercial applications. These compounds can be produced in high quantity by using the specific plants in the hairy root culture system.

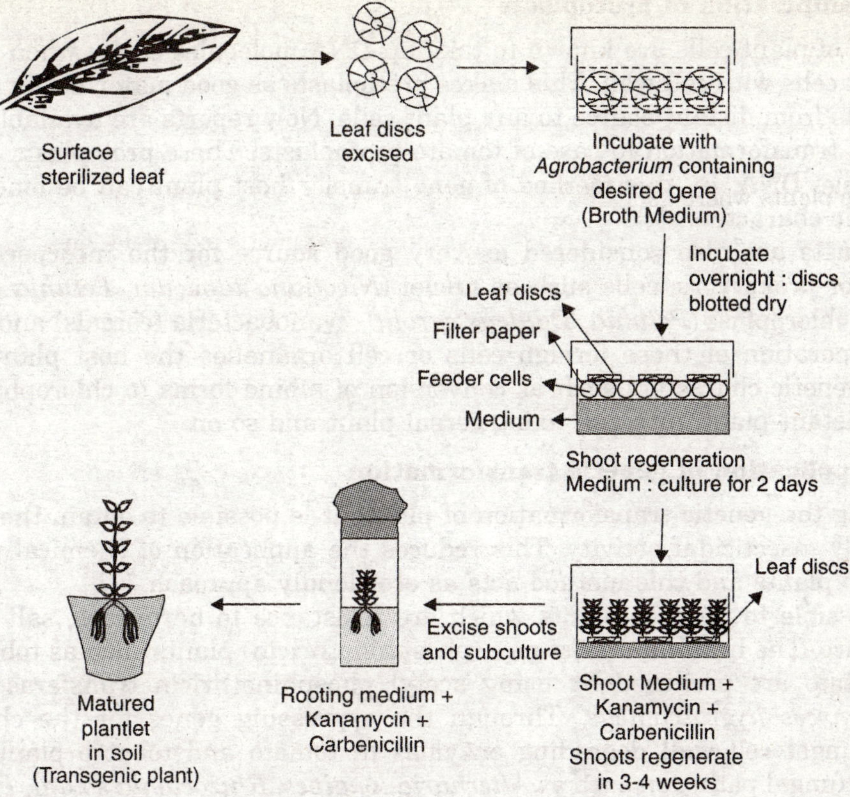

Fig. 20.10. A schematic representation of the procedure for *Agrobacterium-mediated* gene transfer into leaf tissues to obtain genetic transformation.

Fig. 20.11. Hairy roots formation at infected site of the plant.

The hairy root culture of any plant can be started by inoculating thick bacterial suspension by wounding of surface sterilized leaf disc or stem of plantlet or seed or petioles or stem segments from green house plants.

Genetic manipulation of protoplasts

Protoplasts of plant cells are known to take up DNA molecules easily when compared to intact plant cells with cell wall. This makes protoplasts as good materials for introduction of new traits from distant source to any plant cells. Now reports are available to indicate this type of transformation in case of tomato protoplasts. These protoplasts are taken in lambda phage DNA. By this method of gene transfer host plant can be made to change their genetic characters.

Protoplasts are also considered as very good source for the incorporation of cell organelles or prokaryotic cells such as nuclei (*Nicotiana tabacum, Petunia hybrida* and *Zea mays*), chloroplast *(Petunia, Daulcus carota),* cyanobacteria (cereals) and *Rhizobium.* After incorporation of these foreign cells or cell organelles the host plant cells show change of genetic characters such as conversion of albino forms to chlorophyllous plant, disease resistant plant, nitrogen fixing cereal plant and so on.

Practical application of genetic transformation

- By using the genetic transformation of plants it is possible to obtain the plants with generally insecticidal activity. This reduces the application of chemical pesticides to the crop plants and this method acts as ecofriendly approach.
- It is possible to produce plants which are resistance to herbicides, salt and disease resistance. The herbicide resistance (phosphinothricin) plants such as tobacco, tomato and potato are produced by using acetyl phosphinothricin transferase gene from *Streptomyces hygroscopicus.* Through the expressing genes for the chitinase and other fungal cell wall degrading enzymes in tomato and tobacco plants protection against fungal pathogen such as *Alternaria longipes, Rhizoctonia solani, Phytophthora infestans* was achieved.
- Antisense RNA is a RNA molecule having complementary sequences for the mRNA. Thus, it inhibits the expression of such specific proteins from the mRNA. By using this antisense sequence in plants we can control the expression of particular enzymes which is involved in ripening processes. For example polygalacturonase is responsible for the ripening of tomato fruits and leads to spoilage. If we have antisense RNA for this enzyme in the plants the ripening process in prolonged.
- Genetic manipulation also carried out in crops for making desirable storage proteins, vitamins, amino acids and for the suppression of antinutritional proteins synthesis.
- Through the genetic manipulation plants are made to produce the antigenic proteins of the pathogenic organism and thus can be used as edible vaccines. Such efforts are made in number of plants including potato, tomato, tobacco for *Vibrio cholerae,* Hepatitis B virus and others.
- Plants are made to transform into hairy roots characters to produce high concentration of secondary metabolites having high commercial value.

SUMMARY

Plant gene technology made it possible to genetically alter the plants for the crop improvement. Now a large number of agriculture and commercially important plants are genetically modified with respect to several desirable characters; such as to enhance yield, insect resistance, herbicide resistance, microbial resistance and etc. Genetically modified plants are called transgenic plants.

The microorganisms such as *Agrobacterium* species is referred as nature's smallest genetic engineer, because it has natural ability to change the genome composition of the host organism for their own benefit. *Agrobacterium tumefaciens* is a soil borne microorganism causes crown gall disease in dicot plants. The ability to cause crown gall disease is associated with the presence of Ti-plasmid within the bacterial cell. This is greater than 200 Kb in size that carries T-DNA, *vir*-region, ori site, *tra* site, genes for the metabolism of opines, genes for synthesis of auxins and cytokinins. The remarkable feature of Ti-plasmid is that, after infection, a part of Ti-plasmid is integrated into the plant chromosomal DNA. This segment is called T-DNA. T-DNA is responsible for unregulated proliferation of the infected cells.

The T-DNA is about 23 Kb in length and transferred into the plant chromosomal DNA from Ti-plasmid with the help of *vir* genes. It consists of 24 bp direct repeat at terminal region known as TL and TR. The genes present in between direct repeats are responsible for tumor formation and opine synthesis. The *vir* region consists of eight operons (A, B, C, D, E, F and H). Totally there are 25 genes present in this eight operons. It is about 40 Kb in its length.

Any desirable gene of interest can be introduced into plant genome using disharmed Ti-plasmids. The removal of cancerous genes from Ti-plasmid is called disharming and plasmid is called disharmed Ti-plasmid. Several transgenic plants have been produced using *Agrobacterium tumefaciens* mediated gene transfer method. This method was not used for transformation of moncot plants with gene of interest. This is due to lack of specific phenolic signal in monocot plants required for the infection by *Agrobacterium tumefaciens*. However, this mechanism has become successful in transformation of barley, wheat, maize and rice in the presence of excess of acetosyringone (phenolic signaling molecule).

Agrobacterium rhizogenes is another siol-borne bacterium causes hairy root disease in dicots. Similar to *Agrobacterium tumefaciens* it is also having a plasmid called Ri-plasmid, which is responsible for hairy root formation. This plasmid has been exploited in the production of root-derived secondary metabolites such as atropine, tropane alkolids, catharanthin and others.

Plant protoplasts serve as best tool for the transformation of plants with desirable character. Thus by using plant cells, protoplasts, tissues, plants can be manipulated in such a way that, it produces required product throughout the year at optimal quantity.

EXERCISE

1. Write a note on *Agrobacterium tumefaciens.*
2. Which organism is known as nature's genetic engineer?
3. Define Ti-plasmid.
4. Name different types of Ti-plasmid.
5. Explain structure of T-DNA.
6. Describe mechanism involved in T-DNA transfer to plants genome. (BU April 2003)
7. Write a note on the importance of plants genome *Agrobacterium rhizogenes.* (BU Oct. 2002)

8. Give an account of practical application of genetic transformation.
9. Write a short note on Ti plasmid. (BU Oct. 2001)
10. Elucidate practical applications of genetic manipulation.
11. Describe role played by signal molecules in plants infection by *Agrobacterium.*
12. Briefly explain activation of VIR operons in *Agrobacterium.*
13. Name the plants where genetic transformation was done by using *Agrobacterium.*
14. Discuss advantages of tissue culture methods used for genetic manipulations.
15. Write a note on antisense RNA.

21

Arabidopsis thaliana and Plant Genomics

21.1 INTRODUCTION

Arabidopsis thaliana is a small weed commonly called mouse-ear cress or wall cress (Fig. 21.1). It belongs to mustard family Brassicaceae and was discovered by Johannes Thal during the 16th century in Harz mountain. In the 18th century William Curtis designated this plant as 'plant of no use'. Despite the fact that genetics originated with Mendel's classical work on garden pea (*Pisum sativum*), our knowledge of the genetics of higher plants has lagged far behind our knowledge of the genetics of higher animals. The understanding of genetics of the fruit fly, *Drosophila melanogaster* supersedes our understanding of genetics of any other higher organisms, although recent work on the nematode worm, *Counorhabolitis elegans* is rapidly closing the gap.

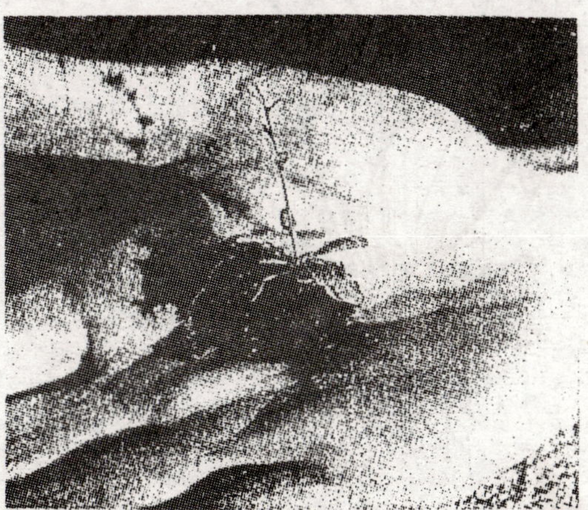

Fig. 21.1. Photograph of *Arabidopsis thaliana*. The plant is relatively small, making it easy to grow many of them in the laboratory and is thus considered as a suitable plant for genomic studies.

Plants are unique in many aspects of development, cell biology, biochemistry and physiological response. Plant cells do not migrate during morphogenesis and totipotency is completely understood/achieved. Many of the most perplexing aspects of plant physiology, development and biochemistry are now coming to be studied by the methods of classical and molecular genetics.

21.2 *ARABIDOPSIS THALIANA* AS A MODEL PLANT

In case of plants, maize is probably the best known genetically. It has been extensively studied for three major reasons: (1) Maize is of major economic importance as one of the world's foremost food crops; (2) As it is diploid, maize is subject to standard genetic analysis, whereas many of the other economically important food crops (oats and wheat)

Fig. 21.2. Sketch of *Arabidopsis thaliana* morphology. It is a small plant of Brassicaceae member and grows over a foot height. The plant also occupies the space of a foot in diameter. This plant also grows in laboratory condition to 1 cm tall, flower and set seeds when restricted by growth in a patri dish (15 × 60 mm). Thus it is considered as a characteristic of a true weed.

are polyploids; (3) The meiotic maize chromosomes are good subjects for cytogenetic studies. Totally, around 200 genetic loci have been mapped on the linkage groups that correspond to the 10 chromosomes of maize. The major disadvantages of maize for genetic studies are (1) its relatively long generation time (four months) and (2) the large size and the complexity of the maize genome. Maize in its large genome has high amount of dispersed repetitive DNA making molecular genetic analysis much more difficult. These disadvantages of maize for molecular work led geneticists to search for a better plant model system.

Arabidopsis thaliana (Fig. 21.2) is a flowering plant, an inconspicuous or small, economically unimportant weed. It belongs to the mustard family Brassicaceae. *A. thaliana* was discovered by Johannes Thal during 16th century in Harz mountain and hence the species name *thaliana*. This weed is commonly called "Mouse-ear cress" or "Wall cress". During the 18th century a botanist William Curtis wrongly said this plant as the plant not having any use for science or society that was found to be wrong. First natural mutant was observed by Braun in 1873. Laibach started working on this plant from 1907 and in 1943 mentioned its character for a model plant in genetical studies. Then the first international conference on *A. thaliana* was initiated.

A. thaliana has been used as a subject for experiments in classical plant genetics for 60 years and during 1980s it was adopted by a number of research laboratories for use in molecular genetic approaches to problems on plant physiology, biochemistry and development. The reason for this interest is that *Arabidopsis thaliana*, though a typical flowering plant in its development, physiology and response to the environment, has been recognized to have properties of life cycle and genomic organization that make it convenient for a number of the methods used in genetic studies. These properties can be divided into those of major importance for classical genetics and molecular genetics.

Table 21.1: Common names of *Arabidopsis thaliana*

Language	Common name
English:	Wall cress; mouse-ear cress
German:	Schmalwand, Gäsekraut, Thal's Gäsekresse
French:	arabette rameuse, arabette des dames
Spanish:	arabide
Dutch:	zandraket
Danish:	gåsemad
Norwegian:	vårskrinneblom
Hungarian:	lúdfü
Polish:	rzodkiewnik
Japanese:	shiro-inu-nazuna

21.3 HISTORY OF *ARABIDOPSIS THALIANA*

Arabidopsis thaliana was discovered by Johannes Thal (hence, *thaliana*) in the Harz mountains in the sixteenth century, though he called it *Pilosella siliquosa* (and it has gone through a number of name changes since). The earliest report of a mutant was in 1873 by

A. Braun. F. Laibach first summarized the potential of *Arabidopsis thaliana* as a model organism for genetics in 1943—he did some work on it much earlier though, publishing its correct chromosome number in 1907. The first collection of induced mutants was made by Laibach's student E. Reinholz. Her thesis was submitted in 1945, the work published in 1947. Langridge played an important role in establishing the properties and utility of

Table 21.2: Time line showing *Arabidopsis thaliana* and its usage in genomic studies

Timeline	Events
2001	Increased emphasis on functional and comparative genomics
2000	Completion of genome sequence
2000	11th International *Arabidopsis* Conference
1999	DNA chips and microarrays available
1999	Chromosomes II and IV sequenced
1998	*Arabidopsis* featured in Science genome issue
1997	Physical maps of all *Arabidopsis* chromosomes completed
1996	*Arabidopsis* Genome Initiative organized
1995	Standard BAC and P1 libraries constructed
1994	cDNA sequencing efforts initiated
1993	High-efficiency transformation established
1992	First chromosome walk published
1991	Stock centers and database established
1990	*Arabidopsis* Genome Project initiated
1989	First T-DNA tagged mutant gene cloned
1988	First RFLP chromosome map published
1988	Electronic *Arabidopsis* newsgroup established
1987	Third International *Arabidopsis* Conference
1986	First *Arabidopsis* gene sequences published
1986	Transformation with *Agrobacterium* reported
1985	First promoted as model for molecular genetics
1984	Genome size and complexity characterized
1983	First detailed genetic map published
1980	Expanded interest in use of *Arabidopsis* to study plant biochemistry, physiology, and development
1976	Second International *Arabidopsis* Conference
1970	First major review article published
1965	First International *Arabidopsis* Conference
1964	First *Arabidopsis* newsletter published
1945	E. Reinholz, a Laibach's student, collected first induced mutants
1943	F. Laibach first considered potential of *Arabidopsis thaliana as* a model organism for genetics
16th Century	Johannes Thal discovered *Arabidopsis thaliana*

the organism for laboratory studies in the 1950s, as did Rédei and others (such as J.H. van der Veen in the Netherlands, J. Veleminsky in Czechoslovakia and G. Röbbelen in Germany) in the 1960s. One of Rédei's many important contributions was to write scholarly reviews on *Arabidopsis*, a particularly thorough one is in Bibliographica Genetica vol. 20, No. 2, 1970, pp. 1–151. He wrote a more easily found one in Ann. Rev. Genet. (1975) vol. 9, pp. 111–127. Both go through some of the early histories of the use of *Arabidopsis* in the laboratory, though the longer 1970 one has all the details".

Arabidopsis thaliana in classical genetics

Laibach in 1943 first categorically pointed out some major features of *A. thaliana* that suit it for laboratory work in classical genetics. They include the small size of the mature plant (less than 30 cm). This makes that several plants can be grown in patri dish or in a square centimeter area (~ 10,000 seeds in single patri dish). *A. thaliana* also produces small seeds that makes more convenient for large-scale mutagenesis or storage of seeds in small volumes. The total of 50,000 seeds will be stored in 1.5 ml small eppendrof tube. The generation time is around 5 to 6 weeks thus permitting the rapid construction of multiply marked strain in less period. The small size of the plant and its ability to grow well in constant fluorescence light at room temperature, either in small soil mixture or in sterile, defined media. This makes indoor maintenance of large populations of *A. thaliana* plants inexpensive and easy.

Rédei (1975), Estelle and Somerville (1986) showed a simple way of large-scale mutagenesis by using seed mutagen such as ethyl methane sulfonate (EMS).

A. thaliana consists of five small chromosomes in haploid genome (2n = 10, n = 5). Koornneef and coworker (1983–1987) found out large number of mutations on each of the chromosomes. Totally more than 80 visible and biochemical mutations were discovered and the list is still growing. Chang *et al.* (1988) found out 90 molecular markers that are used in linkage maps.

***A. thaliana* in Molecular genetics or genomic studies**

In *A. thaliana* it is easily possible to induce new mutations, screen or select for desired phenotypes and determine the location of the genes revealed by mutant phenotypes. All these can be done in less space that can be afforded by a normal laboratory and without any special facilities. Because of these number of useful properties Whyte (1946) referred *A. thaliana* as "botanical *Drosophila*". Its properties for experiments in molecular biology include a very small nuclear genome with very low level of dispersed repetitive DNA. The haploid nuclear genome of *A. thaliana* is approximately 70,000 kilobase pairs (kb). This will be about 15 times more than the prokaryotic *E. coli* and five times more than that of eukaryotic yeast, *Saccharomyces cerevisiae.* Its genome is almost similar in size to *Caenorhabditis elegans.* But *Drosophila melanogaster* contains two times more of DNA and mammals such as mice and humans have 30 to 40 times more of genome. The majority of flowering plants used in genetic studies such as maize (*Zea mays*) are similar to genome size of mammals.

21.4 ADVANTAGES OF *A. THALIANA* USAGE IN GENOMIC STUDIES

Thus *A. thaliana* is now considered to be the best plant genetic model for genetic analysis of plant development and other biological processes unique to plants (e.g. photosynthesis).

The major advantages of using *A. thaliana* in genomic studies are summarized as follow:

1. **Small plant size**—Very large number of plants can be grown in relatively small growth chambers (height usually less than 30 cm).
2. **Short generation time**—Is about 5 weeks.
3. **Higher seed production**—Upto 40,000 seeds per plant are produced; upto 50,000 seeds can be stored in 1.5 ml tube.
4. Very small genome size—Genome contains 7×10^7 base pairs (15 times more than *E. coli* and less than 1/100th size of maize. It contains 25,498 genes.
5. **Very little interspersed repetitive DNA**—Very little or almost absence of interspersed/dispersed DNA. Thus has a unique sequence of DNA on average 120,000 base pair in length (as compared to maize having only 1000 unique sequences).
6. **Natural self-pollination**—Having very low levels of cross-pollination, self-pollination is virtually assured, because pollen is released and fertilization occurs before the flowers even open.

Because of small size of the plant and large number of seeds per plant along with the natural self-pollination *A. thaliana* are very important for mutational studies. Thus a large number of seeds are screened from M_2 (2nd generation after mutagenesis) plants. Somerville and his associates even discovered several mutants that can be identified by using gas chromatography (for change fatty acid profiles) to screen extracts of M_2 plants. This even helped in finding out genes encoding enzymes in all the metabolic pathways of *Arabidopsis*.

The genomes of other plants such as maize, rice are large enough to pose problems for using in genome & genomic studies. The small genome of *Arabidopsis* allows rapid repetitive screening of genomic libraries with minimal effort. And also a lambda or cosmid library representing several genomes of *Arabidopsis* DNA can be plated on a single large patri dish. This character along with absence of repetitive DNA makes chromosome walking much easier.

A major advantage of *Arabidopsis* is that once a mutation has been identified in a gene of interest, the mutant allele can be mapped and used as a tool in cloning the gene. Because of the small genome size and the rare occurrence of interspersed repetitive DNA sequences *in Arabidopsis*, genes can be cloned by a procedure called "*chromosome walking*". The mutant allele of interest is simply mapped relative to the positions of previously mapped molecular markers by standard genetic crosses, and then one "walks" along the chromosome from the closest marker by isolating overlapping (cross-hybridizing) genomic clones until one reaches the gene of interest (see the following discussion for details). Chromosome walking is difficult to impossible in most higher plants because of the huge genomes (one usually has too far to "walk") and abundance of repetitive DNA (each repeated sequence is a potential roadblock to the "walk").

Genetic crosses in *Arabidopsis* are performed by hand under a dissecting scope or magnifying glass by *emasculating flowers* (removing the stamens) prior to the release of pollen and then applying pollen from the male parent of choice to the stigmas of the emasculated female parent. Since the flowers of *Arabidopsis* are quite small, the emasculation procedure is somewhat tedious; however, the other major advantages of *Arabidopsis* far outweigh this minor disadvantage.

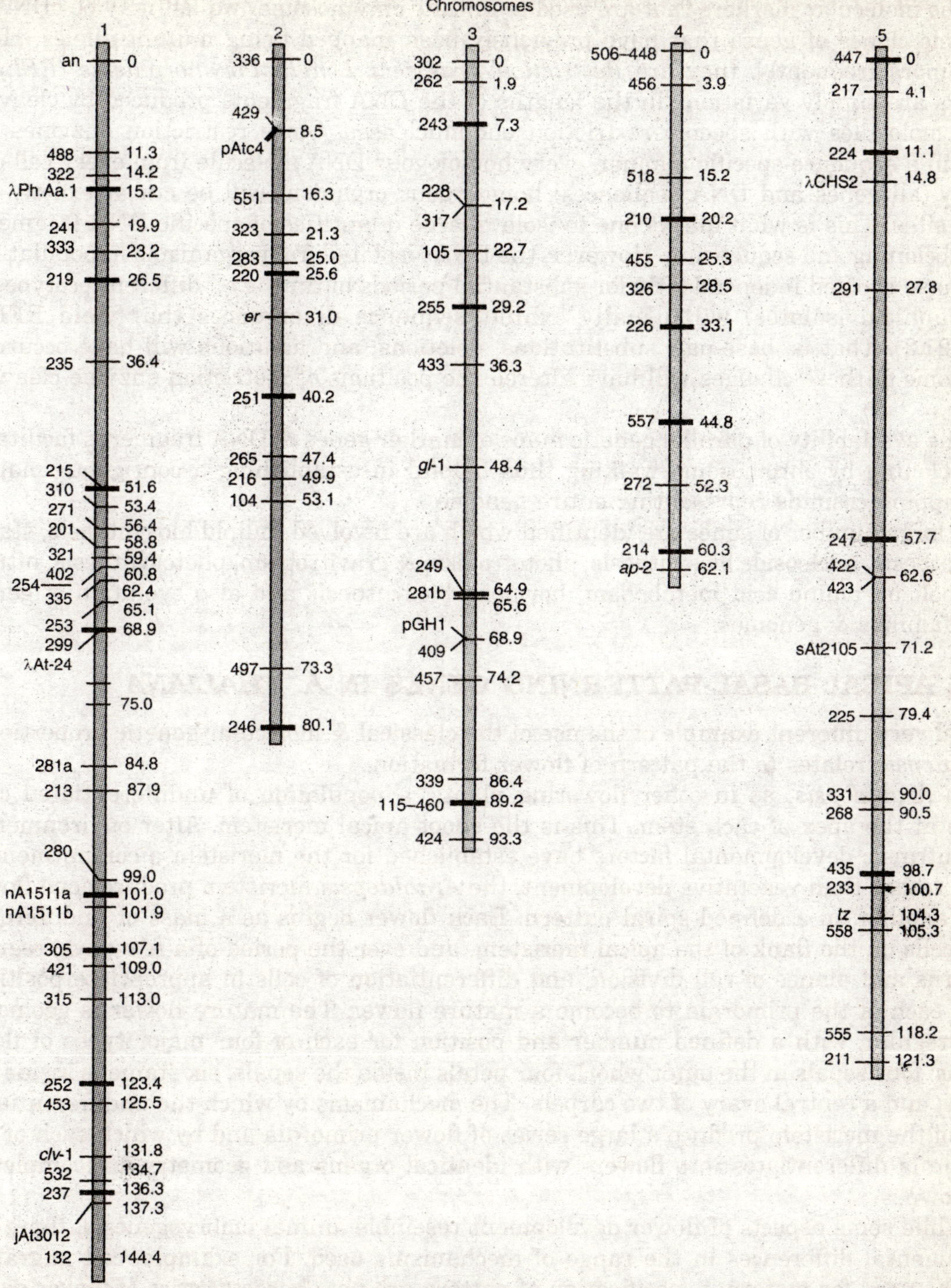

Fig. 21.3. The Restriction Fragment Length Polymorphism Linkage Map of the *Arabidopsis thaliana* genome from chromosome 1 to chromosome 5. Clone names are given on the left side of each linkage group on chromosome. The uppermost marker is named as 0 & then downwards marked on the sight side of the chromosome map.

The molecular markers that are used to initiate chromosome "walks" may be cDNA or genomic clones of genes that have previously been mapped using mutant alleles. However, more frequently, they are *Restriction Fragment Length Polymorphisms (RFLPs).* RFLPs are simply variations in the lengths of the DNA fragments produced by cleaving DNA molecules with specific restriction endonucleases. Since restriction enzymes cut DNA in a sequence-specific manner, every homologous DNA molecule from every cell of a totally (all genes and DNA sequences) homozygous organism will be cut at exactly the same sites. This is what allows one to isolate large quantities of specific DNA fragments for subcloning and sequencing. However, the DNAs isolated from organisms in populations that have evolved independently for substantial periods of time (*e.g.*, different ecotypes or geographical isolates) will usually exhibit sequence divergences that yield RFLPs (Fig. 21.3). That is, base-pair substitutions, deletions, and additions will have occurred, and some of these changes will have altered the positions of restriction enzyme cleavage sites.

The availability of parallel genetic maps of marker genes & DNA fragments facilitates gene cloning by chromosome walking thus helpful in establishing a contiguous map of overlapping cosmids representing entire genome.

A large number of genes are identified which are involved in lipid biosynthesis, starch biosynthesis, nucleoside biosynthesis, phototropism & gravitropism, photosynthesis, nitrate metabolism, amino acid metabolism, hormone biosynthesis and also evolution of genes, gene families & genomes.

21.5 APICAL-BASAL-PATTERNING GENES IN *A. THALIANA*

A final very different example of the use of the classical & molecular genetic properties of *Arabidopsis* relates to the pattern of flower formation.

In *Arabidopsis*, as in other flowering plants, a population of undifferentiated cells exists at the apex of each stem. This is the shoot apical meristem. After environmental and intrinsic developmental factors have established for the meristem a commitment to floral rather than vegetative development, the *Arabidopsis* meristem produces one flower after another, in a defined spiral pattern. Each flower begins as a mass of undifferentiated cells on the flank of the apical meristem, and over the period of a few days, regular patterns and planes of cell division, and differentiation of cells in appropriate positions leads each of the primordia to become a mature flower. The mature flower is geometrically regular, with a defined number and position for each of four major types of floral organs: four sepals in the outer whorl, four petals inside the sepals, six stamens inside the petals, and a central ovary of two carpels. The mechanisms by which the undifferentiated cells of the meristem produce a large series of flower primordia and by which each of the primordia differentiates into flowers with identical organs and geometry are completely unknown.

While some aspects of flower development resemble animal embryogenesis, there are fundamental differences in the range of mechanisms used. For example, cell migration and the need for maternal specification of pattern are not characteristics of flower development. One striking similarity between insect and flower development is the existence of homeotic mutations, which cause one organ to appear in place of another (Bateson,

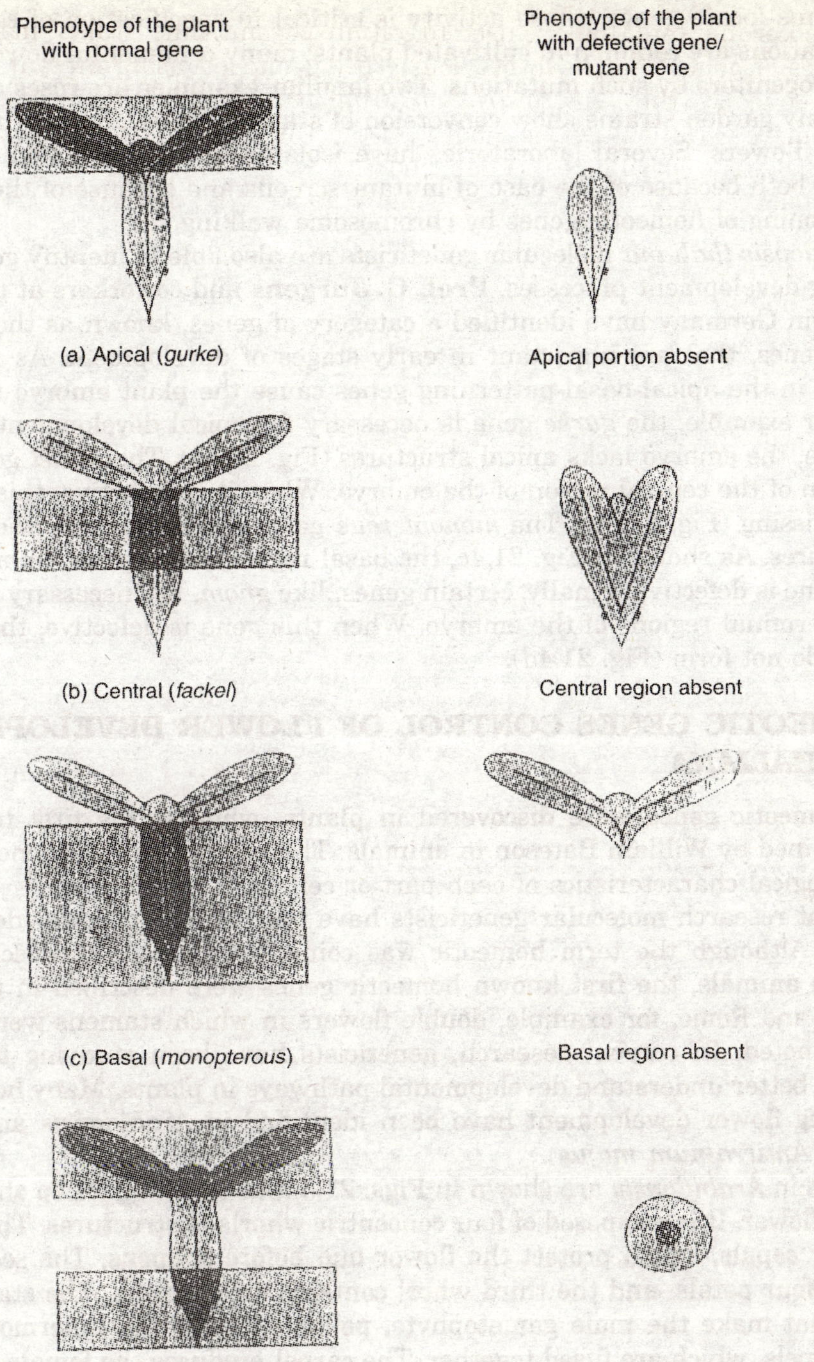

Fig. 21.4. The morphological effects of mutations in the apical-basal-patterning genes of *Arabidopsis thaliana*. Left side pictures are due to the normal genes. When the respective genes are mutated then the right side morphology will be formed in the plant embryo.

1894) and thus locate genes whose activity is critical in specification of cell fate. These types of mutations are common in cultivated plants; many ornamental flowers differ from their wild progenitors by such mutations. Two familiar examples are roses and camellias, in which many garden strains show conversion of stamens to petals and therefore larger and showier flowers. Several laboratories have isolated and studied such mutations in *Arabidopsis*, both because of the ease of mutant screens and because of the potential for molecular cloning of homeotic genes by chromosome walking.

In *Arabidopsis thaliana* molecular geneticists are also able to identify genes that play a role during development processes. **Prof. G. Jurgens** and coworkers at the University of Munchen in Germany have identified a category of genes, known as the apical-basal-patterning genes, that are important in early stages of development. As shown in Fig. 21.4, defects in the apical-basal-patterning genes cause the plant embryo to develop abnormally. For example, the *gurke* gene is necessary for apical development. When it is defective, the embryo lacks apical structures (Fig. 21.4a). The *fackel* gene influences the formation of the central region of the embryo. When it is defective, this region of the embryo is missing (Fig. 21.4b). The *monopterous* gene is necessary for the formation of basal structures. As shown in Fig. 21.4c, the basal region of the plant embryo is missing when this gene is defective. Finally, certain genes, like *gnom,* are necessary for the formation of the terminal regions of the embryo. When this gene is defective, these regions of the embryo do not form (Fig. 21.4d).

21.6 HOMEOTIC GENES CONTROL OF FLOWER DEVELOPMENT IN *A. THALIANA*

The first homeotic genes were discovered in plants, whereas the first term homeotic genes was coined by William Bateson in animals. The role of homeotic genes is to dictate the morphological characteristics of each part or cell.

In current research molecular geneticists have been studying flower development in *A. thaliana*. Although the term homeotic was coined by Bateson to describe homeotic mutations in animals, the first known homeotic genes were described in plants. In ancient Greece and Rome, for example, double flowers in which stamens were replaced by petals were noted. In current research, geneticists have been studying these types of mutations to better understand developmental pathways in plants. Many homeotic mutations affecting flower development have been identified in *Arabidopsis* and also in the snapdragon (*Antirrhinum majus*).

Examples in *Arabidopsis* are shown in Figs. 21.5 and 21.6. Fig. 21.5a shows a normal *Arabidopsis* flower. It is composed of four concentric whorls of structures. The outer whorl contains four sepals, which protect the flower bud before it opens. The second whorl is composed of four petals, and the third whorl contains six stamens. The stamens are the structures that make the male gametophyte, pollen. Finally, the innermost whorl contains two carpels, which are fused together. The carpel produces the female gametophyte. The homeotic mutants shown in Figure 21.5 have undergone transformations of particular whorls. For example, in Figure 21.5b, the sepals have been transformed into carpels, the petals into stamens. The triple mutation in which all of the whorls have been changed into leaves is shown in Figure 21.5c.

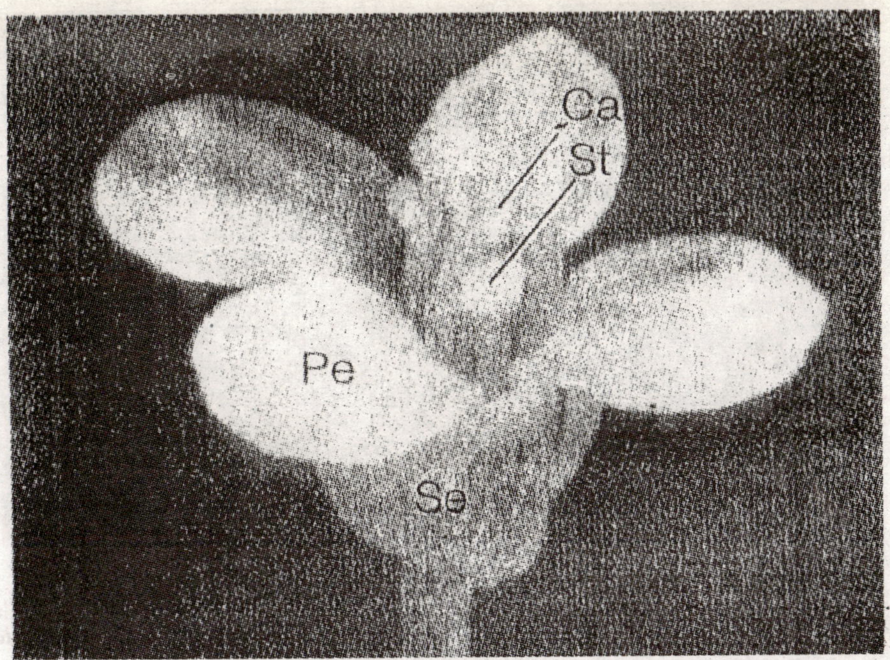

(a) Normal flower

(b) Single homeotic mutant

Fig. 21.5. Normal and different types of homeotic mutations in *Arabidopsis thaliana* flower. (a) Normal flower (b) Single homeotic mutant

(c) Triple homeotic mutant

Fig. 21.5. Normal and different types of homeotic mutations in *Arabidopsis thaliana*.
(c) Triple homeotic mutant

(a)

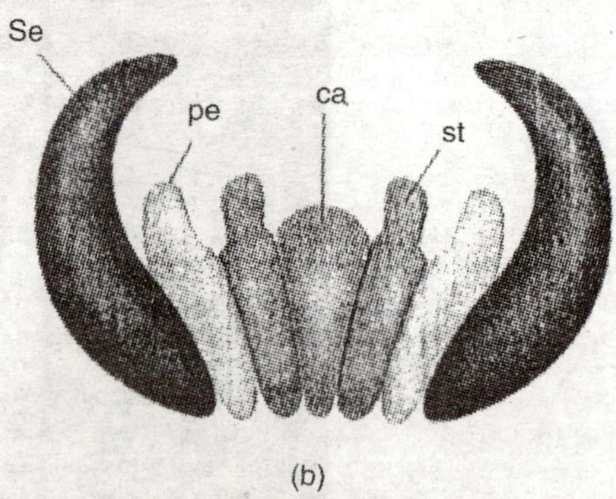

(b)

Fig. 21.6. Flower development in *Arabidopsis thaliana*. (a) The mature products of the four whorls of a flower. (b) A cross-sectional diagram of the developing flower, with the normal fates of the four whorls indicated. From outermost to innermost, they are sepal (se), petal (pe), stamen (st) and carpel (ca).

By analyzing the effects of many different homeotic mutations in *Arabidopsis*, Elliot Meyerowitz at the California Institute of Technology and his colleagues have proposed the *ABC* model for flower development. In this model, three classes of genes (Figs. 21.6 and 21.7), called *A, B,* and *C*, govern the formation of sepals, petals, stamens, and carpels. Figure 21.7 illustrates how these genes affect normal flower development and what happens when the gene products are defective, as in homeotic mutants. In the outermost whorl (whorl 1), the gene *A* products are made. This promotes sepal formation. In whorl 2, both gene *A* and gene *B* products are made, which promote petal formation. In whorl 3, the expression of genes *B* and *C* causes stamens to be made, and in whorl 4, only gene *C* is expressed, which promotes carpel formation.

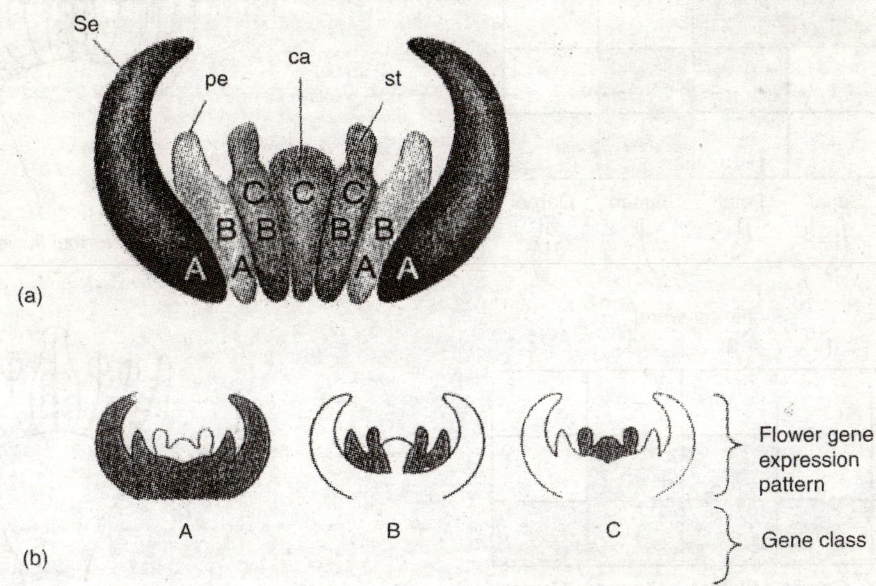

Fig. 21.7. Flower-identity gene expression and the establishment of whorl fate. (a) The patterns of gene expression corresponding to the different whorl fates. (b) The shaded regions of the cross-sectional diagrams of the developing flower indicate the gene expression patterns for genes of the A, B, and C classes. Refer to Figure 21.6 for the normal anatomy of the developing flower.

Now let's take a look at what happens in certain homeotic mutants. In this model, genes *A* and *C* repress each other's expression, and gene *B* functions independently. In a mutant defective in gene *A* expression, gene *C* will also be expressed in whorls 1 and 2. This produces a carpel-stamen-stamen-carpel arrangement. When gene *B* is defective, a flower cannot make petals or stamens. Therefore, a gene *B* defect yields a flower with a sepal-sepal-, carpel-carpel arrangement. Finally, when gene *C* is defective, gene *A* is expressed in all four whorls. This results in a sepal-petal-petal-sepal pattern.

Overall, it appears that the types of genes described in Figs. 21.8 and 21.9 promote cell differentiation that leads to either sepal, petal, stamen or carpel structures. But that happens if all three types of genes are defective. As shown in Figure 21.5c, this produces a "flower" that is composed of leaves. These results indicate that the leaf structure is the default pathway and that the *A, B,* and *C* genes cause development to deviate from a leaf structure in order to make something else. In this regard, the sepals, petals, stamens, and carpels can be viewed as modified leaves.

In *Arabidopsis,* there are two types of gene *A* (*apetala 1* and *apetala 2*), two types of gene *B* (*apetala 3* and *pistillata*), and one type of gene *C* (*agamous*). All these homeotic genes encode transcription factor proteins. The proteins contain a DNA-binding domain and a dimerization domain. However, the *Arabidopsis* homeotic genes do not contain a

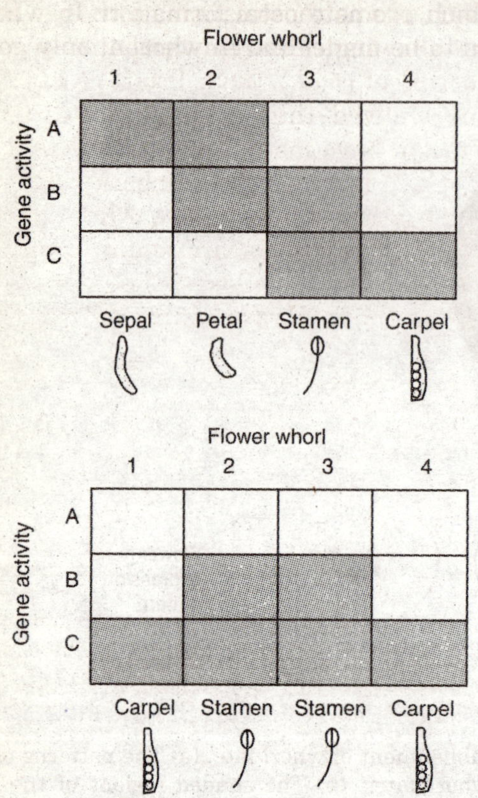

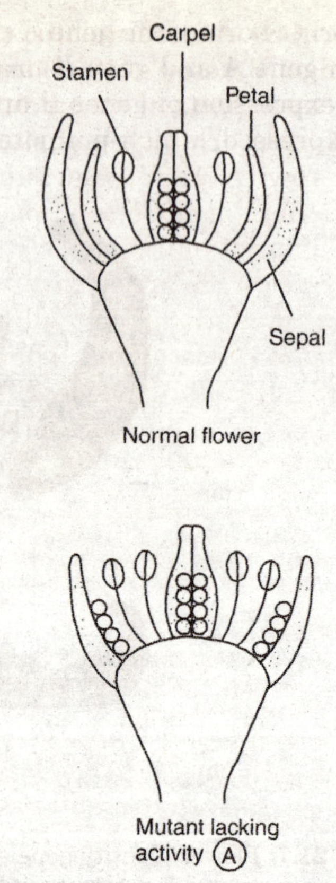

Fig. 21.8. Diagram showing ABC model of homeotic gene action in *Arabidopsis thaliana*. The Top pictures show a normal expression of all genes. Below pictures show the effect when gene *A* is absent.

sequence similar to the homeobox found in animal homeotic genes. Instead, most of them (except for *apetala 2*) contain a MADS ox (an acronym of the first four plant genes of this type that were identified: *ICMI, AG, DEF,* and *SRF*). In the transcription factor proteins, the MADS domain promotes the binding of the transcription factor to specific DNA sequences.

Like the *Drosophila* homeotic genes, plant homeotic genes are thought to be art of a hierarchy of gene regulation. Genes that are expressed within the flower bud primordium produce proteins that activate the expression of these homeotic genes. Once they are transcriptionally activated, the homeotic genes then regulate the expression of realizator genes, the products of which promote the formation of pals, petals, stamens, or carpels.

Arabidopsis homeotic mutations include *apetala 2*, which causes conversion of sepals to leaves and conversion of petals to stamens, *apetala 3*, which converts petals to sepals and stamens to carpels, *pistillata,* which converts petals to sepals, *fl02, fl03, fl04, fl40,* and *fl48*, which convert outer organs to carpels, and *flog,* which may make anthers appear petalloid. This was shown by Koornneef *et al.*, 1983; Pruitt *et al.*, 1987; Meyerowitz, 1987; Haughn and Somerville, 1988; Meyerowitz and Chang, 1988 Komaki *et al.*, 1988; Bowman *et al.*, 1989. Another related class of mutations can be called meristic rather than homeotic, based on Bateson's (1894) definition of meristic variants as those causing a change in the number of parts of an organ or organism without showing clear replacement of one part by another. This type of study was made by Koornneef *et al.*, 1983; Pruitt *et al.*, 1987; Meyerowitz, 1987; Bowman *et al.*, 1988. They showed that meristic mutations in Arabidopsis include *agamous,* which causes flowers to have many sepals and petals, and two different mutations, *clavata l* and *clavata 2,* which increase the number of carpels in the ovary from two to three or four. All of these mutations are recessive. Those that are mapped (which include all but the various *fl0* and *fl* mutations) are found in different chromosomal locations. There is no evidence, therefore, for complexes of related, tightly linked, and interdependent homeotic genes such as in the bithorax complex of Drosophila.

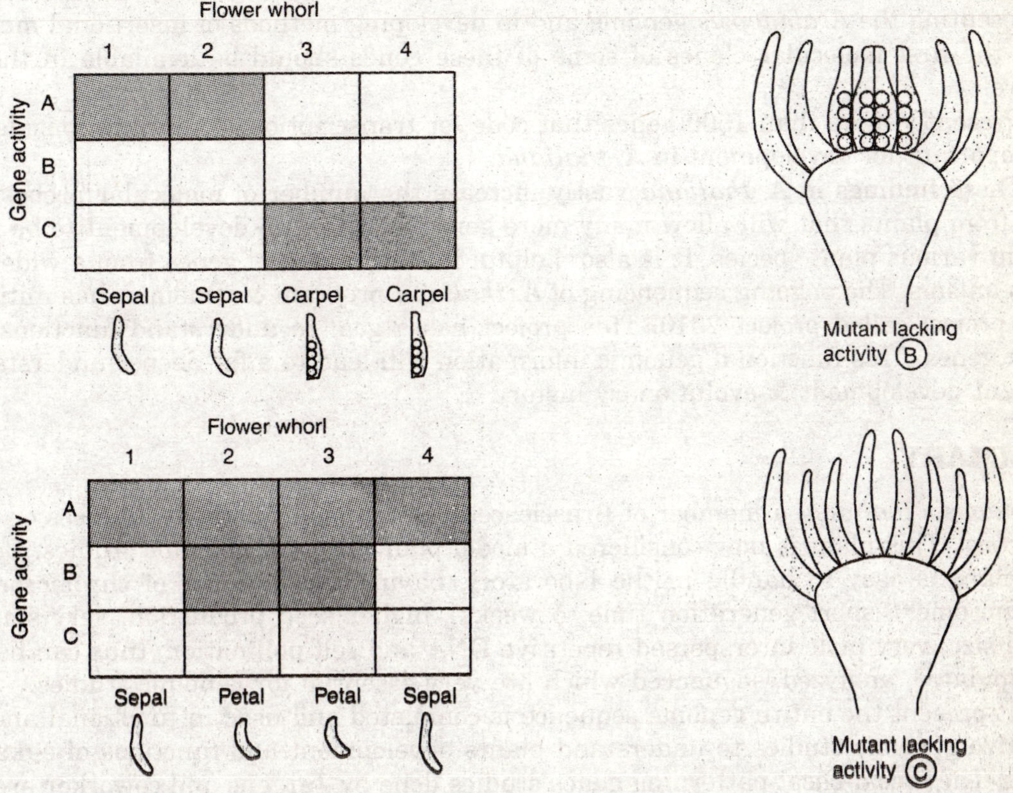

Fig. 21.9. Diagram showing *ABC* model of homeotic gene action in *Arabidopsis thaliana*. Top row shows the morphology of the flower when gene *B* activity is absent. Bottom row is shown when gene *C* activity is absent in flower.

A detailed phenotypic and developmental analysis of a number of these mutants is in progress. Hill *et al.*, Bowman *et al.*, Komaki *et al.*, continuing the study. The strains homozygous for pairs of these mutations have been studied to determine the nature of interactions between the different genes (Bowman *et al.*, 1988, 1989). The phenotypes of double mutant combinations consisting of one homeotic and one meristic mutation are completely additive, with no apparent interaction of the genes. This implies that the two processes, specification of the number and position of organs and specification of the nature of the organs, may be independent. Certain of the homeotic genes are temperature-sensitive, giving a greater degree of homeotic conversion at higher temperatures. Temperature-shift experiments (Bowman *et al.*, 1988, 1989) demonstrate that the temperature-sensitive periods of these mutants occur very early in flower development, before any detectable differentiation of the affected organs. This supports the contention that the wild type products of these genes are necessary for specifying the fate of the cells that will differentiate into specific floral organs.

In the case of these developmental mutants, as in the case of the hormonal mutants, the next stage in understanding the biochemical basis of the phenotypes will involve molecular cloning of the genes. Given the potential for chromosome walking, the availability of a dense RFLP map, and ongoing work in producing a linked collection of cosmids representing the *Arabidopsis* genome and in developing methods of insertional mutagenesis, the first molecular clones of some of these genes should be available in the near future.

Presently more than 1500 genes that code for transcription factors are considered to be important for development in *A. thaliana.*

These findings in *A. thaliana* vastly increase the number of molecular probes available from plants that will allow many more genes that control development to be identified in various plant species. It is also helpful in comparison of genes from a wide range of organisms. The ongoing sequencing of *A. thaliana* program & genomics has initiated a bold project called project 2010. This project has a goal to understand functions of all plant genes. This functional genomic information will lead to a far deeper understanding of plant development & evolutionary history.

SUMMARY

Arabidopsis thaliana, a member of Brassicaceae, is commonly called wall cress or mouse ear cress. This plant is now considered a model plant in plant genomic studies. Because the plant is easy to handle in the laboratory, having less number of chromosomes (5 chromosomes), short generation time (5 weeks), higher seed production, very small genome size, very little interspersed repetitive DNA and self pollination; thus can be used, manipulated, analyzed, sequenced which are very essential for genomic studies.

At present the entire genome sequence is completed and used in functional and comparative genomic studies to understand plants development and functions of eukaryotic genes. The apical-basal-patterning genes studies done by Jurgens and coworker are helpful in understanding how genes play a role in development.

The gene which controls the morphological characteristics of each cell or part are called homeotic genes. Meyerowitz and his associates describe the *ABC* model of flower development by homeotic genes. Now a large number of genes are identified in *A. thaliana* which are involved in starch biosynthesis, lipid biosynthesis, nucleoside biosynthesis, amino acid metabolism, photosynthesis, nitrate metabolism, phototropism and gravitropism and hormone biosynthesis. This clearly is helpful in understanding the complexity and evolution of genes and genomes of wide variety of plants and organisms of other category.

EXERCISE

1. Define genes, genome and genomics.
2. Explain history of *Arabidopsis thaliana* discovery.
3. Describe advantages of *Arabidopsis thaliana* usage in genomic studies.
4. What do you mean by restriction fragment length polymorphism?
5. Name different genes identified in *Arabidopsis thaliana*.
6. Write a note on apical-basal-patterning gene in *A. thaliana* embryo development.
7. Give an account on homeotic genes in *A. thaliana* flower development.
8. Ellucidate advantages of *A. thaliana* use in plant genetic studies.
9. How do *ABC* genes change the different parts of *A. thaliana* flower?
10. Define homeotic genes.
11. What is the function of *gurke, fackel, monopterous* and *gnom* genes in *A. thaliana* embryo?
12. Explain contributions of Meyerowitz to *A. thaliana* genomics.
13. List out discovery steps (time line) from 16th Century to recent times for *A. thaliana* and its usage is genomic studies.
14. Who discovered *A. thaliana* plant?
15. Explain why *Arabidopsis thaliana* plant is considered as botanical *Drosophila* in plant genetic studies.
16. Describe flower parts arrangement in the normal flower of *Arabidopsis thaliana*.
17. What are the resultant flower morphology in the single and triple homeotic mutants in *Arabidopsis thaliana*?
18. Illustrate pattern of gene expression in *ABC* model of *Arabidopsis thaliana* flower having normal expression of *ABC* class genes.
19. Explain in detail flower whorls arrangement in mutant for gene *A* and *C* in case of *Arabidopsis thaliana* flower.
20. What do you mean by carpel, stamen, petals and sepals in a flower?

22

Plants and Industry

Plants and Industry: Production of colouring agents, flavouring agents, biopolymers by genetic recombinant plants.

22.1 INTRODUCTION

Basic understanding and advanced technology with changed technical principles resulted in understanding potentiality of plants in varieties of the field. Now-a-days various industrial products are produced from the cultured plant cells *in vitro* condition. It is gaining importance when compared to other methods because of economic constraints involved in the products' productions. But this can be overcome by the plant culturing system.

With increased health consciousness and knowledge on the harmful effect of the various commercially available, synthetic colouring agents, flavouring agents and biopolymers, made the people to turn into the healthier, safe natural products obtained from plants. The currently available synthetic compounds are having undetermined and unexplained ill-effects on the human and animal health. They are known to have oncogenic (cancer causing), mutagenic (mutation inducing), teratogenic (having effect on developing embryo in uterus) and irreparable effects on the biosphere. Thus now more importance is given to safe, biodegradable and natural products which can substitute the synthetic ones.

Table 22.1: Some of the examples for industrial products produced by the plants

Colouring agent	Plant	Natural Chemical Use
Saffron	*Crocus sativus*	Food, soft drinks
Carotenoids	*Ruta graveolens*	Sweets, foods, cakes
Kokum	*Grasinia indica*	Sweets, drinks,
Anthocyanins		cakes, sweets, cool drinks
Capsaicin	*Capsicum frutescens*	Food stuffs
Anthroquinones	*Morinda citrifolia*	food, sweets, cakes
Yellow colour	*Curcuma longa*	Foods, eatables
Shikonine	*Lithospermun erythrorhizan*	Food stuffs

The development of *Agrobacterium* mediated transformation system made it possible to transfer any genes (prokaryotic or eukaryotic organisms) into plants with stable expression of foreign in the plants. To date, genetically modified (transgenic) plants have been produced in more than 50 species. With other methods such as direct delivery of DNA into protoplasts by electroporation or by treatment with polyethylene glycol or gene gun or viral infection with T-DNA (Agroinfection) made further increase in the integration of DNA in the plants.

Table 22.2: Production of flavouring agents by plants

Flavour	Plant	Use
Flavour of lignin pathway		
Vanilla (Vanillin)	*Vanilla planifolia*	Ice creams, cakes, kheer
Chilli pepper (Capsaicin)	*Capsicum frutescens*	Food stuffs, condiments
Rosmarinic acid	*Coleus blumei* *Anchusa* officinalis	Foods
Flavour from sulfur compounds		
Onion (Alliin)	*Allium cepa*	Foods, curry
Garlic (Alliicin, Alliinin)	*Allium sativum*	Foods, curry
Horse radish (Allyl isothiocyanate)	*Armoracia rusticana*	Foods, curry
Chives (Dipropyl sulphide)	*Allium schoenoprasum*	Condiments, foods, curry, roast
Terpenoids		
Ajwain (thymol)	*Trachyspermum ammi*	Food stuffs
Ajowan (Thymol)	*Carum copsicum*	
Celery (Limonene Seliaene)	*Apium graveolens*	
Cumin (Cuminaldehyde)	*Cuminum cyminum*	Foods, drinks
Pimpinella/Aniseed (Anethol)	*Pimpinella anirum*	Condiments
Lemon grass oil (Citral)	*Cymbopogon citratus*	drinks
Mint oil (Tromenthol)	*Mentha piperita/M. citrata*	drinks, cakes
Origanum (Sesquiterpenoids)	*Origanum vulgare*	drinks, cakes
Basmati (2-acetyl pyroline)	*Oryza sativa*	Foods

Flavouring agents: Flavour production by tissue culture means could be more commercially feasible than colours and sweetness mainly due to the vast applications of flavours in various products of daily use such as high value cosmetics, foods, pharmaceuticals, toiletries and even stationaries. Flavour market amounts to a world market of 3 billion US dollars and the demand for aromas in food and perfume industry is increasing at the rate of 5–7% per annum.

Most of the aromatic amino acids contribute to the formation of several secondary metabolites, some of which contribute for flavour production. Several phenyl propanoids of this pathway are produced in nature as defence compounds which get elicited by pathogen attack.

1. **Vanilla flavour:** Vanilla flavour, from ripe fermented beans of *Vanilla planifolia*, has a great market demand of $100 million/year. The key flavour components in cell cultures of *Vanilla fragrans* and *V. phaentha* were found to be vanillin along with vanillic acid, 4-hydroxybenzaldehyde and 4-hydroxybenzoic acid with some of the other phenolic constituents.

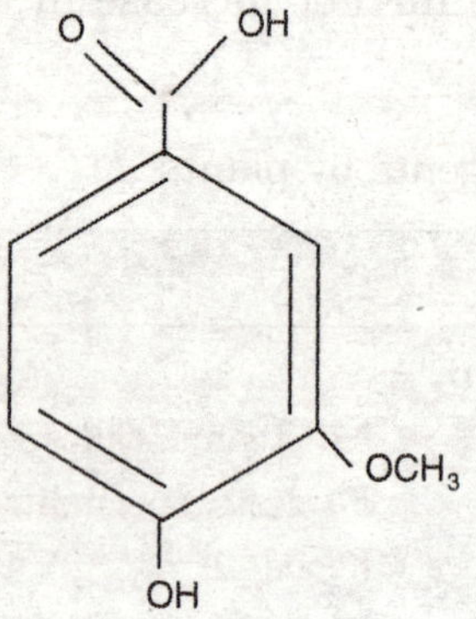

O OH

OCH_3

OH

Vanillic acid

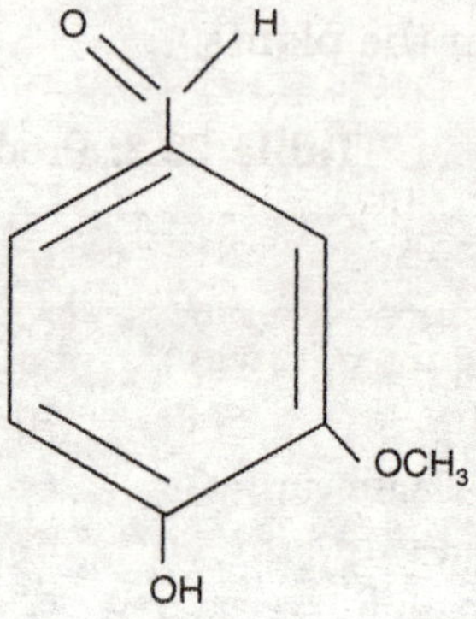

O H

OCH_3

OH

Vanillin

2. **Chilli pepper flavour:** Cell line selection of chilli pepper (*Capsicum annum*) was made on a substrate supplemented with *p*-fluorophenylalanine (PFP), the amino acid analogue. The PFP-resistant cell lines accumulated higher levels of capsaicin than sensitive ones even after prolonged culture on PFP-medium.

Capsaicin production in nonselected cells was only 26.8% of that formed in resistant cells. The capsaicin content in 2 PFP resistant cell lines were 6.7% and 24.9% respectively of that found in intact fruits.

3. **Rosmarinic acid:** Rosmarinic acid (RA) is a caffeic acid ester present in several plants and, in fact, a well studied compound in cell cultures of *Coleus blumei* and *Anchusa officinalis*. In *Coleus*, the level of RA in suspension cultures was highly dependent on the levels of carbohydrates added to the medium. The cell cultures maintained on 2% sucrose produced only 2–3% of rosmarinic acid whereas these on 4% produced 19% of rosmarinic acid on a dry weight basis.

4. **Flavours from sulphur compounds:** Several compounds of sulphur possess distinctive odours which contribute to both pleasant and unpleasant characters of many foods. While fruits produce relatively few sulphur volatiles, the aroma of most of the vegetables is attributed to the presence of a variety of sulphur compounds. The genus *Allium* includes such strongly flavoured vegetables as onions, garlics and chillies, which are used extensively to enhance the flavour of other foods. These vegetables provide the classic examples in which the products formed undergo non-enzymatic degradation to more stable volatile components such as sulphides, disulphides, trisulphides and thisulphonates etc. Garlic varieties contain allyl sulphides and disulphides. Sulphur volatiles in chives are reported to be methyl compounds.

5. **Citrus flavour:** The typical aroma of citrus fruits is attributed to terpenoids, which represent about 5% of citrus oils. Included in this group are the acyclic isoprenoid alcohols, geraniol, nerol, citronellol and farnesol which are all present in citrus essence together with their corresponding aldehydes.

6. **Ajwain: *Trachyspermum ammi*** is an aromatic herb. The seeds yield (2–4%) of essential oil the main compound being thymol constituting upto 60% of the oil.

7. **Aniseed oil: *Pimpinella anisum*** seeds, used as condiments, contain the highly priced fragrant oil — the aniseed oil. The key components of its characteristic flavour were induced in callus and suspension cultures. The accumulation of oil in cultures ranged from 1–4 mg/100 g fresh weight compared to 100–400 mg/100 g fresh weight in intact-plant. The cell cultures produced the main components of the oil in cultures *i.e.*, anethol, *myristicin*, bisabolene and pseudo-isoeugenol-2-methyl-butarate.

8. **Basmati flavour:** Basmati flavours in callus. Cultures of highly scented varieties of basmati rice such as *Oryza sativa* has been reported. 2-acetyl pyrroline having characteristic aroma of most of the basmati varieties was not detected either in rice or callus even though there was perceptible basmati rice flavour in callus.

9. **Saffron:** The saffron of commerce is the dried stigma of *Crocus sativus.* L. saffron is a very highly priced spice commodity with the current international market price of about US $ 2000 per kg. The active principles of saffron *i.e.*, colours (crocin-crocetins) and flavours (picrocrocin and safranal) in pure form (active solids) costs US $ 7,500/kg.

Colouring Agents

Anthocyanins production: Anthocyanins are widely found in several plant species. Since they are natural pigments, are very safe for human health, and have a high potential utility value as a food additive and marker. In cultured plant cells, anthocyanins usually accumulate in only small amounts and their production generally requires light irradiation. Cultures with 1% CO_2 administration showed rapid cell growth and anthocyanin production.

Table 22.3: Production of biopolymers from plants

Compound	Genes donor	Plant host
Biodegradable plastics	*Alcaligenes eutrophus*	*Arabidopsis thaliana*
Polyhydroxy alkonates (PHA)	*Alcaligenes eutrophus*	*Nicotiana tabacum*
Polyhydroxy butyrate (PHB)	*Alcaligenes eutrophus*	*Solanum*
Cyclodextrins	*Klebsiella pneumoniae*	*Solanum tuberosum*

Genetic engineering includes isolation, characterisation, re-arrangement of genetic material and its transfer to plant for obtaining required products. Generally this will be done by keeping in mind to reduce the cost involved in production, high yield, less process for separation and stable compound. As growth of plant is comparatively manageable the required commercial products are easily produced.

With these examples it is clear that genetic engineering and related areas of biotechnology are having a substantial impact on the food, flavour, colour and biopolymer industries. A few reaching commercial realisation from laboratory curiosity, flavour and biopolymer companies are turning to technologies that offer more profit and sale. Biotechnology will continue to capture the attention of industries with rapid advancement in related technologies in recent times. As the consumer becomes more comfortable with r-DNA products, this technology is sure to offer novel products at cost-effective prices.

SUMMARY

With increased environment consciousness of the people throughout the world, now importance is given to biodegradable products from the plant sources. Hence genetic engineering of plants is made use of in the industrial production processes as a result the products can be obtained from plants.

The colouring agents, flavouring agents and biopolymers production are gaining much importance. The synthetic products from the above fields are having large problems which are health-related aspects. Majority of the synthetic products are mutagenic, oncogenic, teratogenic and having residual effect in the environment. Thus, plants are exploited from their natural sources for the industrial production of the above mentioned agents.

EXERCISE

1. What are the advantages of plant derived products?
2. Explain disadvantages of synthetic products.
3. Name different plants used in flavour industry.
4. Elucidate the wide varieties of plants used in colouring agents production.
5. Define biopolymers.
6. Give an account on biopolymer production in plants.
7. Write short notes on:
   (a) Oncogenic effect of colouring agents
   (b) Mutagenic compounds
   (c) Biodegradation
8. What do you mean by teratogenic effects and mention its demerits?
9. What plant derived flavours are preferred in the food industry?
10. In which model organisms the biopolymers are produced in large scale?

Appendix A

COMMONLY USED ABBREVIATIONS

AIDS	Acquired immune deficiency syndrome
ACP	Acyl carrier protein
ADH	Alcohol dehydrogenase
AMP, ADP, ATP	Adenosine 5' -mono di-, and tri-phosphate
APCs	Antigen-presenting cells
cAMP	Cyclic AMP
cDNA	Complementary DNA
COA, COA-SH	Coenzyme A
ConA	Concanavalin A
CR	Complement receptors
DEAE-cellulose	Diethylaminoethyl cellulose
DNA	Deoxyribonucleic acid
DNase	Deoxyribonuclease
DNP	2,4-Dinitrophenol
DTH	Delayed type hypersensitivity
EBV	Epstein-Barr virus
EDTA	Ethylene diamine tetra acetic acid
EPR	Electron paramagnetic resonance
ESR	Electron spin resonance
FAD	Flavin adenine dinucleotide
$FADH_2$	Flavin adenine dinucleotide reduced
Met	N-formyl methionine
FMN	Flavin mononucleotide
$FMNH_2$	Flavin mononucleotide reduced
GDH	Glutamate dehydrogenase
GLC	Gas-liquid chromatography
Glu	Glutamine
cGMP	Cyclic GMP
GS	Glutamine synthetase

HbO_2	Oxyhemoglobin
HEPA	High efficiency particulate air filter
His	Histidine
HnRNA	Heterogeneous nuclear RNA
Ig	Immunoglobulin
Kb	Kilobase
LAF	Laminar airflow
LDH	Lactate dehydrogenase
Leu	Leucine
MAC	Membrane attack complex
MHC	Major Histocompatibility Complex
NAD+	Nicotinamide adenine dinucleotide
NADH	Nicotinamide adenine dinucleotide (reduced)
NADP+	Nicotinamide adenine dinucleotide phosphate
NADPH	Nicotinamide adenine dinucleotide phosphate (reduced)
NMR	Nuclear magnetic resonance
OD	Opticai Density
Pi	Inorganic phosphate
PEP	Phosphoenol pyruvate
RNA	Ribonucleic acid
mRNA	Messenger RNA
rRNA	Ribosomal RNA
sRNA	Soluble RNA
tRNA	Transfer RNA
RNase	Ribonuclease
TCA	Tricarboxylic acid
TLC	Thin layer chromatography
UV	Ultraviolet
UMP	Uridine monophosphate
UDP	Uridine diphosphate
UTP	Uridine triphosphate
Val	Valine
Met	Methionine
Phe	Phenylalanine
PPi	Inorganic pyrophosphate
Ser	Serine
TPP	Thiamine pyrophosphate
Trp	Tryptophan

Appendix B

ALTERNATE NAMES FOR GROWTH FACTORS

Activins (A, AB, B)
- FSH releasing protein (FRP)

Colony-stimulating factor 1 (CSF-1)
- Macrophage colony-stimulating factor (M-CSF)

Epidermal growth factor (EGF)
- Urogastrone
- Erythropoietin (epo; EP)
- Hemopoietine
- Erythrocyte stimulating factor (ESF)

Fibroblast growth factors (FGF)
- Acidic fibroblast growth factor (aFGF)
- Basic fibroblast growth factor (bFGF)
- Brain-derived growth factor (BNDF; BDGF)
- Heparin binding growth factor (HBFG)
- Endothelial cell growth factor (ECGF)
- Retina-derived growth factor (RDGF)
- Eye-derived growth factor (EDGF)
- Kidney angiogenic factor (KAF)
- Adrenal growth factor (AGF)
- Corpus luteum angiogenic factor (CLAF)
- Ovarian growth factor (OGF)
- Placental angiogenic factor (PAG)
- Hepatocyte growth factor (HGF)
- Myogenic growth factor (MGF)
- Cartilage-derived growth factor (CDGF)
- Bone growth factor (BGF)
- Seminiferous growth factor (SGF)
- Prostatropin (PGF)
- Tumor-derived growth factor (TDGF)
- Hepatoma-derived growth factor (HDGF)
- Melanoma-derived growth factor (MDGF)
- Mammary tumor-derived growth factor (MTGF)

Gastrin-releasing peptide (GRP)
- Mammalian bombesin

Granulocyte colony-stimulating factor (G-CSF)
    Macrophagejgranulocyte inducer type 1, granulocyte (MG-1G)
    Pluripotent colony-stimulating factor (pluripoetin)
    Granulocyte-macrophage colony-stimulating factor-p (GM-CSF *P)*
Granulocyte-macrophage colony-stimulating factor (GM-CSF)
    Macrophage-granulocyte inducer
    Colony-stimulating factor 2 (CSF-2)
Inhibin
Insulin-like growth factor I (IGF-I)
    Somatomedin A
    Somatomedin C
    Basic somatomedin
Insulin-like growth factor II (IGF-II)
    Multiplication stimulating activity (MSA)
Interferon-α (IFN-α)
    Leukocyte (Le) interferon
    Type I interferon
Interferon-β (IFN-β)
    Fibroblast (F) interferon
    Type I interferon
Interferon-γ (IFN-γ)
    Immune interferon
    T interferon
    Type II interferon
Interleukin-1 (IL-1)
    Interleukin-l α
    Interleukin-1 β
    Lymphocyte activating factor (LAF)
Interleukin-2 (IL-2)
    T-cell growth factor (TCG F)
Interleukin-3 (IL-3)
    Mast cell growth factor
    P-cell stimulating factor Multi-colony-stimulating factor
    Burst promoting activity
    WEHI-3 hematopoietic growth factor
    Thy-1 inducing factor
    Histamine cell-producing stimulating factor
    20 α -dehydrogenase-inducing factor
Interleukin-4 (IL-4)
    B-cell stimulatory factor 1 (BSF-1)
    T-cell growth factor (TCGF)
    B-cell growth factor I (BCGF-l)
    Mast-cell growth factor (MCGF)
Interleukin-5 (IL-5)
    T-cell replacing factor (TRF-l)
    B-cell growth factor II (BCGF II)
    B-cell differentiation factor *f1* (BCDFf1)

Eosinophil differentiation factor (EDF)
IgA enhancing factor (IgA-EF)
B-cell maturation factor (BMF)
B-cell growth and differentiation factor (BGDF)

Interleukin-6 (IL-6)
B-cell stimulatory factor 2 (BSF-2)
Interferon-β 2
26-kDa protein
Hepatocyte stimulating factor
Hybridoma/plasmacytoma growth factor
Interleukin-HPl
Macrophage granulocyte inducer type 2

Lymphotoxin
Tumor necrosis factor-β (TNF-β)

Mullerian inhibiting substance (MIS)
Anti-MUllerian hormone (AMH)
Mullerian inhibiting factor (MIF)

Nerve growth factor (NGF)

Platelet-derived endothelial cell growth factor (PD-ECGF)

Platelet-derived growth factor (PDGF-AA, PDGF-BB, PDGF-AB)

Transforming growth factor-α: (TGF-α:)
Sarcoma growth factor (SGF)

Transforming growth factor-β (TGF-β)
Cartilage inducing factor-A (CIF-A)
Cartilage inducing factor-B (CIF-B)
BSC-l growth inhibitor (BSC-l GI)
Differentiation inhibitor (DI)
Polyergin

Tumor necrosis factor (TNF) Cachectin

Tumor necrosis factor-α: (TNF-α:)

Appendix C

CHROMOSOMAL LOCATIONS OF HUMAN GROWTH FACTORS/ GROWTH FACTOR RECEPTORS

chromosome 1	Nerve growth factor (p22.1) Transforming growth factor-p2 (q41)
chromosome 2	Inhibin-a (distal portion of long arm) Inhibin-PB (near centromere on short arm) Interleukin-1p (2q13-2q21) Transforming growth factor-a (pl1-p13)
chromosome 4	Basic fibroblast growth factor Epidermal growth factor (q2S-q27)
chromosome 5	Acidic fibroblast growth factor Colony-stimulating factor 1 (q33.1) Colony-stimulating factor 1 receptor (q33.2-33.3) Granulocyte-macrophage colony-stimulating factor (q2331) Interleukin-3 (q23-q31) Interleukin-4 (q23-31) Interleukin-S (q23.3-32) Platelet-derived growth factor receptor (p-subunit)
chromosome 6	Interferon-y receptor
chromosome 7	Epidermal growth factor receptor (p14-p12) Erythropoietin (q11—q22) Inhibin-PA Interleukin-6 (p21) Platelet-derived growth factor A chain
chromosome 9	Interferon-a Interferon-p
chromosome 11	Insulin-like growth factor II
chromosome 12	Insulin-like growth factor I Interferon-y

chromosome 14	Transforming growth factor-p3 (q24)
chromosome 17	Granulocyte colony-stimulating factor (q21-q22) Nerve growth factor receptor (q12-q22)
chromosome 18	Gastrin-releasing peptide (q21)
chromosome 19	Mullerian inhibiting substance Transforming growth factor-p1 (q13)
chromosome 21	Interferon-alpha receptor
chromosome 22	Platelet-derived growth factor B chain

Appendix D

CHROMOSOMAL LOCATIONS OF MOUSE GROWTH FACTORS/ GROWTH FACTOR RECEPTORS

chromosome 1	Transforming growth factor-p2
chromosome 2	Interleukin-1 *a*
	Interleukin-1 *P*
chromosome 3	Epidermal growth factor
	Nerve growth factor
chromosome 4	Interferon-α
	Interferon-β
chromosome 5	Erythropoietin
	Interleukin-6
chromosome 7	Transforming growth factor-p1
chromosome 10	Interferon-y
	Interferon-y receptor
chromosome 11	Epidermal growth factor receptor
	Granulocyte colony-stimulating factor
	Granulocyte-macrophage colony-stimulating factor Interleukin-3
	Interleukin-S
	Nerve growth factor receptor
chromosome 12	Transforming growth factor-β
chromosome 16	Interferon-α receptor
chromosome 18	Colony stimulating growth factor 1 receptor

Appendix E

CD (CLUSTER OF DIFFERENTIATION) MARKERS

CD marker	Mol. Wt. kDa	Cell source
CD1a-1e	49-55	T cells
CD3	50	T cells
CD4	55	T cells
CD8	36	T cells
CD10	100	B cells
CD11a	180	T cells/ B cells
CD25	55	Activated T cells/ B cells
CD35	260	T cells/ B cells
CD114	110	Granulocytes
CD115	150	Macrophages

Appendix F

DECIMAL PREFIXES AND MULTIPLES FOR AREA/ LENGTH, MASS WEIGHT, AND VOLUME

Prefix	Symbol	Multiple	Magnitude	Common Name
yotta	Y	10^{24}	1 000 000 000 000 000 000 000 000	heptillion
zetta	Z	10^{21}	1 000 000 000 000 000 000 000	hexillion
exa	E	10^{18}	1 000 000 000 000 000 000	quintillion
peta	P	10^{15}	1 000 000 000 000 000	quadrillion
tera	T	10^{12}	1 000 000 000 000	trillion
giga	G	10^{9}	1 000 000 000	billion
mega	M	10^{6}	1 000 000	million
kilo	K	10^{3}	1000	thousand
hecto	h	10^{2}	100	hundred
deca	da	10^{1}	10	ten
unit*	-	-	1	one
deci	d	10^{-1}	0.1	tenth
centi	c	10^{-2}	0.01	hundredth
milli	m	10^{-3}	0.001	thousandth
micro	μ(mu)	10^{-6}	0.000 001	millionth
nano	n	10^{-9}	0.000 000 001	billionth
pico	p	10^{-12}	0.000 000 000 001	trillionth
femto	f	10^{-15}	0.000 000 000 000 001	quadrillionth
atto	a	10^{-18}	0.000 000 000 000 000 001	quintillionth
zepto	z	10^{-21}	0.000 000 000 000 000 000 001	hexillionth
yocto	y	10^{-24}	0.000 000 000 000 000 000 000 001	heptillion

* unit = meter (m) for area/length, gram (g) for mass/weight, and liter (L) for volume (capacity)

Appendix G

LENGTH MEASUREMENTS

Kilo meter (km)	=	10^3	meters
Decimeter (dm)	=	10^{-1}	meters
Centimeter (cm)	=	10^{-2}	meters
Millimeter (mm)	=	10^{-3}	meters
Micrometer (μm)	=	10^{-6}	meters
Nanometer (nm)	=	10^{-9}	meters
Angstrom (μm)	=	10^{-10}	meters

MASS/WEIGHT MEASUREMENTS

Kilo gram (kg)	=	10^3	grams
Centigram (cm)	=	10^{-2}	grams
Milligram (mm)	=	10^{-3}	grams
Microgram (μm)	=	10^{-6}	grams
Nanogram (nm)	=	10^{-9}	grams
Picogram (ú)	=	10^{-12}	grams

Appendix H

METRIC MEASUREMENT CONVERSIONS TO ENGLISH SYSTEM

Centimeters (cm)	X	0.40	=	inches
Meters (m)	X	3.28	=	feet
Meters (m)	X	1.09	=	yards
Kilometers (km)	X	0.62	=	miles
Grams (g)	X	0.035	=	ounces
Kilo gram (kg)	X	2.20	=	pounds
Liters (lt)	X	2.10	=	pints
Liters (lt)	X	1.06	=	quarts
Liters (lt)	X	0.264	=	gallons
Milliliters (mlt)	X	0.034	=	fluid ounces

Appendix I

TEMPERATURE CONVERSIONS

Degree Celsius (C) = {F°-32} X 0.555
Degree Fahrenheit (F) = {C° X 1.8} + 32

Fahrenheit Degree F°	Celsius Degree C°
-40°	-40°
0°	-17°
25°	-4°
32°	0°
50°	10°
75°	24°
100°	38°
125°	52°
150°	66°
175°	79°
200°	93°

Glossary

Abacterial Absence of bacteria, or sterile condition.

Abatement Lessening of pain.

Abasia Unable to walk because of loss of motor coordination.

Abdomen Area of body between diaphragm and pelvic bones.

Aberrant A growth that invariably deviates either from the normal or usual type; as exceptional growths (aberrations) which take place in some tissue cultures.

Activated sludge process Aerobic sewage treatment process using aerobic microorganisms present in sewage sludge to break down organic matters in sewage.

ABO Human antigenic system designating the blood groups A, B, AB and O.

Absolute plating efficiency The percentage of cells that give rise to the colonies on the medium.

Accession. Seed or plant sample, strain or population held in either a 'gene bank' or a 'breeding programme' for proper usage and conservation.

Acid phosphatase Enzyme group that catalyzes the hydrolysis of certain organic phosphates. Activator A substance or physical agent that stimulates the transcription of a specific *gene* or *operon*.

Acetyl CoA A condensation product of coenzyme A and acetic acid (attributing acetyl moiety); an intermediate in the transfer of 2-carbon fragments in the process of their gaining entry to tricarboxylic acid (TCA) cycle.

Active site The region of a protein that gets bound to substrate molecules and facilitates a particular 'chemical conversion'. The fundamental underlying *theory of enzyme activity* assumes formation of an enzyme-substrate complex *via* binding at the active site.

Active immunity Immunity acquired because of the individual's own reactions to the corresponding pathogenic microorganisms or their respective *antigens,* attributable to the presence of antibody in response to an antigenic stimulus.

Acquired immunity The ability of an individual to produce specific antibodies in response to antigens to which body has been previously exposed based on the development of a memory response.

Acquired Immune Deficiency Syndrome (AIDS) An infectious disease caused by HIV retrovirus, characterized by the loss of normal functions, followed by various infections.

Adaptive enzymes An enzyme that is duly formed in response to an outside stimulus during adaptation. This terminology has been duly replaced by 'inducible enzymes'.

Aerate To mix or supply with air or gas. The particular process is known as 'aeration'.

Aerobic or Aerobiotic It specifically refers to an *organism* which either lives in or a phenomenon taking place in the presence of molecular oxygen *(i.e.,* O_2-gas).

AFLP Amplified Fragment Length Polymorphism. A highly sensitive method for detecting *polymorphisms* in DNA. Following restriction enzyme digestion of DNA, a subset of DNA fragments are selected for *PCR* amplification and visualisation.

Adjuvants Substances that enhance immunological response to vaccine and may be incorporated to vaccines to slow down absorption and increase effectiveness: substances that enhance the action of an *antigen* or a *drug.*

Affinity It refers to the precise strength of binding of each separate antibody combining site with its corresponding antigenic determinant.

Agglutination Clumping or aggregation of cells due to the reaction of surface bound antigens with homologous antibodies.

Allele A pair of genes that controls a specific character/trait inherited from each parents or Any one of several alternative forms of a *gene*.

Allergen An antigen that induces an allergic responses in the immune system.

Allosteric Refers to a binding site in a protein usually an enzyme. The catalytic function of an enzyme can be duly modified by the introduction of small molecules, not only confined to the 'active site', but also at a 'spatially distinct (allosteric) site' having an altogether different specificity. An *allosteric effector* is a molecule bound at such a site that enhances or lowers the *activity of the enzyme.*

Allosteric Enzymes Enzymes with a binding and catalytic site for the substrate and a different site where a *modulator (i.e., allosteric factor)* usually acts.

Allozyme An *isozyme* whose synthesis is controlled by *codominant* alleles of one gene.

Agrobacterium tumefacnies The bacterium which is responsible for causing crown-gall disease in plants, besides inducing tumour to form. The Ti plasmid of *A. tumefaciens* is known to initiate and cause the disease; and a part of it is invariably utilized as *vector* in the ensuing genetic modification of higher plant cells.

Alkaline Phosphatase An enzyme that catalyzes hydrolyses of monophosphate ester moiety from specific compounds, such as: 3'- and 5'- terminal ends of nucleic acids (DNA and RNA), and nucleotides.

Ammonification The release of ammonia (NH3) gas from the nitrogenous organic matter by microbial action.

Anaphylactic shock Physiological shock resulting from an anaphylactic hypersensitivity reaction, invariably followed by death in extremely severe instances *e.g.,* penicillin-anaphylactic shock.

Anaphylactic Hypersensitivity An observed exaggerated immune response either to foreign protein or other substances involving *degranulation of mast-cells* and subsequent release of histamine.

Anchorage Dependent Cell A cell which will exclusively grow and multiply when in close contact with a *suitable solid support i.e.,* several animal cell-lines.

Anneal The spontaneous pairing of *complementary DNA* or RNA sequences by hydrogen bonding to form a double-stranded polynucleotide.

Anti-idiotype Antibody An antibody having the properties of an *antigen.*

Antibody A glycoprotein molecule produced in the body in response to the introduction of an *antigen* or *hapten* that can react with the antigen. It is also termed as immunoglobulin, that is essentially a part of the serum fraction of the blood formed in response to the antigenic stimulus, and which eventually reacts with antigens with utmost specificity.

Antibody-Mediated Immunity Immunity caused by the activation of the B-lymphocyte population thereby giving rise to the production of *several categories of antibodies.*

Antibody-Dependent Cytotoxic Hypersensitivity It is a Type II hypersensitivity in which an *antigen* present on the surface of a cell combines with an antibody, resulting in the death of that cell by stimulating phagocytic attack or initiating the complement pathway.

Antiserum A serum that specifically contains a mixture of antibodies raised against a particular *antigen* is referred to as antiserum to that *antigen.*

Antitoxin An antibody to a *toxin* capable of reacting with that 'poison' and also neutralizing the toxin produced.

Autoimmunity *Immunity* or *hypersensitivity* to certain constituent in one's own body *i.e.,* immune-reaction with self-antigens.

Antifoam A compound added to bioreactors so as to reduce the production of foam *viz.,* surface active agents.

Antioxidant A substance which is sometimes added to the solutions to either *prevent* or *inhibit* oxidative process.

Aseptic Technique Procedures employed to prevent the introduction of *bacteria, fungi, viruses, mycoplasma* or other organisms into cell tissue and organ cultures.

Aneuploid The condition which exists when the nucleus of a cell does not contain an exact multiple of the haploid number of chromosomes.

AP-PCR Arbitrarily Primed Polymerase Chain Reaction. A technique for amplifying anonymous stretches of DNA using *PCR*. Related to *RAPD*.

Arbitrary primer A short *oligonucleotide* primer used in certain *PCR* methods to initiate DNA synthesis at random locations on the target DNA.

Aspirate To draw something in or out, up or through using suction or a vacuum *(i.e.,* negative pressure); as aspiration (vacuum) may be employed in the 'disinfection process' to draw *disinfectant* right into the *surface layers of plant tissue.*

Assay The 'substance' to be analyzed *quantitatively* or the process of examining or testing it either by chemical means or by other suitable means.

Attenuation A process of minimizing the virulence of a pathogen.

Attenuated Vaccine A vaccine produced by using an attenuated strain or virus or bacterium.

Autocrine This refers to the ability of cytokines to show the effect on the cell that produced it.

Autoradiography A technique where radioactively labelled molecules are visualised through exposure to X-ray film.

Axenic Culture A culture without foreign or undesired life forms. An axenic culture may also include the purposeful co-cultivation of different types of cells, tissues or organisms.

Auxotroph Mutant that differs from the wild type in requiring a nutritional supplement for specific growth *viz.,* as in a deficient mutant.

Avidity Total binding between a heterogeneous antiserum and a multi-valent antigen, regardless of how many different specific antibodies each with a precise affinity, contribute to the total binding. This particular measurement is known as the avidity of the antiserum.

B-Lymphocyte White blood cells (WBC) that are capable of producing specific immunoglobulins. In fact, their surfaces do carry specific *immunoglobulin antigen-binding receptor sites.*

Backcross The cross of a hybrid to either of its parents, the cross of a *heterozygote* with its *homozygous* recessive parent, or the cross of a plant of unknown *genotype* with a homozygous recessive.

Bacteria Single celled, ubiquitous prokaryotic organism having peptidoglycan cell wall.

Bacteriophage A virus infecting bacterium that replicates with the cells of bacteria.

Base The chemical unit which characterises a *nucleotide*. In DNA the bases found are adenine, guanine, thymine, and cytosine. In RNA the bases are adenine, guanine, uracil, or cytosine.

Base pair Two nucleotide *bases* on different strands of a nucleic acid molecule that are held together by hydrogen bonds. Bases can pair in only one way - adenine with thymine (DNA) or uracil (RNA), and guanine with cytosine (DNA).

Base Pairing The particular H-bonding taking place between *purines* and *pyrimidines* in double-stranded nucleic acids *(i.e.,* DNA and RNA).

Base sequence The order of *nucleotide* bases in DNA or RNA.

Batch Culture A cell suspension grown in a liquid medium of a known volume. Organisms in this system invariably exhibit a *'sigmoid'* type growth curve. The *inocula* of successive subcultures are of similar size, and the cultures usually contain almost an equivalent quantum of *'cell-mass'* at the end of each passage. Cultures frequently exhibit five distinct phases per passage, namely: (i) *lag phase; (ii) exponential growth phase; (iii) linear-growth phase; (iv) deceleration phase;* and *(v) stationary phase.*

Bacteriostatic A chemical entity or other substance that does not kill but prevents the possible bacterial growth and multiplication.

Bactericide. An agent or a substance that usually kills bacteria (rather rapidly); and this activity is known as bactericidal activity.

Bioassay A biological assay carried out on the *'living cells'* or on a *'living organism'.*

Biochip A hi-tech device that essentially combines a small scale 'biosensor' with an integrated circuit.

Bioconversion The actual transformation of matter from one form to another by living organisms or enzymes.

Biodegradation A recognized and accepted phenomenon whereby a material is broken down into its smaller chemical fragments by the action of living organisms *e.g.,* biodegradable polymers, biodegradable packing components etc.

Bioethics That refers to the day-to-day moral decisions confronted in handling or caring for living organisms.

Biolistics A technique that injects DNA into the cells using tiny bullets composed of gold or tungsten particles with DNA attached on the surface.

Biomass The cell mass produced during fermentation or the total weight of living matter in a population.

Biopolymer An altogether different type of large molecules duly formed by organisms *i.e.* nucleic acids, proteins, lipids, and polysaccharides.

Biopiracy Illegal commercial development of naturally occurring biological materials such as plant parts or genetic cell lines.

Bioprospecting The legal collection of biological samples that can help medicinal and scientific research with affecting natural sources.

Bioreactor A vessel, preferably made up of Stainless Steel, employed to perform a biological reaction *i.e.,* a *reactor* used for the culture of aerobic cells, or to the columns or packed beds of immobilized cells or enzymes.

Biosensor A highly specific and sophisticated device that usually makes use of an agent either having a biological origin or a biological principle, for the assay of a chemical compound *viz., tissues, immunosystems, isolated enzymes, organelles,* and *whole cells as catalysts.* The 'catalyst' is immobilized and used in conjunction with a physico-chemical device.

Biosynthesis Refers to *'biological synthesis' i.e.,* the building or forming of biological compounds in a living organism.

Biotechnology The meticulous application of *organisms, biological systems,* or *biological phenomena* specifically either to manufacturing units or to service industries. However, this very definition has been logistically extended to include any process or device wherein organisms, tissue cells, organelles, isolated enzymes, design and usage of 'bioreactors', fermentors, downstream processing, and above all the analytical and control equipment associated with biological manufacturing process.

Biotype A certain group of organisms having the same genetic characteristic features.

Biotin [Vitamin H; Co-carboxylase; $C_{10}H_{16}O_3N_2S$; MW 244.31]: It is a component of vitamin B complex. It serves as a prosthetic moiety for carboxylaseenzymes. Usually present in all living cells bound to proteins or polypeptides, and is of utmost importance in carbohydrate, protein, and fat metabolism. It is essentially component of most plant tissue culture media.

Buffer An admixture of various chemical substances (mostly inorganic) in an aqueous medium, or system that prevents/resists change in pH (despite addition of small quantum of acid or base) and withstand shock; as buffered media can appreciably resist pH drift. The extent to which pH change can be resisted logically represents as a measure of the solution's buffering capacity.

Budding A form of asexual reproduction wherein a *daughter cell* develops from a small outgrowth or protrusion of the *mother (parent) cell.*

Callus *(plural-calluses* or calli). An actively growing undifferentiated (parenchymous) tissue formed in higher plants in response to wounded surface or certain infections. It usually represents a disorganized tumour-like masses of plant cells which are formed in a culture medium. It proliferates in an irregular tissue mass which varies extensively in several aspects *e.g.,* appearance, texture, rate of growth. Callogenesis - refers to the phenomenon of callus formation *i.e.,* a function of the tissue type *(species* and *explant)* as well as the composition of the medium.

Capsid A protein coat of a virus enclosing the naked nucleic acid *(i.e.,* DNA and RNA).

Capsomere The individual protein subunits which essentially form the *capsid* of a virus.

Carcinogen An agent capable of initiating development of malignant (cancerous) tumours in tissues or cells.

Carrageenan Sulphated cell-walled polysaccharide found in certain *red algae.* Contains repeating sulphated disaccharides of galactose and anhydrogalactose.

Carrier A matrix strategically placed within a bioreactor or cell-culture system to serve as a support upon which the active biomass or cells grow and get immobilized ultimately.

Carbon Source A source for non-metallic element carbon (C) *e.g.,* as organic substances, sugars, com-steep liquor, maltose, glucose, glucose-syrup, whey powder, molasses (containing 8-10% residual sucrose) etc., that are usually taken up and metabolized by organisms in culture media or by plant-tissue cultures.

Casein Hydrolysate (Edamin) A milk protein (casein) digestive product very much composed of amino acids (casamino acids) and other substances. The resulting complex *(i.e.,* unidentified product) is sometimes used as an additive (0.02 - 2.10%) in nutrient solutions employed in various culture media and plant tissue culture as a non-specific source of organic nitrogen.

Cell Count The number of cells per unit suspension volume or cells weight. Tissue is treated with chromic acid (5-8%) or pectinase (0.25%) for up to 15 minutes normally followed by mechanical dispersion, and subsequently the cell numbers are determined with the help of a haemocytometer in microscope.

Cell Disrupter A procedure employed to liberate the contents of cells.

Cell Hybridisation The formation of synkaryons viable cell hybrids produced through cell fusion. These hybrids may be identified by their enhanced chromosome number in comparison to the parent cells and the possession of typical characters found in one of the parent cells.

Cell Line A cell line is permanently established culture which will proliferate indefinitely giving suitable fresh medium and space. The developmental history or descent through cell division is from a *single original cell.* Any deviation in the culture technique may favour one cell line over another.

Cell-Mediated Hypersensitivity Delayed hypersensitivity reaction involving T lymphocytes, and taking place almost 24-72 hours after due exposure to the antigen.

Cell-Mediated Immunity A highly specific acquired immunity involving T cells, primarily responsible for resistance to infectious diseases caused by certain bacteria and viruses that reproduce within the host cells.

Cell Number The absolute number or approximation of the number of cells per unit area of a culture or medium volume.

Cell Selection Selection within a group of genetically different cells, and usually involves competition between cells often under certain stress. The selection criteria may involve *three* main variants, namely: *(a)* cell viability; *(b)* biochemical activity; *(c)* another basis for choice. Normally, the selected cells or cell lines are carefully relocated to either fresh medium for continued selection process or duly exposed to an enhanced level of the stress agent. However, the ultimate aim is usually to regenerate plants from those select cells with a possibility that the plants may exhibit the specific traits selected for at the cellular level.

Cell Suspension Cells and small aggregates of cells suspended in a liquid medium *e.g.,* cell suspension cultures. Callus or explants derived from cell suspensions are transferred to liquid medium, and the cultures are subsequently agitated to a mechanical shaker. Importantly, the ensuing single cells as well as small cell clusters are employed for a number of purposes in plant tissue culture *viz.,* single-cell cloning.

Cell Bank A facility for preserving and keeping cells frozen at extremely low temperatures. These cells are used for investigating hereditary diseases, human aging, and cancer. Collection of banked cells are kept by the National Institute of Health (the Human Genetic Mutant Cell Repository and the Aging Cell Repository) and at the Cornell Institute for Medical Research.

Cell Counter (Electronic) An electronic instrument employed to count blood cells, employing either an electrical resistance or an optical grating technique *e.g.,* flow-cytometry.

Cell Cycle The series of events that take place during the growth and development of a cell *e.g.,* meiosis and mitosis.

Cell Growth Cycle The order of physical and biochemical events which take place during the growth of cells. In tissue-culture studies the cycle changes are divided into specific periods or phases, namely: *(i)* DNA synthesis (or S period); *(ii)* G_2 period (or gap); *(c)* M or mitotic period; and *(d)* G_1 period.

Cell-kill In antineoplastic therapy, the number of malignant tumour cells destroyed by a treatment process.

Cell Kinetics The study of cells and their subsequent growth and division. Study of these factors has enormously led to much better understanding of cancer cells and has been found to be useful in developing newer chemotherapeutic methods.

Cell Mass In embryology, the mass of cells that develops into an organ or structure.

Cell Organelles Any of the structures in the cytoplasm of a cell *viz.,* mitochondria, endoplasmic reticulum, Golgi complex, ribosomes, lysosomes, and centriole.

Cell Sorting A technique used to separate cells with a surface antigen from those without it.

Classical Competent Pathway Series of reactions initiated by the formation of a complex between an *antigen* and an *antibody* which ultimately lead to the lysis of microbial cells or the enhanced ability of phagocytic blood cells to eliminate such cells.

Chimera (Chimaera) A plant or tissue composed of more than one kind of genetic tissue or genome.

Chimeric Gene An artificial gene produced by combining the DNA sequences from several different sources.

Chromatid Single chromosome containing only one DNA complex.

Chromatins The DNA-protein complex that constitutes a chromosome.

Chromosome The deoxyribonucleic acid (DNA) bearing structure that carries the inheritable characteristics of an organism.

Chelate A chemical compound (liquid) with which metal atoms may be combined in such a manner so as to prevent them (bivalent, trivalent metal ions) from precipitating out of solution, and thus rendering them unavailable to plants, such as: ethylenediamine tetraacetic acid (EDTA), its disodium salt (NaTEDTA), which undergoes complexation to Fe^{2+} (ferrous ion)/Fe^{3+} (ferric ion) present in nutrient solutions employed for plant tissue culture technique. The phenomenon is usually termed as chelation.

Chemotherapeutant A chemical frequently employed to pretreat diseased source plants prior to *excision* or *incorporation* into the media to support certain specific therapeutic objective, for instance malachite green, virazole (ribavirin) are employed to eliminate completely the virus present in the meristem-tip culture. The phenomenon is known as chemotherapy.

Claims The claims are that part of a complete specification which defines the boundaries of the protection sought by the patent. They provide legal protection for the invention. They form a protective boundary line (around the invention) defined by the words and phrasing in the claims. What is not claimed in the 'claims' will be considered as **disclaimer,** even if the matter is disclosed in the description.

Clone A group of genetically identical *cells* or *organisms* descended asexually from a common ancestor. All cells in the clone have the same genetic material and are exact copies of the original.

Clonal Selection Theory The theory accounts for exclusive antibody formation during foetal development wherein complete set of lymphocytes are developed. Each lymphocyte consisting of the genetic information for initiating an immune response to a single specific antigen.

Conjugate A reagent which is formed by covalently coupling two molecules together, such as fluorescein coupled to an immunoglobulin molecule.

Constant regions The relatively invariant parts of immunoglobulin heavy and light chains, and the a, _, y ,md 1) chains of the T-cell reeeptor,

Contact hypersensitivity A delayed intlammatory reaction on the skin seen in type IV hypersensitivity. Co-stimulation. The signals required for the activation of a lymphocyte, in addition to the antigen-specific signal delivered via their antigen receptors. CD28 is an important co-stimulatory molecule for T cells and CD40 for B cells.

Cross reaction The sharing of antigenic determinants of two different antigens thus binds to same antibodies.

***CSFs* (colony stimulating factors)** A group of cytokines which control the ditlerentiation of haemopoietic stem cells.

CTLA-4 (CD152) A down regulatory signaling molecule of T cells that competes with CD28 for ligation by B7 on antigen-presenting cells.

Culture Medium A specifically prepared nutrient solution (substrate) which may be chemically defined for growing plant tissues, and microorganisms *in vitro.*

Cybrid A plant *or* cell cytoplasmic hybrid *(i.e., heteroplast)* having essentially the nucleus *of* one and cytoplasmic organelles *of* another, or *of* both cells or plants.

Cyclophosphamide A cytotoxic drug frequently used as an immunosuppressive chemotherapeutic agent/drug.

Cyclosporin A T-cell suppressive drug that is particularly useful in suppression of graft rejection in transplantation.

Cytolysis Cell dissolution or disintegration.

Cytokines A generic term for soluble molecules which mediate interactions between cells.

Cytotoxic A chemical or other agent which is toxic to cells.

Cytotoxic T Cells A class *of* T lymphocytes that are able to kill cells as part *of* the cellmediated immune response.

D genes Sets of gene segments lying between the V and J genes in the immunoglobulin heavy chain genes, and in the T-cell receptor and J chain genes which arc recombined with V and J genes during ontogeny.

Decay accelerating factor ***(DAF)*** A cell surface molecule on mammalian cells which limits activation and deposition of complement C3b.

Defensins A group of small antibacterial proteins produced by neutrophils,

Degranulation Exocytosis of granules from cells such as mast cells and basophils.

Dendritic cells A set of cells present in tissues, which capture antigens and migrate to the lymph nodes and spleen, where they arc panicularly active in presenting the processed antigen to T cells. Dendritic cells can be derived from either the lymphoid or mononuclear phagocyte lineages.

Delayed Hypersensitivity T Cell A class *of* T -Cells.

Deacceleration Phase The declining growth rate phase following the linear phase and preceeding the stationary phase in most batch-supension cultures.

Decontaminate To free from contamination or surface sterilization.

Deficiency In genetics, a chromosome aberration which comes into being from the actual loss *of* a gene or series *of* genes by deletion.

Deficiency Disease A disease resulting from a diet that invariably lacks essential nutrients *e.g. rickettsia* -from vitamin D deficiency, *vision impairment* - from vitamin A deficiency; *improper blood-coagulation* - from vitamin K deficiency etc.

Defined Culture Medium A culture medium wherein all the components are quantified and known.

Deletion The lost portion either *of* a chromosome or *of* a nucleotide sequence in nucleic acids *e.g.*, DNA, RNA.

De Novo (Latin: *anew, from the very beginning)*. Arising sometimes spontaneously from either unknown or very simple precursors.

Differentiation The resumption ofmeristematic activity mostly by mature cells *via* a reversal *of* the process *of* cell or tissue differentiation.

Dehumidifier A equipment or device or apparatus that aids in the removal *of* moisture from the air.

Deionized Water Water obtained after due passage *via.* an ion exchange device to get rid *of* either soluble inorganic salt (minerals) or certain organic salts. The phenomenon is termed as deionization.

Derepression The mechanism by which repression is invariably eleviated. Whenever a re pressing metabolite is removed it usually results into an enhanced level *of protein* or *enzyme.*

Diffusional Limitation The phenomenon by which the rate *of* diffusion *of* substrate into and product out *of* an *enzyme, cell, tissue* or above all the aggregate *of immobilized cells* becomes limiting.

Diffusion The movement *of* molecules across a concentration gradient specifically from the area *of higher concentration* to the area *of lower concentration.*

Direct Embryogenesis *Embryoid formation* directly upon the outer surface *of* either somatic or zyogtic embroys, or on seeding plant tissues in culture medium without essentially heating an intervening callus phase.

Direct Organogenesis *Organ formation* directly upon the surface *of* comparatively bigger intact explants mostly without an intervening callus phase.

Dismutages An *enzyme* of great value specifically in organ transplantation, and also in the treatment of 'heat attack' where tissues have been adequately deprived of blood for a short duration. Nevertheless, it prevents reperfusion injury by permitting the oxygen deprived tissues so as to enable their due recovery to the normal state in a more defined, elaborated, and orderly manner.

Dilution Rate Refers to the continuous fermentation whereby a measure of the rate at which the prevailing medium is duly replaced with the fresh medium appropriately. The dilution rate is reciprocal of the *hydraulic retention time.*

Diploid Cell Cell having its chromosome in homologous pairs, and thus having two distinct copies of each autosomal genetic locus.

Double Diffusion Test *Immuno diffusion test* wherein *antigen* and *antibody* predominantly diffuse towards each other from separate wells cut into agar gel.

Downstream processing Separation and purification of products from a fermentation process produced in a large scale fermentors.

Dys-functional Immunity An immune response which causes essentially an allergic reaction or the apparent lack of immunse response resulting in failure to protect the body agirinst a host of infections or toxic agents.

DNA Ligases Enzymes that specifically bind together the ensuing DNA fragements.

Dual Culture A specialized culture system which invariably includes plant tissue and one organism *(e.g., nematode species)* or microorganism *(e.g., fungus* or bacteria). In actual practice, dual cultures are mostly employed to investigate and study the host-parasite interactions or the generation of axenic cultures for a variety of purposes. Normally the microorganisms selected is an obligate parasite.

Enzyme Anyone of a number of specialized proteins produced in the living cells which particularly speed-up, augment, and catalyze the 'rate of specific chemical reactions' even at an extremely low concentrations *(e.g.,* organic catalyst), but is not consumed in the reaction.

Enzyme Induction Regulation in the biosynthesis of specific enzymes (largely accelerated under certain conditions) depending exclusively upon the nature of the nutrients.

Enzyme Inhibition. The rate of an *enzyme-catalyzed reaction* may sometimes be controlled in a specific manner by inhibitors as well as inactivators. Substances like acids, alkalis, detergents, urea, proteases etc., serve as enzyme inhibitors and distinctly lower enzyme-catalyzed reaction; however, enzyme inhibition could be either reversible or irreversible. Importantly, the enzyme activation contrarily gives rise to an enhanced rate of an enzyme-catalyzed reaction. Therefore, both inhibition and activation of enzymes by key metabolites predominantly provides the usual means of metabolite control.

Enzyme Activity An expression of the ability of a given enzyme preparation to catalyze a specific reaction effectively.

Enzyme Immobilization It is invariably described as the process whereby an enzyme is converted from a *homogeneous catalyst* into a *heterogeneous catalyst.* It is separated into an altogether distinct and characteristic phase *i.e.,* nonnally water-insoluble and quite often of high molecular weight, that is found to be absolutely separate from the bulk substrate-containing phase. Enzyme immobilization is physically confined and even localized in a particular defined region.

Enzyme-Linked Immunosorbent Assay (ELISA) The excellent technique is invariably employed for detecting and quantifying specific serum antibodies as well as antigens exclusively based upon tagging the antigen-antibody complex with a specific substrate that may be enzymatically converted to readily quantifiable product by a specific enzyme.

Enzyme Immunoassay (EIA) An assay involving the substitution of an enzyme for a radiolabeled product.

Enzymeimmunodiffusion Test A specific test wherein an *antigen* itself is an enzyme and the antigen-antibody complex exhibits distinct enzyme activity. Thus, this test is very useful in detecting invisible or non-stainable antigen-antibody aggregates.

Epigenetic Gene Control It may be regarded as an unique *newer concept.* It essentially describes that the control of gene expression in higher eukaryotic organisms is related to the methylation of *cytosine residues* in gene coding sequence.

Eukaryotes Cellular organisms having a membrane-bound nucleus within which genome of the cell is stored as chromosomes composed of DNA *i.e.,* algae, fungi, protozoa, plants and animals.

Enhancer Element Refers to the short regions of DNA that possess the effect of enhancing the levels of transcription, but unlike promoters, they invariably affect genes from the chromosomalloci at a relatively longer distance.

Environmental Chamber (Incubator) It is a controlled environment cabinet wherein temperature, light quality, intensity, duration, relative humidity and controlled air flow.

Epitopes Specific antibody binding sites on the surface of *antigen,* also antigenic determinants.

Erythroprotein Refers to the glucoprotein usually produced in the kidney, and that regulates the production of RBC present in the bone marrow.

Exoenzymes Enzymes that mostly occur either attached to the outer surface or, in the *peri plasmic space* or, released into the medium surrounding a cell.

Exponential Phase Refers to the specific phase in culture wherein the cells usually undergo the maximum rate of cell division.

Etiology The elaborated study of the particular causation of a 'disease'.

Fab (Antigen Binding Fragment) Represents either of the two identical fragments generated in the event when an immunoglobin (Ig) is cleaved by papain; and the *antigen-binding-portion* of an antibody, including the hypervariable region.

Fc (Crystallisable Fragment) The remainder of the molecule when an immunoglobulin (Ig) undergoes cleavage and the Fab fragment gets separated. The crystallisable portion of an *Ig molecule* made up of the *constant region,* and the end of an Ig that in turn binds with the complement of serum.

F-factor Refers to the plasmid that confers the ability to conjugate upon the bacterial cells, and carries the transfer genes.

Fab Batch System An anaerobic bioreactor operated in a batch mode to which substrate either in *solid* or in concentrated *liquid* form is incorporated several times in the course of the run.

Facultative Anaerobe Usually refers to an *organism,* or a *bacterium,* or a *fungus* that may adapt its metabolism to survive and grow in the presence or the absence of oxygen.

Fibroblast Resident cell of connective tissue that are derived from mesoderm.

Fixative A compound which precisely fixes or sets or stabilizes other compounds or structures securely so as to retain their structural integrity. The phenomenon of fixation makes use of chemical agents to prepare tissues or cells for microscopy rather permanently.

Floccule The coalescence (aggregation) of microorganisms or colloidal particles floating in or on a liquid. Flocculation may be seen in certain contaminated liquid media showing up as a cloud.

Fouling The deposition of the material both in the *'wrong place'* and at the *'wrong time'*. In bioreactors and cooling systems the fouling of heat-exchanger surfaces by *wrong biomass, protein, polysaccharides* or other similar materials invariably occurs.

Frame Shift Mutation Deletion or insertion of a number of bases not divisible by three in an open reading frame of a DNA-sequence.

Freeze Drying (Lypholization) The process of drying in a frozen state under vaccum *e.g.,* tissues are freeze-dried to obtain a dry weight or to preserve them for analysis. Lypholization of a specimen is often carried out with liquid N2, and subsequently causing sublimation of water from the specimen under vacuum. Thus, the proteins are maintained in a reasonably native form, and can easily be dehydrated *viz.,* snake venom, sperm, biological fluids etc.

Fusogen A fusion-inducing agent employed for *protoplast agglutination* in carrying out the somatic hybridization studies *e.g.,* polyethylene glycol (PEG).

Fusion The combination of two distinct cells or macromolecules into a single integrated unit (or entity).

Folic Acid (Vitamin M) [$C_{19}H_{19}N_7O_6$; 441.40] A member of the vitamin B group, and is found in green leaves which also exhibit certain extent of coenzyme activities. It is occasionally included as an ingredient for the *plant tissue culture media.*

Gelation A gelatinous and proteinaceous material produced by boiling bones (animal carcas) with water, which partially hydrolyzes the collagen of animal connective tissues. It is employed skill fully to get or solidify nutrient solutions required for plant-tissue culture.

Gelrite Refers to the brand name of a synthetic refined polysaccharide *(Pseudomonas-de*rived) invariably employed either as a gelling agent or as an agar substitute.

Generation Time The time between successive generations of *cells* or *organisms* within a population.

Gene A sequence of nucleotide that specifies a particular polypeptide chain or RNA-sequence or The functional unit of heredity - a locus on a *chromosome* which encodes a specific functional product..

Gene Amplification Selective replication of DNA-sequence within a cell, producing a multiple extra copies of that sequence.

Gene Cloning The insertion of a DNA-sequence into a vector that may eventually be propagated in a host organism thereby producing a huge number of copies of the sequence.

Gene Expression The effective utilization of information in a gene *via* transcription and translation ultimately giving rise to the generation of a *'protein';* and, therefore, the appearance of the phenotype determined by that gene.

Gene Isolation The meticulous removal of genetic information, in the form of a DNA-sequence, from a selected organism in order to study its structure or insert it into a vector in the course of 'gene manipulation'.

Genetic linkage The proximity of two or more *genes* on a *chromosome* such that they tend to be inherited together.

Gene magnification Phenomenon in which the number of copies of the ribosomal genes is increased in mutants of Drosophila. Such mutants are termed Bobbed, since they also have reduced number of bristles.

Gene map Order of nucleotides and genes that is deduced from genetic recombination experiments. Individual genes can be assigned to specific loci within a genome or a chromosome as a result of gene mapping.

Gene mapping The determination of the relative positions of genes on a *chromosome* or *plasmid* and the distance between them.

Gene pool The total of all genes in an interbreeding population at a given time. Gene redundancy. Presence in a cell of many copies of a single gene.

Gene regulatory protein Protein that functions by altering the transcriptional state of a specific gene. The repressor protein that regulates the activity of the lac operon in *E. coli* is one such protein, and they are widely believed to be an important aspect of eukaryotic gene regulation also.

Gene sequencing A technique used to determine the sequence of bases in the length of DNA coding for particular polypeptide.

Gene splicing The enzymatic attachment of one gene or part of a gene to another; also removal of introns and splicing of exons during mRNA synthesis.

Genetic Engineering (Recombinant DNA Technology) It refers to the specific technology involving man-made changes in the genetic constitution of cells (apart from some selective breeding). Genetic engineering invariably uses a vector *(e.g.,* Ti plasmid of *Agrobacterium tumefaciens)* for transferring useful genetic information usually from a donor organism into *a cell* or *organism* which does not possess it.

Genetic Selection Refers to selection of genes, clones etc., by man either within populations or between populations or species. The useful purpose is to cause a change in the various specific phenotypic character. Genetic selection normally results in differential success rates of the various genotypes thereby reflecting several variables including selection pressure along with the genetic variability in populations.

Genetic Transformation The precise transfer of extracellular DNA *(genetic information)* among and between species *viz.* usage of viral or bacterial vectors.

Genetic Variation The actual differences in individuals derived from the same genotype in distinction to differences caused by the environment. Genetic variance designates the very proportion of phenotypic variance caused by differences in the genetic make-up of an individual.

Genome The complete genetic constitution of an organism or individual or A collective term for the total hereditary material of a cell found in the *chromosome* set.

Genotoxic Carcinogenic and toxic to the chromosomes.

Genotype The genetic constitution of an organism.

Germplasm Refers to the reproductive body tissues distinct from somatic *i.e.,* non-productive, tissues. The basis of heredity of an organism *i.e.,* the *'genetic material'* is usually passed on through the previous generation(s) or The total genetic variability available to a population of organisms as represented by germ cells or seeds etc.

Gibberellic Acid [Gibberellin; GA_3; $C_{19}H_{22}O_6$; MW 346.37]. It represents one of the gibberellins, a group of plant-growth hormones essentially promoting *cell division, elongation,* and *germination.* It is isolated from the fungal pathogen *Giberella jujikuroi,* and is extensively used in plant-tissue culture.

Glycosylation Transfer of glucose residue invariably from nucleotide sugar derivative. Glycolysis. The conversion of monosaccharide to pyruvate *via* the glycolytic pathway *[i.e.,* Embden-Meyerhoff Pathway (EMP-Cycle) in carbohydrate metabolism] in the cytosol.

Glyoxylate Cycle Metabolic pathway present in bacteria and in the glyoxysome 4-C-decarboxylic acid.

HEPA Filter [High Efficiency Particulate Air Filter] A filter capable of screening out particles larger than 0.3 µm. They are frequently employed in Laminar Air Flow Cabinets (Hoods) as well as sterile zones for carrying out transfers/operations in highly aseptic conditions.

Hairpin Loop A region of double helix formed by base pairing very much within a single strand of DNA or RNA that has precisely folded back on itself.

Hapten A substance that elicits antibody formation only when combined with other molecules or particles which may specifically react with performed antibodies.

Haploid A single set of homologous chromosomes that have half of the normal diploid number of chromosome.

HAT Medium A highly selective growth medium for animal tissue cells that essentially comprise of *hypoxanthine, aminopterin* (a folate antagonist), and *thymidine.* It is exclusively used for the selection of hybrid somatic cell lines, such as: production of monoclonal antibodies.

Haemeagglutination The agglutination or clumping of blood cells by an antibodies *e.g., erythrocytes.*

Haemeagglutination Inhibition. Refers to the inhibition of haemeagglutination *(i.e.,* antibody mediated clumpting of RBC), invariably by means of specific immunoglobulins or enzymes, employed to establish/determine whether a patient has been exposed to a specific virus.

Hepatitis B Viron [Dane Particle] It has 42 nm diameter having an outer wall enclosing inner 27 nm core particle comprising of the circular DNA. Aggregates of the envelope proteins are usually found in plasma and are referred to as hepatitis B surface antigen (HBsAg).

Heterokaryon A cell containing genetically different nuclear. Found naturally in several fungal species, and can be produced by cell-fusion technique experimentally.

Histones Proteins found in the nuclei of most-eukaryotic cells where they are eventually complexed to DNA in chromatin and chromosomes.

Histocompatibility Antigens Genetically determined antigens located on the lipoprotein membranes of the nucleated cells of most tissues that cause an immune response when grafted on to a genetically different individual and thus determine the capability of tissues in transplantation.

Hypertrophy Refers to an abnormal increase in cell size thereby leading to irregular *growth* or *swelling* of body parts or organs. It is invariably regarded as a disease or other stress-induced response.

Hypersensitivity An exaggerated immunological response upon exposure to a particular antigen or allergen.

Hypervariable Region A region of immunoglobulin (Ig) which essentially accounts for the specificity of antigen-antibody reactions, *e.g.,* specified terminal regions of the Fab (antigen binding) fragments.

Hybridoma A cell hybrid wherein a tumour cell forms one of the original source cells.

Idiotopes The specific short segments of amino acids that are present either in the variable region *(i.e.,* region not actually involved in antigen binding) or in the combining-site itself, may act like antigenic determinants, and possess the ability to combine with *self-antibodies* or the *receptors* on certain lymphocytes. These *self-determinants* are known as idiotopes.

Idiotype Immunoglobulin (Ig) molecules having distinct variable regions determining the precise specificity of the ensuing antigen-antibody reactions.

Immobilization A physical or chemical phenomenon used to fix microorganisms, enzymes, and cultures of animal and plant cells, or organelles derived from these sources either upon a solid support or have them trapped in a solid matrix.

Immunodiagnostic An analytical test method exclusively based upon the highly specific interaction between the *antibody* and an *antigen.* The interaction is linked to a method of quantifying the result

Insulin A polypeptide hormone usually found in both *vertebrates* and *invertebrates* secreted by B-lymphocytes (cells) of the endocrine pancreas in response to high blood sugar levels, it gives rise to the causation of hypoglycaemia.

Incubator A sophisticated apparatus providing absolute identically controlled environmental conditions *(e.g.,* light, temperature, humidity, photoperiod) most suitable for the growth of microorganisms, yeasts, plant cells or plant cultures (incubation), and animal cells *viz.,* BOD-Incubator.

Infringement An invasion of an exclusive right of intellectual property. Infringement of a utility patent involves making, using, or selling a patented product or process without permission. Infringement of a design patent involves fabrication of a design that, to the ordinary observer, is substantially the same as an existing design, where the resemblance is intended to induce the observer to purchase one thing supposing it to be another. Infringement of a trademark consists of the unauthorized use or imitation of a mark that is the property of another in order to deceive, confuse, or mislead others. Infringement of a copyright involves reproducing, adapting, distributing, performing in public, or displaying in public the copyrighted work of someone else.

Indirect Embryogenesis Refers to the formation upon the callus tissues meticulously derived from somatic or zygotic embryos, seeding plants or other tissues in culture.

Indirect Organogenesis. The organ formation upon callus tissue derived from explants in tissue culture.

Induction Media Refers to the media which may afford variation or mutation in the tissues exposed to it.

Interleukins (ILs) A particular variety of substances (entities) that are produced by leucocytes. Their activity usually get triggered off specifically during inflammatory responses. Interleukins (ILs) represent the larger class of lymphokines.

Intellectual property (includes patent-trademark-copyright-trade secret) Creative ideas and expressions of the human mind that have commercial value and receive the legal protection of a property right. The major legal mechanisms for protecting intellectual property rights are copyrights, patents, and trademarks. Intellectual property rights enable owners to select who may access and use their property and to protect it from unauthorized use.

Ionizing Radiation Refers to such radiations as: γ-radiation and X-ray radiation that specifically gives rise to toxic free radicals which mostly attribute chemical reactions critically disruptive to the *biochemical organization of microorganisms.*

***In situ*:** In the original, natural or intact position *viz.,* location of the explants on the mother plant prior to excision.

***In vitro*:** [Latin = in glass]. Experimentation on organisms or portions thereof in a test tube or any other glassware, *viz.,* growing under artificial conditions as in a tissue culture of plants and animals.

***In vivo*:** [Lain = in life]. Experimentation on organisms under absolute natural parameters very much within intact living organisms, *viz.,* laboratory animals, voluntary humans etc.

Isotypes Antibodies differing in heavy chain constant regions intimately associated with different class and subclasses of immunoglobulins (Ig).

Jumping Gene A populist terminology for 'transposon' *i.e.,* a genetic unit such as DNA sequence that is transferred from one cell's genetic material to another.

Jiffy Pots A brand name for peat pots that are sometimes employed for *in vitro* transplantation purpose.

Karyotype A complete set of metaphase chromosomes of a cell or organism arranged in decreasing order of chromosomal length.

Karyotyping The meticulous identification and classification of organisms, cells or tissues invariably based on their actual chromosome content.

Kilobase (Kb) A sequence of 1000 bases or base-pairs of DNA, other nucleic acids or nucleotides.

Kinin The original class name for substances promoting cell-division to which the prefix *cyto* has been added duly *(e.g.,* cytokinins) to distinguish them clearly from kinins found in animal systems.

Knock-out animals Transgenic animals in which an original replaced or "knocked-out" of action by a new gene that doesn't function.

Krebs Cycle The *citric-acid cycle* or *carboxylic-acid cycle* occurring in 'carbohydrate metabolism'.

Kinases Enzymes which specifically catalyze the transfer of the a-phosphate of a nucleotide 5' - triphosphate to the 5' -OH terminus of a DNA or RNA molecule.

Killer T Cells A class ofT lymphocytes (cells) that actually participate in killer activity during cell-mediated immune (CMI) response.

Laminar flow. Layered monodirectional airflow. In a laminar airflow (LAF) cabinet air is forced through filters and specially designed ducting to produce an airflow with a minimal lateral disturbance. This has the effect of producing 'layers' of air where the adjacent layers do not mix except on the molecular scale. In the vertical LAF cabinet, this airflow is in the form of a narrow 'curtain' of air passing down the inner front surface of the cabinet: in a horizontal LAP cabinet the air forms a 'block' the size of the front opening blown towards the operator.

Lag Phase Mostly used to describe initial phase of the growth of most batch propagated cell suspension culture wherein the inoculated cells in fresh medium get adapted to the altogether new-environment and get ready to undergo division.

Ligase An enzyme which specifically catalyzes the condensation phenomenon of two molecules coupled to the breakdown of a pyridine triphosphate.

Ligation The process of having direct linkage of the nucleic acid fragments.

Layering *In vitro* layering essentially involves the horizontal placement on agar of cultured shoots (with or without) leaves or nodal segments to promote auxillary bud proliferation.

Lecithin A naturally occurring, choline-containing phospholipid present in animal and plant tissues. Chemically, lecithins are very much akin to fat, but they are found to be rich in phosphorus and nitrogen *e.g.,* yellow yok of an egg; soyalecithin. These are mostly used in food products as an 'emulsifier'.

Limiting Factor An environmental variable whose absolute level at a given time limits the growth or other activity of an organism.

Linear Phase The constant increase in 'cell numbers' following the exponential growth phase.:'inkers. These are short DNA double strands *(i.e.,* decameric oligonucleotides), which contain sites for the action of one or more restriction enzymes.

Liposome A spontaneously formed layered lipid vesicle in an aqueous medium used as a DNA rector in cell hybridization research activities.

Lyase An enzyme that catalyzes the addition of groups to double bonds or the formation of double bond-lyases as one of the major moieties.

Lymphocyte White cells of the blood, mediator of specific immunity, derived from the stem:ells of the lymphoid series. Two main categories are: T - and B-Lymphocytes have been duly recognized; the former subdivided into subsets *(viz.,* helpers, suppressors, and cytotoxic T::ells), responsible both for CMI as well as for regulation of the activation of B-cells, whereas the latter solely responsible for antibody production.

Lyophilize or Freeze Dry To freeze rapidly then dehydrate under high vacuum; and the process s termed as 'lyophilization'.

Lysis Cell rupture or destruction *via* enzymatic action.

Lysosomes Organelles essentially containing hydrolytic enzymes involved in autolytic and digestive processes.

Lysogeny The ability of certain specific *'phages'* to survive in a bacterium as a result of the integration of their respective DNA into the corresponding host chromosome.

Lymphokines A group of biologically active extracellular proteins duly generated by activated T lymphocytes involved in CMI; the chemical mediators of cellular immunity.

Macerase A brand name for *Pectinase,* which is used mostly in the isolation of intact cells and protoplasts of higher plants.

Macerate To disintegrate or separate tissues *via* cutting, soaking, enzymatic or other action, resulting in cell dissociation.

Macrolides A group of antibiotics having macrolytic structures from actinomycetis *e.g.,* erythromycin, clarithromycin.

Macromolecules Biological terminology referring to proteins, nucleic acids, and carbohydrates. Maternal Inheritance. Inheritance controlled by extra chromosomal (cytoplasmic) hereditary determinants.

Maintenance Energy Energy needed by an organism or a cell culture to survive in a viable condition.

MDA Microdroplet array technique.

Malignant A primary tumour essentially having the inherited potential to invade locally and also to cause metastasis, *i.e.,* movement of bacteria or body cells (esp. cancer cells) from one part of the body to another.

Marker An 'allele', *i.e.,* one of the two or more different genes containing specific inheritable characteristics that occupy corresponding positions (loci) on paired chromosomes, whose actual inheritance is under observation in a cross.

Matrix Ground substance wherein *'things'* are duly embedded. A common usage is for a loose meshwork within which the cells are usually embedded.

Maximum Oxygen Transfer Rate It refers to a measure of the fastest possible rate of uptake of oxygen (O_2) by a growing culture expressed in terms of the volumetric mass transfer coefficient times the oxygen solubility.

Maximum Specific Growth Rate (m_{max}) The theoretical maximum rate of growth of a culture under optimal conditions of nutrient supply.

Meristem A localized region of continuing mitotic cell divisions *(meristametic cells)* obtained due to protoplasmic synthesis and tissue initiation. Interestingly, from these undifferentiable tissue new cells arise which eventually differentiates into the .specialized tissues. Metabolite. A substance that acts as a substrate for or is produced by a metabolic process or enzyme reaction.

Macrophages: (or Mononuclear Phagocytes) These refer to large, active phagocytic cells found in spleen, liver, lymph nodes, and blood. They represent important factors in non-specific immunity.

Major Histocompatibility Complex (MHC). The genetic region in human beings and mice that controls invariably not only tissue compatibility but also the development and activation of part of the prevailing immune system.

Mast Cells Cells that contain granules of heparin, histamine, and serotonin especially in connective tissues involved in hypersensitivity reactions.

Memory Cells. Clones of lymphocytes having receptors of high affinity for a particular antigenic molecule; cells that are sensitised duly during the secondary response.

Microdroplet Array [or Multiple Drop Array (MDA) or Hanging Droplet Technique (HDT)] Introduced by Potrykus *et at.* (1979), this technique is commonly employed to evaluated sufficient numbers of 'media modifications', employing small amounts of medium into which are placed small numbers of cells. Droplets of liquid culture *(i.e.,* medium and suspended cells or protoplasts) are arranged carefully on the lid of a petri-dish, inverted over the bottom half of the dish containing a solution with a lower osmotic pressure, and the dish is now sealed. The cells or protoplasts form a monolayer at the droplet meniscus and can easily be examined.

Michaelis Constant (Km) A kinetic parameter used to characterize an enzyme defined as the concentration of a substrate that allows half-maximal rate of reaction.

Micropropogation Refers to propagation in culture by axillary or adventitious means. It is a general terminology used for vegetative *(i.e., asexual) in vitro* propagation.

Micro injection The specific insertion of a substance into a cell through a microelectrode. To extrude the substances *via* the very five electrode tips, either using hydrostatic pressure or electric current.

Microsatellite DNA Very short DNA motifs (1-10 base pairs) which occur as *tandem repeats* at numerous *loci* throughout the *genome*. Also known as simple sequence repeats (SSR), simple tandem repeats or simple repetitive sequences.

Minisatellite DNA Short DNA motifs (10-60 base pairs) which occur as *tandem repeats* at many *loci* on the *genome.*

Minimum Inoculation Size. The smallest inoculums that may be used successfully for subculture.

Monoclonal Derived exclusively from a single cell; pertaining to a single clone. Monoclonal Antibodies (MABs). Produced from hybrid cells using hybridoma technology. In other words, an antibody produced from hybrid cells employing hybridoma technology.

Mutagen A chemical or physical treatment or agent capable of producing genetic mutation in an organisms. It may cause insertion, deletion, or alteration of a base or part of the nucleic acid chain (DNA).

Mycelia. The interwoven mass of discrete fungal hyphae.

Mutability The propensity of an individual's genes or genotype to undergo heritable mutation. Nick. A point specific in a double-stranded DNA molecule where there is no phosphodiester bond existing between the adjacent nucleotides of one strand.

Nick Closing Enzyme. A form of DNA polymerase which is capable of restoring base sequences. A nick is strategically introduced adjacent to an incorrect base-pair. The base is subsequently removed and the correct base in inserted.

Natural Complex A complicated, non-synthetic, and invariably unidentified addendum in plant tissue culture media, *viz.,* orange or tomato juice or coconut milk.

Neoplasm. Localized cell-multiplication or tumour, a collection of cells that have undergone genetic transformation. These cells usually differ in structure and function from the original cell type.

Nif -Genes The complex of genes in nitrogen fixing bacteria, which code for the proteins required for the nitrogen fixation, specifically the enzyme nitrogenase.

Nitrocellulose Paper. Paper having a high non-specific absorbing power for biological macromolecule.

Nonsense Codon (or Nonsense Triplet) The three codons, UAA (ochre), UAG (amber), and UGA (opal), which do not code for an amino acid but serve as signals for the termination of protein synthesis.

N-Terminal The end of a protein or polypeptide chain that contains a free amino group (-NH2); therefore, the abbreviation N.

Nucleases Enzymes that catalyze hydrolysis of the phosphodiester bonds of polynucleotides *i.e.,* DNA or RNA.

Nuclease SI A very specific nuclease which essentially degrades single-stranded nucleic acids or splits short single-stranded stretches in a DNA sequence, but fails to attack any double stranded structure. It is used invariably in gene manipulation for converting sticky ends of duplex DNA to form blunt ends or to trim off single-stranded ends of nucleic acids after due conversion of single-stranded cDNA to the double-stranded form.

Nucleic Acids Refer to linear polymer of nucleotides, usually linked together by 3', 5'phospodiester bonds. In deoxyribonucleic acid (DNA), the sugar moiety is deoxyribose, and the corresponding bases of the nucleotides are invariably: *adenine, cytosine, guanine,* and *thymine.* The ribonucleic acid (RNA) has ribose as the sugar, and uracil replaces *thymine* amongst the four bases stated earlier.

Nucleoprotein A conjugated protein closely associated with nucleic acid such as histone protein.

Nucleotide Transferase These enzymes do catalyze attachment of the mononucleotide triphosphate to the 3' -OH function of the initiator polymer, with the release of pyrophosphate.

Obviousness (in patent) A condition in which an invention cannot receive a valid patent because a person with ordinary skill in that technology can readily deduce it from publicly available information (prior art).

Open Continuous Culture. A cell suspension culture with a continuous influx of fresh medium, maintained at constant volume by the efflux of cells and spent medium.

Operator The specific regulatory section present in a nucleic acid having binding capacity for a repressor protein that categorically blocks the synthesis of mRNAs.

Organ Culture The growth in an aseptic culture of plant organs *e.g.,* roots or shoots, starting with organ segments and maintaining the characteristic features of the specific organ(s).

Organized Growth The *in vitro* development of organized explants *viz.,* shoot tips or meristem tips or floral buds or organ segments or their *de novo* formation *from* unorganized tissues in culture.

Organogenesis The initiation *(de novo)* and growth of organs (invariably roots and shoots) *form* cells or tissues; as in organ culture system. Organs may be formed upon the very surface of the explants (direct organogenesis) or upon an intervening callus phase (indirect organogenesis) in tissue culture.

Organoid A anomalous *(i.e.,* unusual) organ-like structural growth formed in culture *e.g.,* as could be seen on leaves, roots, or callus.

Ortet The original *'mother plant'* or *'donor plant'* from which vegetatively propagated plants are derived meticulously. **Operon** A cluster or group of structural gene whose coordinated expression is duly controlled by a regulator gene.

Osmolarity The overall molar concentration of the solutes affecting the osmotic potential of a solution or nutrient medium.

Oxidative Phosphorylation A metabolic sequence of reactions taking place within a membrane wherein an electron is transferred *from* a reduced coenzyme by a series of electron carriers thereby strategically establishing an electrochemical gradient across the membrane that virtually drives the formation of ATP from ADP and nitrogen phosphate by chemiosmosis.

Opsonins Refer to intrinsic blood factors that help in establishing a link between particles to be ingested and the macrophage.

Opsonisation The process whereby a cell becomes more susceptible to phagocytosis and lytic digestion when a surface antigen gets combined with an antibody or other serum component.

Oxygenase An enzyme enabling either a system or an organism able to utilize atmospheric oxygen.

Paratope An important site or part in the antibody molecule which essentially combines with the antigen.

Parahormone A substance with hormone-like properties which is not a secretory product *e.g.,* ethylene, CO2.

Parasexual Hybridization Refers to genetic recombination by means other than *via* fertilization of germ cells (parasexual) leading to the formation of individuals or hybrid cells; as could be seen in hybrid cells or plants derived from somatic cell fusion.

Particle Radiation Refers to α-particles (+ vely charged) and β-particles (-vely charged), electrons, protons, and neutrons. These particles are usually employed to produce mutant cells or organisms in plant tissue culture.

Parts Per Million (ppm) This terminology has largely been replaced by equivalent terms, such as : for solids: mg per liter; and for liquids and gases: micro liter per liter.

Patents The patents are granted by the Government, for the commercial exploitation of an invention for a specific period of time in consideration of the disclosure of the invention so that on expiry of the terms of the patent the information can benefit the public at large.

Packed Cell Volume (PCV). Refers to a 'quantitative method' of estimating cell growth. It is based on the total cell volume in an aliquot of cell culture. The aliquot is duly centrifuged for 5 min. at 2000 rpm after which the packed cell volume (PCV) is expressed as a percentage (%) of the aliquot volume.

PCR Polymerase Chain Reaction A method for amplifying a DNA sequence in large amounts using a heat-stable *polymerase* and suitable *primers* to direct the amplification of the desired region of DNA.

Peptide A compound or a *'chemical entity'* existing of two or more amino acids covalently linked by a 'peptide bond' (-CO-NH-) peptide, that are formed with three or more amino acids are usually termed as 'polypeptides'.

Pharming A term derived from pharmaceutical and farming that refers to the growing of genetically engineered organisms for the production of pharmaceutical compounds.

Plasmid An autonomously replicating, extrachromosomal, double stranded, circular DNA present in bacterial cells. **Precipitin Test (or Microprecipitin Test)** A 'serological assay' wherein visible particulate matters *(i.e.,* precipitates) are duly formed by the interaction of soluble *antigen* and *antibody.* The precipitin test detects and identifies 'antigens'.

Prefilter A *coarse filter (i.e.,* Furnace Filter) *e.g.,* those used in a laminar air-flow cabinet to screen large particles before air is forced *via* a much *finer filter (i.e.,* HEPA-Filter).

Protein Digest The enzyme hydrolysis of proteins to give rise to their building block components, amino acids and short-chain peptides *(i.e.,* chains comprising of two to several amino acids normally without enzymatic function).

Protoclone Distinct phenotypic regenerates from a plant protoplast. A clone initiated from a protoplast or protoplast-fusion product.

Protocol A sequence of events, activities, techniques or procedures linked to accomplish a definite purpose and objective (goal); as a schedule or protoplast-fusion product.

Protoplast Fusion The coalescence of plasmalemma and cytoplasm of two or more protoplasts in contact with one another. Initial adhesion is indeed a random process but coalesence may be promoted *via* various means *(e.g.,* induced fusion). Nevertheless, when adhesion takes place between adjacent protoplasts during enzymatic wall-degradation or between freshly isolated protoplasts in the absence of a fusion agent, it is known as 'spontaneous fusion'.

Pure Culture A culture that contains cells of one type only; the progeny of a single cell i.e. genetically identical.

Pure Line All cells or individual members that are essentially *homozygous* for one or more characters or genes and will yield more cells or organisms with the character(s) under consideration and are genetically pure.

Primary Immune Response The first ever immune response to a particular antigen that has a characteristically long lag period and a low liter of antibody production.

Quiescent Quiet, at rest, not necessarily dormant and having the potential for the resumed activity; may also apply to cells unlikely to divide the non-meristematic cells.

Radiation Rays of heat, light or particles represented in the wave form.

Ramet An individual of a clone. It may also refer to the vegetatively propagated offspring of an ortet.

Radioimmunoassay (RIA) A highly sensitive serological technique invariably employed to assay specific antigens or antibodies, using a specific radioactive label.

Radial Immunodiffusion Radial diffusion is solely based upon the principle that a quantitative relationship prevails between the amount of nitrogen placed in a well where the antibody is already incorporated in the gel, and a ring of precipitation results; generally employed in the assay of serum proteins.

Random Amplified Polymorphic DNA (RAPD) A widely used technique for amplifying anonymous stretches of DNA using *PCR* with *arbitrary primers*.

Recombinant Any organism whose genotype has come into being (arisen) as a consequence of recombination; besides, any nucleic acid that has arisen as a result of recombination. Recombinant DNA. The *hybrid DNA* generated specifically by joining pieces of DNA from various sources, invariably designated as rDNA; and the phenomenon is termed as 'recombinant DNA technology' (or Genetic Engineering).

Recon or Reconstructed Cell A viable cell hybrid, cybrid or a transformed cell accomplished as a result of 'genetic engineering'.

Reculture The aseptic transfer of a pure culture to another fresh sterilized nutrient medium: also referred to as a 'subculture'.

Regime or Regimen Refers to a systematic treatment.

Redifferentiation Cell or tissue reversal in differentiation from one specific type to another type of cell or tissue.

Rejuvenation Refers to the treatment which leads to *culture invogoration (viz.,* subculture) or *revival (viz.,* dormancy termination).

Regulatory Genes Genes that serve as a regulatory function; genes that do not code for specific peptides but instead regulate the expression of structural genes.

Replica Plating A technique whereby various types of mutants may be isolated from a population of bacteria grown under nonselective conditions, based upon plating cells from each colony onto multiple plates and thereby noting the exact positions of inoculation.

Replication Multiplication of microorganisms, duplication of nucleic acid (DNA or RNA) from a template strand.

Repression Altered gene expression causing the failure of a particular protein synthesis (anabolism) in the organisms.

Resistance Transfer Factor A plasmid, present in certain types of organism, for instance: *E. coli,* which may impart specific resistance to antibiotics or toxic compounds in animals that are eventually exposed to them.

Restriction Endonucleases Enzymes that specifically cleave DNA at a definite information bearing sequences of nucleotides, and thereby allow an insertion of foreign sequences. Restriction Sites. Specific sites representing categorically the loci at which the 'foreign DNA' gets integrated.

Restriction Fragment Length Polymorphism (RFLP) Fragments of differing lengths of DNA that are produced by cutting with restriction endonucleases enzymes.

Rhizogenesis Refers to the formation of roots and their growth as one may come across in root development *de novo* from callus.

Ribosomal RNA (rRNA) RNA of various sizes that make up part of the ribosomes, constituting up to 90% of the total RNA of a cell; single-strand RNA having essentially helical regions duly formed by base-pairing between complementary regions within the strand.

Rotator A wheel-like device for slowly *(ca.* 1 rpm) rotating and gently agitating cultures usually in a vertical plane.

Rotary Shaker A platform shaker having a *'circular motion'* employed particularly for shaking culture flasks at variable speeds.

S-Phase The cell cycle phase during which RNA synthesis takes place.

Sarcoma A solid tumor growing from derivatives of tissue such. as embryonal connective tissue, bone, muscle and fat.

Satellite DNA Very high repetitions (from 1,000 to over 100,000 copies) of a basic motif or repeat unit (commonly 100 - 300 base pairs) which occur at a few *loci* on the *genome.*

Scale-up Expansion of laboratory experiments to full-sized industrial processes.

Schiff's Reagent An admixture of pararosaniline hydrochloride *(i.e.,* an aniline dye) and sodium bisulphite [$NaHSO_3$] invariably employed in staining chromosomes and other nuclear substances present for detection and identification respectively.

Selection Culture Makes use of difference(s) in the environmental parameters or more commonly in culture medium composition so that preferred variant cells or cell-lines are predominantly favoured over other variants or the wild-type.

Selection Pressure It is a measure of the effectiveness of natural or experimental selection in altering the genetic composition of a population.

Selection Unit. Single cells or small clusters, units of optimum size for isolating and regenerating variants or mutants; the minimum number of cells effective in the screening process.

Selectable Markers Refer to nucleic acid sequences that are phenotypically easy to recognize *i.e.,* antibiotic resistance genes of *E. coli.* plasmids, auxotrophic markers etc.

Semicontinuous Culture The maintenance of cells in a culture vessel in an actively dividing state by draining periodically the medium and replenishing by adding fresh medium.

Serial Float Culture Sunderland's technique of floating anthers *(i.e,* part of a flower's stamen containing pollen) on liquid medium and subculturing them to a new medium at an interval of several days, whereby anther dehiscence, pollen release, and development take place-ultimately enhancing the anther productivity.

Shake Culture An agitated suspension culture, usually an Erlenmeyer flask containing the culture is attached to a horizontal or platform shaker or agitated with a magnetic stirrer to provide adequate aeration for cells in the liquid medium.

Secondary Immune Response The response of an individual to the second or subsequent contact with a specific antigen, characterized by a short lag-period, and the production of a high antibody titre.

Sequence Characterised Amplified Regions (SCAR) A SCAR is a locus representing a single *RAPD* fragment which has been sequenced. Primers specific to the locus can be designed and used in *PCR* amplification.

Somaclone A plant regenerated from a tissue culture originating from somatic tissue of the plants.

Somatic Cell Variant (or Embryoid) An organized embryonic structure morphologically similar to a zygotic embryo but initiated from somatic (non-zygotic) cells. Ultimately, these develop into plantlets *in vitro via* developmental processes which are very much akin to zygotic embryos sexual embryos.

Somatic Hyrbid Refers to a cell or plant-product of somatic cell fusion; caused as a result of cell or protoplast fusion and implying genomic integration. The phenomenon is termed as s0matic hybridization.

Somatic Organogenesis The production of roots, shoots or other organs upon the somatic tissues of explants (direct organogenesis) or by induction on callus generated by explants (indirect organogenesis).

Spent Medium Medium discarded when a culture or organism is subcultured. The implication is that the medium has been duly depleted of nutrients, dehydrated or accumulated toxic metabolic products.

Spontaneous Fusion Uninduced protoplast fusion that may occur between freshly isolated protoplasts or following adhesion of adjacent cells during enzymatic cell wall degradation of plant cells.

Spontaneous Variation Refers to the variation in plant populations derived from respective tissue cultures not previously exposed to mutagens but taking place as a consequence of the culture parameters.

Stationary Culture A non-agitated culture *i.e.,* an 'antonym' of shake-culture.

Stages of Culture (I-IV) Refer to: Stage-I: Aseptic explanation or establishment of the explant in culture; State-II: Multiplication of the propagates; State-III: Rooting of the propagates and preparation for transplant to soil; State- IV: Establishment of State II or III propagates *in vitro* in soil.

Structural Gene A gene whose product is an enzyme, structural protein, tRNA, or rRNA as opposed to regulator gene whose product prominently regulates the transcription of structural gene.

Stock Solution A solution invariably concentrated (10 to 100 times the final medium concentration) of select medium constituents that may be mixed together for attaining perfect compatibility and to avoid precipitation; and usually prepared before hand to save time during preparation of medium. In usual practice, the stock solutions are either frozen (in deep-freezer) or stored in the refrigerator to avoid deterioration in strength and quality. Portions of stock solutions are used conveniently as and when required for preparation of media.

Subculture (Passage) A culture derived from another culture or the aseptic division and transfer of a culture or a portion of that culture (inoculation) to a fresh nutrient medium. Invariably, sub-culturing is Gamed out at a predetermined time intervals, the length of duration is known as the *subculture interval* or *passage time.*

Subline A cell line regenerated from a unique cell line of a hybrid callus colony.

Suspension Culture Cells and group of cells (aggregates) that are adequately dispersed in an aerated, usually agitated, liquid culture medium.

Supraoptimum An amount greater than required as an inhibitory concentration of an exogenous growth factor.

Surface Sterilization. The removal of plant surface micro-flora prior to aseptic excision of explants. Surface sterilization is accomplished by immersing of tissue in one of several sterilants *e.g.*, calcium hypochlorite, sodium hypochlorite, hydrogen peroxide, mercuric chloride, silver nitrate, bromine water for an empirically determined period of time.

Synchronized Cells Synchronized mitosis in a group of cells in culture by natural or artificial means.

Synchronous Culture A microbial or plant-cell culture treated in such a manner so as to have most cells or individuals in the same stage of development or mitosis. It may be accomplished in several ways including: *via temperature variation* and *nutrient limitation.*

Tandem repeats Multiple copies of the same *base sequence* on a *chromosome*. Tandem repeats are found in *satellite*, *minisatellite* and *microsatellite* DNA.

Ti Plasmid A portion of the genome of the bacterium *Agrobacterium tumefaciens i.e.,* the agent directly involved in crown gall disease. This specific plasmid happens to be an useful (experimental) vector for the transfer of genetic information into the plant cells.

Tissue Explant An excised section of plant tissue usually employed to initiate a culture in the laboratory.

T-Cells. T lymphocytes that are differentiated in the thymus and are virtually important in cell mediated immunity (CMI); besides, in the regulation of antibody-mediated immunity (AMI).

Totipotency The potential or inherent capacity of a plant cell or tissue to develop into an entire plant when stimulated appropriately. It critically implies that all the information necessary for growth and reproduction of the organism is very much contained in the cell.

Transcription Refers to conversion of a DNA (gene) information sequence into a mRNA sequence.

Transduction Introduction of foreign genetic material into a cell by the help of **virus** or, in the particular instance of bacteria by means of **phages.**

Transformation Uptake of free nucleic acids or plasmids into a cell.

Transgenic Organisms that consists of (foreign) genetic materials from a different species or genus.

Translation Refers to the synthesis of a protein upon mRNA.

Toxic Poisonous in nature; as are certain chemical substances (toxicants) or any specific substance present in excess which tends to be detrimental to the normal plant or animals function or growth.

T Suppressor Cells A class of T cells that usually suppress the activates of B cells in the antibody mediated immunity.

Triplet Code Refers to the 'genetic code' *i.e.,* three sequential nucleotides in mRNA are required to code for a specific amino acid.

Tumour Inducing Principle (TIP) The plasmid carried by *Agrobacterium tumefaciens* the crown gall organism. Through incorporation into the host genome the host tissue is duly transformed into the tumour tissue.

Turbidostat An open continuous culture system wherein the inflow of fresh medium is controlled by the turbidity of the culture, a function of the amount of cell growth. Thus, balancing the fresh medium inflow which being a regulated outflow of cells and spent medium thereby restoring the original turbidity level.

Ultrasonic Cleaner A device to include high frequency vibration of materials, removing adhering substances from surfaces by mechanical action. This specific device is useful for cleaning glassware and for also disinfecting plant materials.

Ultraviolet (UV) Radiation The short wavelength of electromagnetic radiations having 10 to 400 nm with high energy.

Unorganized Growth *In vitro* formation of tissues with a few differentiated cell types and lacking recognizable structure.

Unwinding Enzymes Refer to such enzymes that specifically catalyze separation of complementary strands of DNA.

Vaccine Any antigenic preparation administered to stimulate the recipients immune defence mechanisms with respect to a given pathogen or toxic agent.

Variable Number of Tandem Repeats (VNTRs) A class of polymorphism characterised by the highly variable copy number of identical or closely related sequences. See also *tandem repeat*, *minisatellite* and *microsatellite* DNA.

Vector Plasmid or virus after ligation with a foreign DNA is employed to introduce new information into the cell.

Vegetative Cells Refers to such cells that are intimately engaged in nutrition and growth; they do not act as specialized reproductive or dormant forms.

Virus Elimination Chemotherapy, thermotherapy and meristem or meristem tip culture, used alone or in combination have been used successfully for the elimination of systemic viruses from plants.

Virus-Free (or Virus Tested) A plant that appears healthy and repeatedly gives negative tests for the presence of one or more identifiable viruses. Such a plant may subsequently employed as a stack or donor plant (explant source) solely for propagation purposes, which may be duly certified as virus tested (or certified virus tested).

Wild Types The *genotype* or *phenotype* of an organism predominating in the control (standard or wild) population, in its natural environment.

Water of Hydration The quantum of water chemically bound to a substance. The amount could be variable and should be taken into account when solutions of salts are prepared; as in the instance of medium preparation.

Xylem The vascular tissue of plants that conduct water and dissolved minerals from roots to the aerial parts of plants.

Yeast Extract An unidentified complex Vitamin B complex used as an addendum to certain plant tissue-culture media.

Yield Refers usually to the 'productivity assessment'.

Zygote The fertilized egg or ovum.

Index

A

B

C

H

I

N

O

P